I0042734

SCIENCE & SOCIETY

SCIENCE & SOCIETY

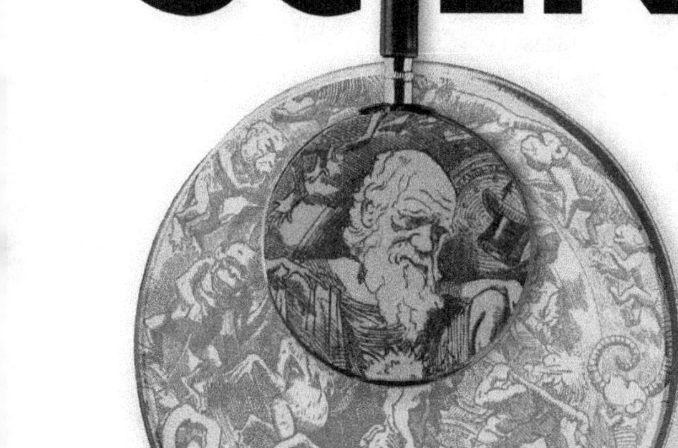

John Scales Avery

University of Copenhagen, Denmark

World Scientific

NEW JERSEY • LONDON • SINGAPORE • BEIJING • SHANGHAI • HONG KONG • TAIPEI • CHENNAI • TOKYO

Published by

World Scientific Publishing Co. Pte. Ltd.
5 Toh Tuck Link, Singapore 596224
USA office: 27 Warren Street, Suite 401-402, Hackensack, NJ 07601
UK office: 57 Shelton Street, Covent Garden, London WC2H 9HE

Library of Congress Cataloging-in-Publication Data
Names: Avery, John, 1933–
Title: Science and society / John Scales Avery, University of Copenhagen, Denmark.
Description: New Jersey : World Scientific, 2016. | Includes index.
Identifiers: LCCN 2016026718| ISBN 9789813147706 (hardcover : alk. paper) |
ISBN 9789813147713 (pbk. : alk. paper)
Subjects: LCSH: Science--Social aspects--History. | Science and civilization--History.
Classification: LCC Q175.5 .A94 2016 | DDC 303.48/3--dc23
LC record available at https://lccn.loc.gov/2016026718

British Library Cataloguing-in-Publication Data
A catalogue record for this book is available from the British Library.

Contents

Preface

This book is a history of the social impact of science and technology, from earliest times up to the present. Looking at this epic story, we can see how the key inventions, agriculture, writing and printing with movable type, initiated an explosive growth of knowledge and human power over the environment. We also see how the Industrial Revolution changed the relationship between humans and nature, and initiated a massive use of fossil fuels. In the modern world, problems related to nuclear power, nuclear weapons, genetic engineering, information technology, exhaustion of non-renewable resources, use of fossil fuels and climate change are examined. Finally, the need for ethical maturity to match our scientific progress is discussed.

The history of science and technology and their social impact ought to be part of the education of both scientists and non-scientists. In a democratic society, the general public is often required to make decisions about how our constantly-increasing power over nature ought to be used. *Science and Society* was written to fulfill the need for a textbook in this field. An earlier and smaller version of the book was printed locally and used as a text for courses on the history and social impact of science, given in Denmark, Sweden, Switzerland and England. The present edition has been greatly enlarged and brought up to date.

Most of this book can be understood by readers without a scientific background. However, some parts, especially Chapter 17 (Genesplicing) and Chapter 18 (Artificial Intelligence), are more demanding. I apologize to general readers for the inclusion of advanced material, and I hope that they will just skip over any parts that give difficulty. At the same time, I hope that the advanced material that is included, for example in Chapters 17 and 18, will be useful to those readers who have scientific training.

I have tried to tell the story of the rise of our scientific civilization in a way that shows the hand-prints of individuals, and for this reason the book is not a comprehensive history but a chronological sequence of pictures of the accumulation of the treasures of knowledge that we possess today.

In many respects, our cultural evolution can be regarded as an enormous success. However, at the start of the 21st century, most thoughtful observers agree that civilization is entering a period of crisis. As all curves move exponentially upward, population, production, consumption, rates of scientific discovery, and so on, one can observe signs of increasing environmental stress, while the continued existence and spread of nuclear weapons threaten civilization with destruction. Thus, while the explosive growth of knowledge has brought many benefits, the problem of achieving a stable, peaceful and sustainable world remains serious, challenging and unsolved.

Our modern civilization has been built up by means of a worldwide exchange of ideas and inventions. It is built on the achievements of many ancient cultures. China, Japan, India, Mesopotamia, Egypt, Greece, the Islamic world, Christian Europe, and the Jewish intellectual traditions, all have contributed. Potatoes, corn, squash, vanilla, chocolate, chili peppers, and quinine are gifts from the American Indians.

The sharing of scientific and technological knowledge is essential to modern civilization. The great power of science is derived from an enormous concentration of attention and resources on the understanding of a tiny fragment of nature. It would make no sense to proceed in this way if knowledge were not permanent, and if it were not shared by the entire world.

Science is not competitive. It is cooperative. It is a great monument built by many thousands of hands, each adding a stone to the cairn. This is true not only of scientific knowledge but also of every aspect of our culture, history, art and literature, as well as the skills that produce everyday objects upon which our lives depend. Civilization is cooperative. It is not competitive.

Our cultural heritage is not only immensely valuable; it is also so great that no individual comprehends all of it. We are all specialists, who understand only a tiny fragment of the enormous edifice. The great, complex and fragile edifice of human civilization is far too precious to be risked in a thermonuclear war. It has been built by all humans, working together. And by working together, we must now ensure that it is handed on intact to our children and grandchildren.

Chapter 1

THE BEGINNINGS OF CIVILIZATION

Early ancestors of man

Almost three million years ago, manlike creatures lived on the shores of Lake Rudolf in Kenya. The skull of one of these early "homenoids" was found in 1972 by Richard E. Leakey. Pouring fine sand into the reconstructed skull, Dr. Leakey and his associates measured the brain capacity as 800 c.c. - considerably less than the modern brain volume of 1400 c.c., but still remarkably large considering the early date of the skull. Potassium-argon dating of the volcanic ashes in which the skull was found established its age as approximately 2.8 million years.

At the Oldavai Gorge in Tanzania, not far from Lake Rudolf, Louis and Mary Leakey (Richard Leakey's father and mother) discovered many remains of a somewhat more advanced homenoid which they called *Homo habilis*. Among these remains, which were shown to be 1.8 million years old, Louis and Mary Leakey found many chipped stones, probably representing tools and weapons used by *Homo habilis*. The discoveries of the Leakey family, as well as those of Raymond Dart and Robert Broom, indicate that the early evolution of the human race probably took place in Africa. The early ancestors of man seem to have been hunter-gatherers living in small bands on the East African grasslands.

Terra Amata

We catch another glimpse of early man at the Terra Amata site at Nice in Southern France, where 300,000 years ago, in a warm period between the Mindel and Riss glacial eras, a small tribe came every summer to spend a few weeks hunting and food-gathering on the shore of the Mediterranian.

The huts which these early people built on their brief summer visits to the beach are among the earliest man-made dwellings ever discovered. They were between 26 and 29 feet long, and were built in an oval shape out of leafy saplings leaned against a central ridge pole. The central ridge pole in each hut was supported by vertical tree trunks embedded in the sand. Around the oval perimeter of the huts were walls of large stones for protection against the wind, and inside the huts were hearths on which small fires were built. (This is almost the earliest use of fire known, although earlier hearths have been found in strata of the Mindel ice age at Verteszölos in Hungary.)

Water for the camp came from a nearby spring. The level of the Mediterranian Sea was then 85 feet higher than it is today. It covered most of the plane of Nice, and near the camp it had cut a small cove with a sandy pebble-strewn beach into the western slope of Mount Boron. On the slopes of the mountain grew heather, sea pine, Aleppo pine and holm oak. A human footprint nine and one-half inches long is preserved in the sand of the ancient dune. Evidence shows that these summer visitors of 300,000 years ago spent their time gathering shellfish, hunting and making tools. Among the animals which they hunted were stag, an extinct elephant, wild boar, ibex, rhinoceros and wild ox. They stayed at Terra Amata only a few weeks each year, and then continued their travels, following the migrations of the animals which they hunted.

The Soultrian and Magdalenian cultures

In the caves of Spain and Southern France, not far from the Terra Amata site, are the remains of vigorous hunting cultures which flourished at a much later period, between 30,000 and 10,000 years ago. The people of these upper paleolithic cultures lived on the abundant cold-weather game which roamed the southern edge of the ice sheets during the Wurm glacial period: huge herds of reindeer, horses and wild cattle, as well as mammoths and wooly rhinos. The paintings found in the Dordogne region of France, for example, combine decorative and representational elements in a manner which contemporary artists might envy. Sometimes among the paintings are stylized symbols which can be thought of as the first steps towards writing.

In this period, not only painting, but also tool-making and weapon-making were highly-developed arts. For example, the Soultrian culture, which flourished in Spain and southern France about 20,000 years ago, produced beautifully worked stone lance points in the shape of laurel leaves and willow leaves. The appeal of these exquisitely pressure-flaked blades must have been aesthetic as well as functional. The people of the Soultrian culture had fine bone needles with eyes, bone and ivory pendants, beads and bracelets, and long bone pins with notches for arranging the hair. They also had red, yellow and black pigments for painting their bodies.

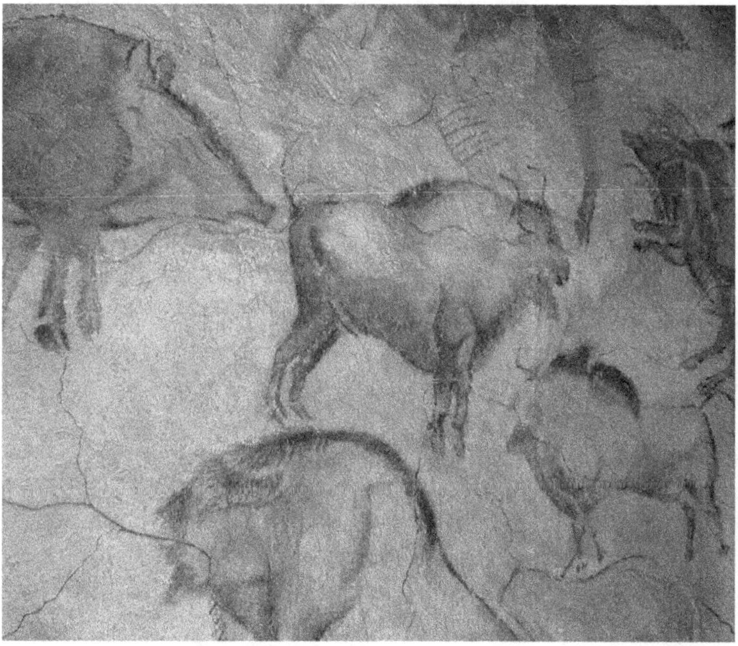

Fig. 1.1 *Cave painting from the cave of Altamira. Public domain, Wikimedia Commons*

The Soultrian culture lasted for 4,000 years. It ended in about 17,000 B.C. when it was succeeded by the Magdalenian culture. Whether the Soultrian people were conquered by another migrating group of hunters, or whether they themselves developed the Magdalenian culture we do not know.

The agricultural revolution

Beginning about 9,000 B.C., the way of life of the hunters was swept aside by a great cultural revolution: the invention of agriculture. Starting in western Asia, the neolithic agricultural revolution swept westward into Europe, and eastward into the regions which are now Iran and India.

By neolithic times, farming and stock breeding were well established in the Near East. Radio-carbon dating shows that by 8,500 B.C., people living in the caves of Shanidar in the foothills of the Zagros mountains in Iran had domesticated sheep. By 7,000 B.C., the village farming community at Jarmo in Iraq had domesticated goats, together with barley and two different kinds of wheat.

At Jerico, in the Dead Sea valley, excavations have revealed a prepottery neolithic settlement surrounded by an impressive stone wall, six feet wide and twelve feet high. Radio-carbon dating shows that the defenses of the town were built about 7,000 B.C.. Probably they represent the attempts of a settled agricultural people to defend themselves from the plundering raids of less advanced nomadic tribes.

By 4,300 B.C., the agricultural revolution had spread southwest to the Nile valley, where excavations along the shore of Lake Fayum have revealed the remains of grain bins and silos. The Nile carried farming and stock-breeding techniques slowly southward, and wherever they arrived, they swept away the hunting and food-gathering cultures. By 3,200 B.C. the agricultural revolution had reached the Hyrax Hill site in Kenya. At this point the southward movement of agriculture was stopped by the great swamps at the headwaters of the Nile. Meanwhile, the Mediterranian Sea and the Danube carried the revolution westward into Europe. Between 4,500 and 2,000 B.C. it spread across Europe as far as the British Isles and Scandanavia.

Mesopotamia; the invention of writing

In Mesopotamia (which in Greek means "between the rivers"), the settled agricultural people of the Tigris and Euphraties valleys evolved a form of writing. Among the earliest Mesopotamian writings are a set of clay tablets found at Tepe Yahya in southern Iran, the site of an ancient Elamite trading community halfway between Mesopotamia and India.

The Elamite trade supplied the Sumarian civilization of Mesopotamia with silver, copper, tin, lead, precious gems, horses, timber, obsidian, alabaster and soapstone. The practical Sumerians and Elamites probably invented writing as a means of keeping accounts.

Fig. 1.2 *Cuneiform tablet of the old Babylonain period (c. 1800 B.C.) with its container printed with cylinder seals. Uploaded by Einsamer Schütz, [CC BY-SA 3.0], Wikimedia Commons*

The tablets found at Tepe Yahya are inscribed in proto-Elamite, and radio-carbon dating of organic remains associated with the tablets shows them to be from about 3,600 B.C.. The inscriptions on these tablets were made by pressing the blunt and sharp ends of a stylus into soft clay. Similar tablets have been found at the Sumarian city of Susa at the head of the Tigris River.

In about 3,100 B.C. the cuneiform script was developed, and later Mesopotamian tablets are written in cuneiform, which is a phonetic script where the symbols stand for syllables.

Mesopotamian science

In the imagination of the Mesopotamians (the Sumerians, Elamites, Babylonians and Assyrians), the earth was a flat disc, surrounded by a rim of mountains and floating on an ocean of sweet water. Resting on these mountains was the hemispherical vault of the sky, across which moved the stars, the planets, the sun and the moon. Under the earth was another hemisphere containing the spirits of the dead. The Mesopotamians visualized the whole spherical world-universe as being immersed like a bubble in a limitless ocean of salt water.

By contrast with their somewhat primitive cosmology, both the mathematics and astronomy of the Mesopotamians were startlingly advanced. Their number system was positional, like ours, and was based on six and sixty. We can still see traces of it in our present method of measuring angles in degrees and minutes, and also in our method of measuring time in hours, minutes and seconds.

The Mesopotamians were acquainted with square roots and cube roots, and they could solve quadratic equations. They also were aware of exponential and logarithmic relationships. They seemed to value mathematics for its own sake, for the sake of enjoyment and recreation, as much as for its practical applications. On the whole, their algebra was more advanced than their geometry. They knew some of the properties of triangles and circles, but did not prove them in a systematic way.

Although the astronomy of the Mesopotamians was motivated largely by their astrological superstitions, it was nevertheless amazingly precise. For example, in the beginning of the fourth century B.C., incredibly accurate tables of new moons, full moons and eclipses were drawn up by Nabu-rimani; and about 375 B.C. Kidinnu, the greatest of the Babylonian astronomers, gave the exact duration of the solar year with an accuracy of only 4 minutes and 32.65 seconds. (This figure was found by observing the accumulated error in the calendar over a long period of time.) The error made by Kidinnu in his estimation of the motion of the sun from the node was smaller than the error made by the modern astronomer Oppolzer in 1887.

In medicine, the Mesopotamians believed that disease was a punishment inflicted by the gods on men, both for their crimes and for their errors and omissions in the performance of religious duties. They believed that the cure for disease involved magical and religious treatment, and the diseased person was thought to be morally tainted. However, in spite of this background of superstition, Mesopotamian medicine also contained some practical remedies. For example, the prescription for urinary retention was as follows: "Crush poppy seeds in beer and make the patient drink it. Grind some myrrh, mix it with oil and blow it into his urethra with a tube of bronze. Give the patient anemone crushed in alppanu-beer."

Until recently it was believed that the Mesopotamians had no idea of hygiene and preventive medicine. However, the following remarkable text was published recently. It is a letter, written by Zimri-Lim, King of Mari, who lived about 1780 B.C., to his wife Shibtu: "I have heard that Lady Nanname has been taken ill. She has many contacts with the people of the palace. She meets many ladies in her house. Now then, give severe orders that no one should drink in the cup where she drinks. No on should sit on the seat where she sits. No one should sleep in the bed where she sleeps. She should no longer meet many ladies in her house. This disease is contagious."

We can guess that the Mesopotamians were aware of some of the laws of physics, since they were able to lift huge stones and to construct long aquaducts. Also to their civilization must be credited a great cultural advance: the invention of the wheel. This great invention, which eluded the civilizations of the western hemisphere, was made in Mesopotamia in about 3600 B.C..

The early Hebrew culture was closely related to that of the Mesopotamian region, and a vivid picture of the period which we have been describing can be obtained by reading the Old Testament.

It may seem surprising that so many of the early steps in the cultural evolution of mankind were taken in a region much of which is now an almost uninhabitable desert. However, we should remember that in those days the climate of the Near East was very different - very much wetter and cooler than it is now. Even today, the process of drying up after the last ice age is not yet complete, and every year the Sahara extends further southward.

Early metallurgy in Asia Minor

Whatever the ancient civilizations of the Near East knew about chemistry and metallurgy, they probably learned as "spin-off" from their pottery industry. In the paleolithic and neolithic phases of their culture, like people everywhere in the world, they found lumps of native gold, native copper and meteoric iron, which they hammered into necklaces, bracelets, rings, implements and weapons. In the course of time, however, after settled communities had been established in the Near East for several thousand years, it became much more rare to find a nugget of gold or metallic copper.

Although the exact date and place are uncertain, it is likely that the first true metallurgy, the production of metallic copper from copper oxide and copper carbonate ores, began about 3,500 B.C. in a region of eastern Anatolia rich in deposits of these ores. It is very probable that the discovery was made because colored stones were sometimes used to decorate pottery. When stones consisting of copper oxide or copper carbonate are heated to the very high temperatures of a stone-ware pottery kiln in a reducing atmosphere, metallic copper is produced.

Imagine a potter who has made this discovery - who has found that he can produce a very rare and valuable metal from an abundant colored stone: He will abandon pottery and go into full-scale production as a metallurgist. He will try all sorts of other colored stones to see what he can make from them. He will also try to keep his methods secret, exaggerating their miraculous character, and he will try to keep a monopoly on the process. Such was probably the beginning of metallurgy!

However, it is impossible to keep a good thing secret for long. Knowledge of smelting and refining copper spread eastward along the mountain chain to Khorassan and Bukhara, and from there southward to Baluchistan, whose mines supplied copper to the peoples of the Indus valley. Also, from Bukhara, metallurgy spread northeast through the Kizal Kum desert to the ancestors of the Shang tribe inhabiting the Yellow River valley in China.

By 3,000 B.C., Sumer, Egypt and Cyprus also had adopted metallurgy and had even discovered secret methods of their own. Egypt obtained its copper ores from mines in Sinai, while Sumer imported ore from Oman. The use of the Oman copper ores was fortunate for the Sumerians, because these ores contain as much as fourteen percent tin and two percent nickel, so that the metal produced by reducing them is natural bronze, whose properties are much more desirable than those of copper. The demand for bronze continued even after the Oman ores were exhausted, and eventually it was discovered that bronze could be produced artificially by adding tin and nickel to copper.

The Egyptian civilization

The prosperity of ancient Egypt was based partly on its rich agriculture, nourished by the Nile, and partly on gold. Egypt possessed by far the richest gold deposits of the Middle East. They extended the whole length of the eastern desert, where more than a hundred ancient mines have been found; and in the south, Nubia was particularly rich in gold. The astonishing treasure found in the tomb of Tutankhamen, who was certainly not the most powerful of the pharaohs, gives us a pale idea of what the tombs of greater rulers must have been like before they were plundered.

In the religion of ancient Egypt, the distinction between the gods and the pharaohs was never very clear. Living pharaohs were considered to be gods, and they traced their ancestry back to the sun-god, Ra. Since all of the pharaohs were thought to be gods, and since, before the unification of Egypt, there were very many local gods, the Egyptian religion was excessively complicated. A list of gods found in the tomb of Thuthmosis III enumerates no fewer than seven hundred and forty! The extreme conservatism of Egyptian art (which maintained a consistent style for several thousand years) derives from the religious function played by painting and sculpture.

The famous gods, Osiris, Isis, Horus and Set probably began their existence as real people, and their story, which we know both from hieroglyphic texts and from Pliney, depicts an actual historical event - the first unification of Egypt: Osiris, the good ruler of the lower Nile, was murdered and cut to pieces by his jealous brother Set; but the pieces of Osiris' body were collected by his faithful wife Isis, who performed the first mummification and thus made Osiris immortal. Then Horus, the son of Osiris and Isis, like an Egyptian Hamlet, avenged the murder of his father by tracking down his wicked uncle Set, who attempted to escape by turning into various animals. However, in the end Horus killed Set, and thus Horus became the ruler of all of Egypt, both the lower Nile and the upper Nile.

This first prehistoric unification of Egypt left such a strong impression on the national consciousness that when a later pharaoh named Menes reunified Egypt in 3,200 B.C., he did so in the name of Horus. Like the Mesopotamian story of the flood, and like the epics of Homer, the story of the unification of Egypt by Horus probably contains a core of historical fact, blended with imaginative poetry. At certain points in the story, the characters seem to be real historical people - for example, when Osiris is described as being "handsome, dark-skinned and taller than other men". At

Fig. 1.3 *Facsimile of a vignette from the Book of the Dead of Ani. The deceased Ani kneels before Osiris, judge of the dead. Behind Osiris stand his sisters Isis and Nephthys, and in front of him is a lotus on which stand the four sons of Horus. Public domain, Wikimedia Commons*

other times, imagination seems to predominate. For example, the goddess Nut, who was the mother of Osiris, was thought to be the sky, while her husband Geb was the earth. The long curved body of Nut was imagined to be arched over the world so that only the tips of her toes and fingers touched the earth, while the stars and moon moved across her belly. Meanwhile her husband Geb lay prostrate, with all the vegetation of the earth growing out of his back.

The idea of the resurrection and immortality of Osiris had a strong hold on the ancient Egyptian imagination. At first only the pharaohs were allowed to imitate Osiris and become immortal like him through a magical ceremony of mummification and entombment. As part of the ceremony, the following words were spoken: "Horus opens the mouth and eyes of the deceased, as he opened the mouth and eyes of his father. He walks! He speaks! He has become immortal! ... As Osiris lives, the king lives; as he does not die, the king does not die; as he does not perish, the king does not perish!" Later the policy became more democratic, and ordinary citizens were allowed mummification.

Fig. 1.4 *Old Egyptian hieroglyphic painting showing an early instance of a domesticated animal (cow being milked). Public domain, Wikimedia Commons*

Fig. 1.5 *A view of the well preserved and beautifully painted Tomb TT3 from Deir el-Medina on the West Bank of Luxor. This scene depicts the god Osiris with the Mountains of the West behind him. It belonged to Pashedu who served as an Ancient Egyptian artist and foreman at Deir el-Medina under pharaoh Seti I. Uploaded by BetacommandBot, [CC BY-SA 2.0], Wikimedia Commons*

Imhotep

The tradition of careful mummification and preservation of the pharaohs led to the most impressive and characteristic expression of Egyptian civilization: the construction of colossal stone temples, tombs and pyramids. Ordinary houses in Egypt were made of brick, but since the tombs, in theory, had to last forever, they could not use brick or even the finest imported ceder wood. They had to be made entirely of stone.

The advanced use of stone in architecture began quite suddenly during the reign of Zoser in the Third Dynasty, in about 2,950 B.C.. During the Second Dynasty, a few tentative and crude attempts had been made to use stone in building, but these can hardly be thought of as leading to the revolutionary breakthrough in technique which can be seen in the great step pyramid of Zoser, surrounded by an amazing series of stone temples, and enclosed by a wall 33 feet high and nearly a mile long.

It is tempting to believe that this sudden leap forward in architectural technique was due to the genius of a single man, the first scientist whose name we know: Imhotep. The ancient Egyptians certainly believed that the whole technique of cutting and laying massive blocks of stone was invented entirely by Imhotep, and they raised him to the status of a god. Besides being King Zoser's chief architect, Imhotep was also a physician credited with miraculous cures. After his deification, he became the god of medicine, and his tomb became a place of pilgrimage for sick people seeking to be cured, more or less in the manner of Lourdes.

The craftsmanship of the pyramid builders has never been surpassed in any country. No scholar has been able to explain fully the methods by which they were able to fit enormous blocks of stone together with such astonishing accuracy. However, it is known that their method of quarrying was as follows: Along the line where a limestone block was to be split away from a cliff, a V-shaped groove was cut with copper tools. Along the bottom of the groove, wedge-holes were drilled, and wooden wedges were hammered into the holes. The wedges were soaked in water, and the force of expansion split the block away from the cliff face. Obviously, this is a slow and laborious method of quarrying, and therefore from the standpoint of economy it was better to cut one huge block rather than a hundred small ones. Also, from the standpoint of achieving enormous size and permanence in the finished structure, large blocks were by far the best.

In building the great pyramid of Cheops (c. 2,600 B.C.), on which 100,000 men were said to have worked 30 years, 2,300,000 blocks were used.

The average weight of the stones was two and one half tons, but many of them weighed as much as fifteen tons, and the enormous slabs of granite which form the roof of the king's chamber weigh almost fifty tons apiece.

Fig. 1.6 *Statuette of Imhotep, chancellor to the pharaoh, priest of Ra and architect. Uploaded by Hu Totya, [CC BY-SA 3.0], Wikimedia Commons*

The blocks were dragged from the quarries on sleds pulled with ropes by teams of men. On the front of each sled stood a man, pouring water in front of the runners, so that the clay on which they slid would be made slippery. Also standing on the sled, was a foreman who clapped his hands rhythmically to coordinate the movements of the workmen. His clapping was amplified by a second foreman, who banged two blocks of wood together in the same rhythm.

Hieroglyphic writing

The Egyptian hieroglyphic (priest writing) system began its development in about 4,000 B.C.. At that time, it was pictorial rather than phonetic. However, the Egyptians were in contact with the Sumerian civilization of Mesopotamia, and when the Sumerians developed a phonetic system of writing in about 3,100 B.C., the Egyptians were quick to adopt the idea. In the cuneiform writing of the Sumerians, a character stood for a syllable. In the Egyptian adaptation of this idea, most of the symbols stood for combinations of two consonants, and there were no symbols for vowels. However, a few symbols were purely alphabetic, i.e. they stood for sounds which we would now represent by a single letter. This was important from the standpoint of cultural history, since it suggested to the Phoenicians the idea of an alphabet of the modern type.

In Sumer, the pictorial quality of the symbols was lost at a very early stage, so that in the cuneiform script the symbols are completely abstract. By contrast, the Egyptian system of writing was designed to decorate monuments and to be impressive even to an illiterate viewer; and this purpose was best served by retaining the elaborate pictographic form of the symbols.

The invention of paper

The ancient Egyptians were the first to make books. As early as 4,000 B.C., they began to make books in the form of scrolls by cutting papyrus reeds into thin strips and pasting them into sheets of double thickness. The sheets were glued together end to end, so that they formed a long roll. The rolls were sometimes very long indeed. For example, one roll, which is now in the British Museum, is 17 inches wide and 135 feet long.

(Paper of the type which we use today was not invented until 105 A.D.. This enormously important invention was made by a Chinese eunuch named

Tsai Lun. The kind of paper invented by Tsai Lun could be made from many things: for example, bark, wood, hemp, rags, etc.. The starting material was made into a pulp, mixed together with water and binder, spread out on a cloth to partially dry, and finally heated and pressed into thin sheets. The art of paper-making spread slowly westward from China, reaching Baghdad in 800 A.D.. It was brought to Europe by the crusaders returning from the Middle East. Thus paper reached Europe just in time to join with Gütenberg's printing press to form the basis for the information explosion which has had such a decisive effect on human history.)

The flooding of the Nile

The date of the flooding of the Nile was predicted each year by the priests, so that the farmers could move their families and possessions in time. The Egyptian calendar contained 365 days, 360 of which were ordinary days and five of which were holidays on which the birthdays of the principal gods were celebrated. The 360 ordinary days of the calendar were divided into 36 weeks of ten days. Three weeks formed a month, so that the year consisted of twelve months, each with approximately the same number of days as the moon's period. However, the exact number of days in a year is not 365 but 365.2422..., and therefore the Egyptian calendar gradually got out of phase. The priests then found that the most invariant method of predicting the flooding of the Nile was by observing the return of the star Sirius.

The periodic flooding of the Nile meant that each year the land had to be surveyed and boundary lines redrawn. Thus the flooding of the Nile, with its surveying problems, together with the engineering problems of pyramid building, led the Egyptians to develop the science of geometry (which in Greek means "earth measurement").

An ancient Egyptian papyrus book on mathematics was found in the nineteenth century and is now in the British Museum. It was copied by the scribe Ahmose in c. 1,650 B.C., but the mathematical knowledge which it contains is probably much older. The papyrus is entitled *"Directions for Attaining Knowledge of All Dark Things"*, and it deals with simple equations, fractions, and methods for calculating areas, volumes, etc..

The Egyptians knew, for example, that a triangle whose sides are three units, four units, and five units long is a right triangle. They knew many special right triangles of this kind, and they knew that in these special cases the sum of the areas of the squares formed on the two short sides is equal to the area of the square formed on the longest side. However, there is no evidence that they knew that the relationship holds for every right triangle. It was left to Pythagoras to discover and prove this great theorem in its full generality.

Suggestions for further reading

(1) Jaquetta Hawkes and Sir Leonard Wooley, *Prehistory and the Beginnings of Civilization*, George Allen and Unwin (1966).
(2) Luigi Pareti, Paolo Brezzi and Luciano Petech, *The Ancient World*, George Allen and Unwin Ltd., London (1996).
(3) James and Janet MacLean Todd, *Peoples of the Past*, Arrow Books Ltd., London (1963).
(4) Georges Roux, *Ancient Iraq*, Penguin Books Ltd. (1966).
(5) R. Ghirshman, *Iran*, Penguin Books Ltd. (1954).
(6) Francesco Abbate, *Egyptian Art*, Octopus Books, London (1972).

Chapter 2

ANCIENT GREECE

The Minoans

Histories of the development of western civilization usually begin with the Greeks, but it is important to remember that the Greek culture was based on the much earlier civilizations of Mesopotamia and Egypt. The cultural achievements of these very early civilizations were transmitted to the Greeks in part through direct contact, and in part through the Minoan and Mycenaean civilizations.

The Minoan civilization on Crete is the civilization which is familiar to us through the legends of Thesius, the Minotaur and the Labyrinth, and the legend of Daedalus and Icarus. Apart from the Greek legends, whose truth was doubted, nothing was known about the Minoan civilization until 1900. In that year, the English archaeologist, Sir Arthur Evans, began to dig in a large mound at Knossos on Crete. What he uncovered was a palace of great beauty which, to his astonishment, seemed once to have boasted such conveniences as hot and cold running water and doors with metal locks and keys. Sir Arthur Evans considered this to represent the palace of the legendary King Minos.

The Minoan civilization seems to have been based not on agriculture, but on manufacture and on control of the Mediterranian sea trade. It flourished between 2,600 B.C. and 1,400 B.C.. In that year, the palace at Knossos was destroyed, and there is evidence of scattered looting. Other evidence shows that in about 1,400 B.C., a nearby island called Theria exploded in a volcanic eruption of tremendous violence; and probably this explosion, combined with an invasion of Mycenaeans, caused the end of the Minoan civilization. The palace at Knossos was inhabited later than 1,400 B.C., but the later people spoke Greek.

The Minoan civilization, as shown in the graceful works of art found at Knossos, seems to have been light-hearted and happy. The palace at Knossos was not fortified and was apparently protected by sea power. Women's dresses on ancient Crete looked a bit like the dresses which were popular in Europe during the 1900's, except that they left the breasts bare. Some of the wall paintings at Knossos show dances and bull-fights. In the bull-fights, the bull was not killed. The bull-fighter was an acrobat, often a girl, who seized the lowered horns of the charging bull and was tossed in a somersault over its back.

The Mycenaean civilization

The Mycenaean civilization developed at Troy, Mycenae (the home of the legendary Agamemnon), and other sites around the Aegean Sea. It is the civilization familiar to us through the stories of Ulysses, Priam, Ajax, Agamemnon, Paris and Helen. Like the Minoan civilization, the Mycenaean culture was thought to be purely legendary until quite recent times. We now know that the Homeric epics have apdflatex scibk basis in fact, and this surprising revelation is mainly due to the work of a brilliant businessman-turned-archaeologist named Heinrich Schliemann.

As a young (and poor) boy, Schliemann was inspired by reading Homer's Iliad, and he decided that when he grew up he would find the site of ancient Troy, which most people considered to be a figment of Homer's imagination. To do this, he first had to become very rich, a task which he accomplished during the first 45 years of his life.

At last he had accumulated a huge fortune, and he could follow the dream of his boyhood. Arriving in Greece, Schliemann put an advertisement into a newspaper describing himself and saying that he needed a wife. This was answered by a beautiful and intelligent Greek girl, whom he promptly married.

Aided by armies of excavators, his beautiful wife, his brilliant intellect and a copy of Homer, Schliemann actually succeeded in unearthing ancient Troy at a site in Asia Minor! At this site, he uncovered not one, but nine ancient cities, each built on the ruins of the last. He also found beneath the walls of Troy a treasure containing 8,750 pieces of gold jewelry, which he considered to be King Priam's treasure. He went on to uncover many other remains of the Mycenaean civilization at sites around the Aegean.

Fig. 2.1 *A vase depicting a scene from the Trojan wars. Upper frieze: the marriage of Helen and Paris; sirens under the handles facing toward the front of the vessel. Lower frieze with animals: goats and panthers. Source: Rosemania Flickr, [CC BY 2.0], Wikimedia Commons*

Schliemann's discoveries show the Mycenaeans to have been both technically and artistically accomplished. They spoke an Indo-European language (a form of Greek), and they were thus linguistically related to the tribes which conquered Persia, India and Europe.

The Mycenaean civilization lasted until about 1,075 B.C.. Between that date and 850 B.C., the Greek-speaking peoples of the Aegean entered a dark age. Probably the civilized Mycenaeans were conquered by fresh waves of semi-primitive Greek-speaking tribes from the north.

It is known that the Greeks arrived in the Aegean region in three waves. The first to come were the Ionians. Next came the Achaeans, and finally the Dorians. Warfare between the Achaeans and the Ionians weakened both groups, and finally they both were conquered by the Dorians. This conquest by the semi-primitive Dorians was probably the event which brought the Mycenaean civilization to an end. At any rate, during the dark ages between 1,075 B.C. and 850 B.C., the art of writing was lost to the Greeks, and the level of artistic and cultural achievement deteriorated.

Thales of Miletus

Beginning in about 850 B.C., there was a rebirth of Greek culture. This cultural renaissance began in Ionia on the west coast of present-day Turkey, where the Greeks were in close contact with the Babylonian civilization. Probably the Homeric epics were written in Miletus, a city on the coast of Asia Minor, in about 700 B.C.. The first three philosophers of the Greek world, Thales, Anaximander and Anaximenes, were also natives of Miletus.

Thales was born in 624 B.C. and died in 546 B.C.. The later Greeks considered him to have been the founder of almost every branch of knowledge. Whenever the wise men of ancient times were listed, Thales was invariably mentioned first. However, most of the achievements for which the Greeks admired Thales were probably not invented by him. He is supposed to have been born of a Phoenecian mother, and to have travelled extensively in Egypt and Babylonia, and he probably picked up most of his knowledge of science from these ancient civilizations.

One of the achievements which made Thales famous was his prediction of a solar eclipse which (according to modern astronomers) occurred on May 28, 585 B.C.. On the day of the eclipse, the Medes and the Lydians were about to begin a battle, but the eclipse convinced them that they ought instead to make peace and return home. Thales predicted, not the exact day, but only the year in which the eclipse would occur, but nevertheless the Greeks were impressed. The astronomical knowledge which allowed him to make this prediction was undoubtedly learned from the Babylonians, who had developed a system for the accurate prediction of lunar eclipses two centuries earlier.

Thales brought Egyptian geometry to Greece, and he also made some original contributions to this field. He changed geometry from a set of *ad hoc* rules into an abstract and deductive science. He was the first to think of geometry as dealing not with real lines of finite thickness and imperfect straightness, but with lines of infinitessimal thickness and perfect straightness. (Echoes of this point of view are found in Plato's philosophy).

Thales speculated on the composition of matter, and decided that the fundamental element is water. He thought this because animals can live by eating plants, and plants (Thales mistakenly believed) can live on water without any other nourishment.

Many stories are told about Thales. For example, Aristotle says that someone asked Thales, "If you're so wise, why aren't you rich?" Thales was offended by this question, and in order to prove a point, he quietly

Fig. 2.2 *Thales of Miletus. Public Domain, Wikimedia Commons*

bought up all the olive presses of the city during the winter of a year when his knowledge of weather told him that the olive harvest would be exceptionally large. When summer came, the harvest was enormous, and he was able to rent the presses at any price he liked to charge. He made himself rich in one season, and then went back to philosophy, having shown that philosophers could easily be rich if they liked, but they have higher ambitions than wealth.

Another story is told about Thales by Plato. According to Plato, Thales was so interested in some astronomical observations which he was making that he failed to look where he was going and fell into a well. He was helped out by a pretty and clever serving maid from Thrace who laughed at him because he was so interested in the stars that he could not see things that were right under his feet!

Thales had a student named Anaximander (610 B.C. - 546 B.C.) who also helped to bring Egyptian and Babylonian science to Greece. He imported the sundial from Egypt, and he was the first to try to draw a map of the entire world. He pictured the sky as a sphere, with the earth floating in space at its center. The sphere of the sky rotated once each day about an axis passing through the polar star. Anaximander knew that the surface of the earth is curved. He deduced this from the fact that as one travels northward, some stars disappear below the southern horizon, while others appear in the north. However, Anaximander thought that a north-south curvature was sufficient. He imagined the earth to be cylindrical rather than spherical in shape. The idea of a spherical earth had to wait for Pythagoras.

The third philosopher in the school of Militus was Anaximenes (570 B.C. - 500 B.C.), a pupil of Anaximander. He was the first of the Greeks to distinguish clearly between the planets and the stars. Like Thales, he speculated about the composition of matter, and he concluded that the fundamental element was air. This (he thought) could be compressed to form water, and still further compressed to form earth. Thus Anaximenes conceived in principle the modern idea of the three states of matter: gas, liquid and solid, which change into one another as the pressure and temperature are changed.

Pythagoras

Pythagoras, who lived from 582 B.C. to 497 B.C., is one of the most important and interesting figures in the history of European culture. It is hard to decide whether he was a religious leader or a scientist. Certainly, in order to describe him, one has to say a little about the religion of ancient Greece.

Besides the official religion, the worship of the Olympian gods, there were also other cults which existed simultaneously, and among these the worship of Bacchus or Dionysos was the most important. Bacchus, Dionysos and Bromios were all names of a many-named Thracian god who represented the forces of nature. The worshippers of Dionysos tried to return to nature, gaining release from the tensions generated by civilization by casting off all civilized constraints and returning temporarily to an animal-like state, reviving long-suppressed instincts. Often the worshippers were women, young girls and slaves, who gathered on the mountain slopes on certain evenings and began to dance. The dancing and drinking of wine continued throughout the whole night, becoming progressively wilder and more primitive.

Intoxicated by wine (the blood of Bacchus) and by the wild rhythm of the drums and pipes, the Bacchae would gradually reach a state of primitive frenzy in which they would tear living animals to bits and eat their raw flesh. By these acts, the Bacchae were re-enacting the legend of Dionysos. According to legend, Dionysos, the beautiful son of Zeus and Persephone, was torn to pieces by the Titans and eaten, all except for his heart, which was returned to Zeus. Dionysos was then reborn, and the Titans were killed by the thunderbolts of Zeus. From the ashes of the Titans mankind was created, and thus the human race contains not only the evil of the Titans, but also the divinity of Dionysos

The legend of Orpheus contains a parallel to the legend of Dionysos. In grief over his lost wife, Orpheus decides to give up sex forever, and this angers the women of Thrace. As Orpheus sings a last beautiful melody, the women of Thrace tear him to pieces, and his head, still singing, floats down the river Hebrus.

In Orphism, which was a reformed version of the cult of Dionysos, the idea of the simultaneously divine and evil nature of the human race is stressed. Followers of the Orphic religion believed that because of the element of evil and original sin in the human soul, it was doomed to a cycle of death and rebirth. However, the soul could be released from the cycle of reincarnation, and it could regain its divinity and immortality. The methods which the Orphists used to purge the soul included both Bacchic catharsis and asceticism. Also, Orphism included primitive tabus. For example, the followers of the cult were forbidden to eat beans, to touch a white cock, too stir the fire with an iron, to eat from a whole loaf, etc..

Pythagoras, who was a student of Anaximander, became a leader and reformer of the Orphic religion. He was born on the island of Samos, near the Asian mainland, and like other early Ionian philosophers, he is said to have travelled extensively in Egypt and Babylonia. In 529 B.C., he left Samos for Croton, a large Greek colony in southern Italy. When he arrived in Croton, his reputation had preceded him, and a great crowd of people came out of the city to meet him. After Pythagoras had spoken to this crowd, six hundred of them left their homes to join the Pythagorean brotherhood without even saying goodbye to their families.

For a period of about twenty years, the Pythagoreans gained political power in Croton, and they also had political influence in the other Greek colonies of the western Mediterranian. However, when Pythagoras was an old man, the brotherhood which he founded fell from power, their temples at Croton were burned, and Pythagoras himself moved to Metapontion, another Greek city in southern Italy.

Although it was never again politically influential, the Pythagorean brotherhood survived for more than a hundred years, and the ideas of the Pythagoreans became one of the foundations on which western civilization ultimately was built. Together with Thales, Pythagoras was the founder

Fig. 2.3 *Detail of Pythagoras with a tablet of ratios, numbers sacred to the Pythagoreans, from The School of Athens by Raphael. Vatican Palace, Rome, 1509. Public Domain, Wikimedia Commons*

of western philosophy; and the ideas of Pythagoras have an astonishing breadth and originality which is not found in Thales.

The Pythagorean brotherhood admitted women on equal terms, and all its members held their property in common. Even the scientific discoveries of the brotherhood were considered to have been made in common by all its members.

Pythagorean harmony

The Pythagoreans practiced medicine, and also a form of psychotherapy. According to Aristoxenius, a philosopher who studied under the Pythagoreans, "They used medicine to purge the body, and music to purge the soul". Music was of great importance to the Pythagoreans, as it was also to the original followers of Dionysos and Orpheus.

Both in music and in medicine, the concept of harmony was very important. Here Pythagoras made a remarkable discovery which united music and mathematics. He discovered that the harmonics which are pleasing to the human ear can be produced by dividing a lyre string into lengths which are expressible as simple ratios of whole numbers. For example, if we divide the string in half by clamping it at the center, (keeping the tension constant), the pitch of its note rises by an octave. If the length is reduced to 2/3 of the basic length, then the note is raised from the fundamental tone by the musical interval which we call a major fifth, and so on.

Having discovered that musical harmonics are governed by mathematics, Pythagoras fitted this discovery into the framework of Orphism. According to the Orphic religion, the soul may be reincarnated in a succession of bodies. In a similar way (according to Pythagoras), the "soul" of the music is the mathematical structure of its harmony, and the "body" through which it is expressed is the gross physical instrument. Just as the soul can be reincarnated in many bodies, the mathematical idea of the music can be expressed through many particular instruments; and just as the soul is immortal, the idea of the music exists eternally, although the instruments through which it is expressed may decay.

In distinguishing very clearly between mathematical ideas and their physical expression, Pythagoras was building on the earlier work of Thales, who thought of geometry as dealing with dimensionless points and lines of perfect straightness, rather than with real physical objects. The teachings of Pythagoras and his followers served in turn as an inspiration for Plato's idealistic philosophy.

Pythagoras also extended the idea of harmony to astronomy. He was the first person we know of who recognized that the earth is spherical in shape. He was also the first person to point out that the plane of the orbit of the moon is inclined with respect to the plane of the earth's equator, and the first Greek to recognize that the morning star (Phosphorus) and the evening star (Hesperus) are the same planet. After his time it was called Aphrodite by the Greeks, and later Venus by the Romans.

Pythagoras pointed out that the sun and the planets do not have the same apparent motion as the sphere of the stars. Each has its own motion. This led him to introduce into his cosmology an independently revolving sphere for each of the planets and for the sun. Pythagoras imagined these spheres to be concentric and transparent, and to revolve about the spherical earth.

The idea of spheres carrying the planets was developed further by later Greek astronomers, the greatest of whom was Hipparchus (190 B.C. - 120 B.C.), and it was incorporated into a famous book by Ptolemy (75 B.C. - 10 B.C.). After the fall of Rome, Ptolemy's book, the *Almagest*, survived in the highly civilized Arab world. It was translated into Latin in 1175 A.D., and it dominated astronomical thinking until the Renaissance. Thus the celestial spheres of Anaximander, Pythagoras, Hipparchus and Ptolemy had a long period of influence, and even some calculational usefulness, before they were replaced by the very much better sun-centered cosmology of Copernicus, Tycho Brahe, Kepler, Galileo and Newton.

Pythagoras searched for mathematical harmony in the motions of the planets. He thought that, just as the notes of the musical scale are connected by simple mathematical relationships, so the motions of the planets should obey a simple mathematical law. The Pythagoreans even imagined that as the celestial spheres turned, they produced a kind of cosmic music which only the most highly initiated could hear. The Pythagorean vision of mathematical harmony in the motion of the planets was laughed at by Aristotle, but in the end, after two thousand years, the dream was fulfilled in the laws Newton.

Having found mathematical harmony in the world of sound, and having searched for it in astronomy, Pythagoras tried to find mathematical relationships in the visual world. Among other things, he discovered the five possible regular polyhedra. However, his greatest contribution to geometry is the famous Pythagorean theorem, which is considered to be the most important single theorem in the whole of mathematics.

The Babylonians and the Egyptians knew that for many special right triangles, the sum of the squares formed on the two shorter sides is equal to the square formed on the long side. For example, Egyptian surveyors used a triangle with sides of lengths 3, 4 and 5 units. They knew that between the two shorter sides, a right angle is formed, and that for this particular right triangle, the sum of the squares of the two shorter sides is equal to the square of the longer side. Pythagoras proved that this relationship holds for every right triangle.

In exploring the consequences of his great theorem, Pythagoras and his followers discovered that the square root of 2 is an irrational number. (In other words, it cannot be expressed as the ratio of two integers.) The discovery of irrationals upset them so much that they abandoned algebra. They concentrated entirely on geometry, and for the next two thousand years geometrical ideas dominated science and philosophy.

The Pythagorean ideal

According to the Pythagoreans, the mind can be out of tune, just as a musical instrument can be out of tune. In medicine and psychiatry, they aimed at achieving harmony in the bodily organs and in the mind. When we speak of "muscle tone" or a "tonic" or "temperence", we are using words which have a Pythagorean origin. The word "philosophy", ("love of wisdom"), was also coined by the Pythagoreans.

In psychiatry, the Pythagoreans used various methods to free the mind from the tyrannical passions and tensions of the body. These methods were graded according to the degree of initiation of the patient. At the lowest level was the catharsis of a Bacchic orgy, followed by a long tranquilizing sleep and then an ascetic regimen to develop self-control. At the highest level of liberation, the mind was drawn away from preoccupation with self by the study of the eternal truths of nature as revealed by mathematics. According to Plutarch, "The function of geometry in Pythagorism is to draw us away from the world of the senses to the world of the intellect and the eternal".

The Orphic religion in some ways resembles the Buddhist and Hindu religions. It is not inconceivable that they have a common origin, since the Greeks were linguistically related to the Indo-European-speaking peoples who conquered India in the first millenium B.C.. In Buddhism, as in Orphism, one aims at release from the wheel of death and rebirth by mastery over self. However, the Pythagorean modification of Orphism introduces an element which is not found in Buddhism. In Pythagorism, the highest level of release and purification is achieved by contemplation of the structure of the universe; and the key to this structure is mathematics.

Pythagoras was the first person to maintain that mathematics is the key to the understanding of nature. In this belief he was completely correct. In the Pythagorean view of nature, mathematical harmony governs the fundamental laws of the universe. In the Pythagorean ethic, the highest vocation is that of the philosopher, and the aim of philosophy is to understand nature through the discovery of the mathematical relationships which govern the universe.

Much of what Pythagoras hoped to achieve in mathematics has been achieved today. For example, quantum theory has shown that the inner structure of an atom is governed by mathematical relationships closely analogous to those governing the harmonics of a lyre string. We have indeed found mathematical harmony in the fundamental laws of nature; but one can ask whether philosophy has brought harmony to human relations, as Pythagoras would have hoped!

We mentioned that the word "philosophy" was invented by the Pythagoreans. The word "theory" in its modern sense is also due to them. The word is derived from the Greek word "thea", meaning "spectacle", (as in the English word "theater"). In Greek, there is a related word, "theorio", meaning "to behold" or "to contemplate". In the Pythagorean ethic, contemplation held the highest place. The Pythagoreans believed that "The greatest purification of all is disinterested science; and it is the man who devotes himself to that, the true philosopher, who has most effectively released himself from the wheel of birth."

One of the Pythagorean mottos was: "A diagram and a step, not a diagram and a penny". Euclid, who belonged to the Pythagorean tradition, once rebuked a student who asked what profit could be gained from a knowledge of geometry. Euclid called a slave and said (pointing at the student): "He wants to profit from geometry. Give him a penny." The student was then dismissed from Euclid's school.

The Greeks of the classical age could afford to ignore practical matters, since their ordinary work was performed for them by slaves. It is unfortunate that the craftsmen and metallurgists of ancient Greece were slaves, while the philosophers were gentlemen who refused to get their hands dirty. An unbridgable social gap separated the philosophers from the craftsmen; and the empirical knowledge of chemistry and physics, which the craftsmen had gained over the centuries, was never incorporated into Greek philosophy.

The idealism of Pythagoras was further developed and exaggerated by Plato, the most famous student of the Pythagorean school. Plato considered the real world, as revealed by the senses, to be an imperfect expression of the world of ideas; and he thought that philosophers should not concern themselves with the real world.

The factors mentioned above prevented the classical Greeks from making use of observation and induction; and for this reason they were far better in mathematics than in other branches of science. In mathematics, one proceeds by pure deduction from a set of axioms. This insistence on pure deduction gives mathematics its great power and certainty; but in other branches of science, deduction alone is sterile. To be fruitful, deduction must be combined with observation and induction.

The Pythagorean preoccupation with harmony and with ideal proportion was reflected in Greek art. The classical Greeks felt that, just as harmony in music is governed by ideal ratios, so also harmony in architecture and in sculpture is governed by ideal proportions. All Greek temples of the classical period exhibited certain ratios which were considered to be ideal; and Greek sculpture showed, not real individuals, involved in emotions of the moment, but calm ideal figures.

Greek drama did not represent the peculiarities of particular individuals, but rather searched for universal truths concerning human nature. In classical Greek drama, one can even see a reflection of the deductive method which characterized Greek philosophy: In the beginning of a play, the characters are faced with a set of circumstances from which the action inevitably follows, just as the theorems of Euclid inevitably follow from his axioms.

The golden age of Athens

Between 478 B.C. and 431 B.C. Athens enjoyed a golden age. Their victory in the Persian war gave great prestige to Athens and Sparta, and these two cities became the leaders of the other Greek city states. Athens was the leader of the Delian league, while Sparta was the leader of the Peloponesian League. The Greek world was divided into two blocks, and although Athens and Sparta had been allies during the Persian war, they soon became political and commercial rivals.

Aided by her large navy, Athens pursued a very aggressive commercial policy aimed at monopolistic control of the Mediterranian sea trade. This brought great prosperity to Athens, but it also brought the Delian League into conflict with the Peloponesian League, a conflict which ultimately led to the downfall of Athens. However, during the period between 478 B.C. and 431 B.C., Athens enjoyed enormous prosperity. Refugees from the Ionian cities on the Asian mainland flocked to Athens, bringing with them their sophisticated culture. These refugees greatly enriched the cultural life of Athens, and their arrival marked the beginning of Athenian intellectual leadership.

The Athenians decided to use the surplus from the treasury of the Delian League to rebuild the Acropolis, which had been destroyed by the Persians. Pericles, the leader of Athens, put his friend, the sculptor Pheidias, in charge of the project. The new Acropolis was dominated by the Parthenon, which was built between 447 B.C. and 432 B.C.. Most of the sculptures of the Parthenon were brought to England in the nineteenth century by Lord Elgin, and they are now in the British Museum. The famous "Elgin marbles", together with the ruins of the Parthenon in Athens, symbolize the genius of the age of Pericles.

Wealthy, full of self-confidence, proud of their victory in the Persian war, and proud of their democratic constitution, the Athenians expressed the spirit of their age in sculpture, architecture, drama, poetry and philosophy which shine like beacons across the centuries.

Anaxagoras

One of the close friends of Pericles was the philosopher Anaxagoras (500 B.C. - 428 B.C.), who came to Athens from Ionia when he was 38 years old. This move by Anaxagoras was important, because it brought to Athens the philosophic tradition of the Ionian cities of Asia Minor. (In a similar way, a century earlier, Pythagoras had carried Ionian philosophy to the Greek colonies of the western Mediterranian.)

Fig. 2.4 *The Parthenon as it looks today. Public domain, Wikimedia Commons*

Anaxagoras was a rationalist and probably also an atheist (unlike the Pythagoreans). He believed that the stars and planets had been brought into existence by the same forces which formed the earth, and that the laws of nature are the same for celestial bodies as they are for objects on the earth. He thought that the sun and stars were molten rocks, and that the sun was about the same size as the Greek peninsula. (A large meteor which fell on Greece during the lifetime of Anaxagoras may have caused him to form this opinion).

Anaxagoras knew that the moon shines by reflected light, and that there are mountains on the moon. In fact, he believed that the moon is very much like the earth, and he thought that it might possibly be inhabited. He explained correctly the cause of both solar and lunar eclipses, and the phases of the moon.

Even the cultured Athenians found these views a bit too advanced. Anaxagoras was thrown into prison, accused (probably correctly) of atheism. The fact that he was a close friend of Pericles did not help him. The political enemies of Pericles, not daring to attack the great leader directly, chose to embarrass him by attacking his friends.

Pericles used his eloquence to defend Anaxagoras, and he succeeded in having his friend released from prison. However, Anaxagoras felt that it was not safe to remain in Athens. In 434 B.C. he retired to the little town of Lampsacus on the Hellespont, where he spent the remainder of his life.

The atomists

In the 5th century B.C. there was a great deal of discussion among the Greek philosophers about whether there is anything permanent in the universe. Heraclitus (540 B.C. - 475 B.C.) maintained that *everything* is in a state of flux. Parmenides (540 B.C. - c. 470 B.C.) maintained that on the contrary *nothing* changes - that all change is illusory. Leucippus (490 B.C. - c. 420 B.C.) and his student Democritus (470 B.C. - c. 380 B.C.), by a lucky chance, hit on what a modern scientist would regard as very nearly the correct answer.

According to Democritus, if we cut an apple in half, and then cut the half into parts, and keep on in this way for long enough, we will eventually come down to pieces which cannot be further subdivided. Democritus called these ultimate building blocks of matter "atoms", which means "indivisible". He visualized the spaces between the atoms as being empty, and he thought that when a knife cuts an apple, the sharp edge of the blade fits into the empty spaces between the atoms and forces them apart.

Democritus believed that each atom is unchanged in the processes which we observe with our senses, where matter seems to change its form. However, he believed that the atoms are in a state of constant motion, and that they can combine with each other in various ways, thus producing the physical and chemical changes which we observe in nature. In other words, each atom is in itself eternal, but the way in which the atoms combine with each other is in a state of constant flux because of the motion of the atoms.

This is very nearly the same answer which we would give today to the question of which things in the universe are permanent and which change. Of course, the objects which we call "atoms" *can* be further subdivided, but if Democritus were living today he would say that we have merely made the mistake of calling the wrong things "atoms". We should really apply the word to fundamental particles such as quarks, which cannot be further subdivided.

In discussing which things in the universe are permanent and which change, we would also add, from our modern point of view, that the fundamental laws of the universe are permanent. In following these unchanging laws, matter and energy constantly alter their configuration, but the basic laws of nature remain invariant. For example, the configuration of the planets changes constantly, but these constant changes are governed by Newton's laws of motion, which are eternal.

Of the various ancient philosophers, Democritus is the one who comes closest to our modern viewpoint. However, the ideas of Democritus, like those of Anaxagoras, were too advanced for his contemporaries. Although Democritus was not actually thrown into prison for his beliefs, they aroused considerable hostility. According to Diogenes Laertius, Plato dislike the ideas of Democritus so much that he wished that all of his books could be burned. (Plato had his wish! None of the seventy-two books of Democritus has survived.) Aristotle also argued against atomism, and because of the enormous authority which was attached to Aristotle's opinions, atomism almost disappeared from western thought until the time of John Dalton (1766 - 1844).

That the ideas of Democritus did not disappear entirely was due to the influence of Epicurus (341 B.C. - 270 B.C.), who made mechanism and atomism the cornerstones of his philosophy. The Roman poet Lucretius (95 B.C. - 55 B.C.) expounded the philosophy of Epicurus in a long poem called *De Natura Rerum* (On the Nature of Things). During the middle ages, this poem disappeared completely, but in 1417, a single surviving manuscript was discovered. The poem was then published, using Gutenberg's newly-invented printing press, and it became extremely popular. Thus, the idea of atoms was not entirely lost, and after being revived by John Dalton, it became one of the cornerstones of modern science.

Hippocrates

The physician Hippocrates was born in about 460 B.C. on the island of Kos. According to tradition, he visited Egypt during the early part of his life. There he studied medicine, especially the medical works of Imhotep. He is also said to have studied under Democritus. Returning to the island of Kos, he founded the most rational school of medicine of the ancient world. He had many students, among whom were his sons and his sons-in law. During the later part of his life, he also taught and practiced in Thrace and Athens.

The medical school founded by Hippocrates was famous for its rationality and for its high ethical standard. The medical ethics of Hippocrates live on today in the oath taken by physicians. The rationality of Hippocrates is evident in all the writings of his school. For example, a book on epilepsy, called *The Sacred Disease*, contains the following passage:

Fig. 2.5 *Hippocrates. Uploaded by Fæ [CC BY 4.0], Wikimedia Commons*

"As for this disease called divine, surely it has its nature and causes, as have other diseases. It arises, like them, from things which enter and leave the body... Such things are divine or not - as you will, for the distinction matters not, and there is no need to make such a division anywhere in nature; for all alike are divine, or all are natural. All have their antecedent causes, which can be found by those who seek them."

More than fifty books of Hippocrates' school were collected in Alexandria in the 3rd century B.C.. All of them were attributed by the Alexandrians to Hippocrates himself, but undoubtedly many of the books were written by his students. The physicians of the school of Hippocrates believed that cleanliness and rest are important for a sick or wounded patient, and that the physician should interfere as little as possible with the natural healing processes of the body. The books of the school contain much careful observation of disease. Hippocrates and his school resisted the temptation to theorize without a basis of carefully observed facts, just as they also resisted the temptation to introduce supernatural causes into medicine.

Hippocrates is said to have died in his hundredth year. According to tradition, he was humane, observant, learned, orderly and calm, with a grave and thoughtful attitude, a complete mastery of his own passions and a profound sympathy for the sufferings of his patients. We feel his influence today, both as one of the great founders of rational medicine, and as a pioneer of observation and inductive reasoning in science.

The Sophists and Socrates

Since Athens was a democracy, the citizens often found themselves speaking at public meetings. Eloquence could be turned into influence, and the wealthy Athenians imported teachers to help them master the art of rhetoric. These teachers, called "Sophists" (literally "wisdomists"), besides teaching rhetoric, also taught a form of philosophy which denied the existence of absolute truth, absolute beauty and absolute justice. According to the Sophists, "man is the measure of all things", all truths are relative, "beauty is in the eye of the beholder", and justice is not divine or absolute but is a human institution.

Opposed to the Sophists was the philosopher Socrates, who believed passionately in the existence of the absolutes which the Sophists denied. According to Socrates, a beautiful object would be beautiful whether or not there were any humans to observe it. Socrates adopted from the Sophists a method of conducting arguments by asking questions which made people see for themselves the things which Socrates wanted them to see.

The Sophists talked about moral and political questions, rather than about the nature of the universe. Socrates was an opponent of the Sophists, but like them he also neglected the study of nature and concentrated on the moral and political problems of man, "the measure of all things".The Sophists, together with Socrates and his pupil Plato, exerted a great influence in causing a split between moral philosophy and natural philosophy.

The beginning of the end of classical Greek civilization came in 431 B.C., when Athens, pushing her aggressive commercial policy to an extreme, began to expel Corinthian merchants from markets around the Aegean. Corinth reacted by persuading the Peloponesian League to declare war on Athens. This was the beginning of a long war which ruined Greece.

Realizing that they could not resist the Spartan land forces, the Athenians abandoned the farmland outside their city, and took refuge inside the walls. The Athenians continued their prosperous foreign trade, and they fed their population with grain imported from the east. Ships bringing grain also brought the plague. A large part of the population of Athens died of the plague, including the city's great leader, Pericles. No leader of equal stature was found to replace him, and the democratic Athenian government degenerated into mob rule.

In 404 B.C., when the fleet of Athens was destroyed in a disastrous battle, the city surrendered to the Spartans. However, the Spartans remembered that without Athens, they would be unable to resist the Persian Empire. Therefore they did not destroy Athens totally, but were content to destroy the walls of Athens, reducing the city to the status of a satellite of Sparta.

Looking for scapegoats on whom to blame this disaster, the Athenian mobs seized Socrates (one of the few intellectuals who remained alive after the Peloponesian War), and they condemned him to death for failing to believe in the gods of the city.

For a short period, Sparta dominated the Greek world; but soon war broke out again, and the political scene degenerated into a chaos of wars between the city states.

Fig. 2.6 *Head of Socrates in Palazzo Massimo alle Terme (Rome). Uploaded by Livoandronico2013, [CC BY-SA 4.0], Wikimedia Commons*

Plato

Darkness was falling on the classical Greek world, but the light of civilization had not quite gone out. Socrates was dead, but Plato, the student of Socrates, kept his memory alive by writing dialogues in which Socrates appeared as a character.

Plato (427 B.C. - 317 B.C.) was an Athenian aristocrat, descended from the early kings of Athens. His real name was Aristocles, but he was called by his nickname, Platon (meaning "broad") because of his broad shoulders. After the death of Socrates, Plato left Athens, saying that the troubles of the city would never end until a philosopher became king. (He may have had himself in mind!) He travelled to Italy and studied under the Pythagoreans. In 387 he returned to Athens and founded a school, which was called the Academy because it stood on ground which had once belonged to a Greek named Academus.

Plato developed a philosophy which was based on the idealism of the Pythagoreans. In Pythagorean philosophy, a clear distinction was made between mathematical ideas and their physical expression. For example, geometry was considered to deal, not with real physical objects, but with idealized figures, constructed from lines of perfect straightness and infinite thinness. Plato developed and exaggerated the idealism of Pythagoras. In Plato's philosophy, the real world is corruptible and base, but the world of ideas is divine and eternal. A real table, for example, is an imperfect expression of the idea of a table. Therefore we ought to turn our eyes away from the real world and live in the world of ideas.

Plato's philosophy was just what the Athenians wanted! All around them, their world was crumbling. They gladly turned their backs on the unpleasantness of the real world, and accepted Plato's invitation to live in the world of ideas, where nothing decays and where the golden laws of mathematics rule eternally.

By all accounts, Plato was an excellent mathematician, and through his influence mathematics obtained a permanent place in education.

Fig. 2.7 *Plato and Aristotle by Raphael. Public domain, Wikimedia Commons*

Aristotle

Plato's favorite student was a young man from Macedon named Aristotle. Plato called him "the intelligence of the school". He was born in 381 B.C., the son of the court physician of the king of Macedon, and at the age of seventeen he went to Athens to study. He joined Plato's Academy and worked there for twenty years until Plato died. Aristotle then left the Academy, saying that he disapproved of the emphasis on mathematics and theory and the decline of natural science.

Aristotle traveled throughout the Greek world and married the sister of the ruler of one of the cities which he visited. In 312 B.C., Philip II, who had just become king of Macedon, sent for Aristotle and asked him to become the tutor of his fourteen-year-old son, Alexander. Aristotle accepted this post and continued in it for a number of years. During this period, the Macedonians, under Philip, conquered most of the Greek city-states. Philip then planned to lead a joint Macedonian and Greek force in an attack on the Persian Empire. However, in 336 B.C., before he could begin his invasion of Persia, he was murdered (probably by an agent of his wife, Olympia, who was jealous because Philip had taken a second wife). Alexander then succeeded to his father's throne, and, at the head of the Macedonian and Greek army, he invaded Persia.

Aristotle, no longer needed as a royal tutor, returned to Athens and founded a school of his own called the Lyceum. At the Lyceum he built up a collection of manuscripts which resembled the library of a modern university.

Aristotle was a very great organizer of knowledge, and his writings almost form a one-man encyclopedia. His best work was in biology, where he studied and classified more than five hundred animal species, many of which he also dissected. In Aristotle's classification of living things, he shows an awareness of the interrelatedness of species. This interrelatedness was later brought forward by Darwin as evidence for the theory of evolution. One cannot really say that Aristotle proposed a theory of evolution, but he was groping towards the idea. In his history of animals, he writes:

"Nature proceeds little by little from lifeless things to animal life, so that it is impossible to determine either the exact line of demarcation, or on which side of the line an intermediate form should lie. Thus, next after lifeless things in the upward scale comes the plant. Of plants, one will differ from another as to its apparent amount of vitality. In a word, the whole plant kingdom, whilst devoid of life as compared with the animal, is yet endowed with life as compared with other corporial entities. Indeed, there is observed in plants a continuous scale of ascent towards the animal."

Aristotle's classification of living things, starting at the bottom of the scale and going upward, is as follows: Inanimate matter, lower plants and sponges, higher plants, jellyfish, zoophytes and ascidians, molluscs, insects, jointed shellfish, octopuses and squids, fish and reptiles, whales, land mammals and man. The acuteness of Aristotle's observation and analysis can be seen from the fact that he classified whales and dolphins as mammals (where they belong) rather than as fish (where they superficially seem to belong).

One of Aristotle's important biological studies was his embryological investigation of the developing chick. Ever since his time, the chick has been the classical object for embryological studies. He also studied the four-chambered stomach of the ruminants and the detailed anatomy of the mammalian reproductive system. He used diagrams to illustrate complex anatomical relationships - an important innovation in teaching technique.

Aristotle's physics and astronomy were far less successful than his biology. In these fields, he did not contribute with his own observations. On the whole, he merely repeated the often-mistaken ideas of his teacher, Plato. In his book *On The Heavens*, Aristotle writes:

"As the ancients attributed heaven and the space above it to the gods, so our reasoning shows that it is incorruptable and uncreated and untouched by mortal troubles. No force is needed to keep the heaven moving, or to prevent it from moving in another manner. Nor need we suppose that its stability depends on its support by a certain giant, Atlas, as in the ancient fable; as though all bodies on high possessed gravity and an earthly nature. Not so has it been preserved for so long, nor yet, as Empedocles asserts, by whirling around faster than its natural motion downward."

Empedocles (490 B.C. - 430 B.C.) was a Pythagorean philosopher who studied, among other things, centrifugal forces. For example, he experimented with buckets of water which he whirled about his head, and he knew that the water does not run out. The passage which we have just quoted shows that Empedocles had suggested the correct explanation for the stability of the moon's orbit. The moon is constantly falling towards the earth, but at the same time it is moving rapidly in a direction perpendicular to the line connecting it with the earth. The combination of the two motions gives the moon's orbit its nearly-circular shape.

Empedocles had thus hit on the germ of the idea which Newton later developed into his great theory of universal gravitation and planetary motion. In the above passage, however, Aristotle rejects the hypothesis of Empedocles. He asserts instead that the heavens are essentially different from the earth, and not subject to the same laws.

Aristotle believed celestial bodies to be composed of a fifth element - ether. This, he thought, was why the heavens were not subject to the laws which apply to earthly matter. He thought that for earthly bodies, the natural motion was a straight line, but for celestial bodies the natural motion was circular because "one kind of motion is divine and immortal, having no end, but being in itself the end of other motions"; and motion in a circle is "perfect, having no beginning or end, nor ceasing in infinite time."

This doctrine, that the motion of celestial bodies must be uniform and circular, was a legacy from Plato. In fact, Plato had placed before his Academy the problem of reconciling the apparently irregular motion of the planets with the uniform circular motion which Plato believed they *had* to have. In a famous phrase, Plato said that the problem was to "save the appearances".

The problem of "saving the appearances" was solved in a certain approximation by Eudoxis, one of Plato's students. He imagined a system of concentric spheres, attached to one another by axes. In this picture, each sphere rotates uniformly about its own axis, but since the spheres are attached to each other in a complex way, the resulting motion duplicates the complex apparent motion of the planets.

Aristotle accepted the system of Eudoxis, and even added a few more spheres of his own to make the system more accurate. In making a distinction between the heavens and the earth, Aristotle gave still another answer to the question of which things in the universe change and which are permanent: According to Aristotle, the region beneath the sphere of the moon is corrupt and changeable, but above that sphere, everything is eternal and divine. Change is bad, permanence is good - that is the emotional content of the teaching of Plato and Aristotle, the two great philosophers of the rapidly-decaying 4th century B.C. Greek civilization.

Besides writing on biology, physics and astronomy, Aristotle also discussed ethics, politics and literary criticism, and he made a great contribution to western thought by inventing a formal theory of logic. His writings on logic were made popular by St. Thomas Aquinas (1225-1274), and during the period between Aquinas and the Renaissance, Aristotle's logic dominated theology and philosophy. In fact, through his work on logic, Aristotle became so important to scholastic philosophy that his opinions on other subjects were accepted as absolute authority. Unfortunately, Aristotle's magnificent work in biology was forgotten, and it was his misguided writings on physics and astronomy which were influential. Thus, for the experimental scientists of the 16th and 17th centuries, Aristotle eventually became the symbol of wrongness, and many of their struggles and victories have to do with the overthrow of Aristotle's doctrines.

Even after it had lost every vestige of political power, Athens continued to be a university town, like Oxford or Cambridge. Plato's Academy continued to teach students for almost a thousand years. It was finally closed in 529 A.D. by the Emperor Justinian, who feared its influence as a stronghold of "pagan philosophy".

Aristotle's Lyceum continued for some time as an active institution, but it soon declined, because although Athens remained a center of moral philosophy, the center of scientific activity had shifted to Alexandria. The collection of manuscripts which Aristotle had built up at the Lyceum became the nucleus of the great library at Alexandria.

The books of Plato and Aristotle survived better than the books of other ancient philosophers, perhaps because Plato and Aristotle founded schools. Plato's authenticated dialogues form a book as long as the Bible, covering all fields of knowledge. Aristotle's lectures were collected into 150 volumes. (Of course, each individual volume was not as long as a modern printed book.) Of these, 50 have survived. Some of them were found in a pit in Asia Minor by soldiers of the Roman general Sulla in 80 A.D., and they were brought to Rome to be recopied.

Some of the works of Aristotle were lost in the West, but survived during the dark ages in Arabic translations. In the 12th and 13th centuries, these works were translated into Latin by European scholars who were in contact with the Arab civilization. Through these translations, Europe enthusiastically rediscovered Aristotle, and until the 17th century, he replaced Plato as *the* philosopher.

The influence of Plato and Aristotle was very great (perhaps greater than they deserved), because of their literary skill, because so many of their books survived, because of the schools which they founded, and because Plato and Aristotle wrote about all of knowledge and wrapped it up so neatly that they seemed to have said the last word.

Suggestions for further reading

(1) Roger Ling, *The Greek World*, Elsevier-Phaidon, Oxford (1976).
(2) Bertrand Russell, *History of Western Philosophy*, George Allen and Unwin Ltd., London (1946).
(3) Michael Grant (editor), *Greek Literature*, Penguin Books Ltd. (1976).
(4) George Sarton, *History of Science*, Oxford University Press (1959).
(5) Morris Kline, *Mathematics in Western Culture*, Penguin Books Ltd. (1977).

(6) E.T. Bell, *Men of Mathematics*, Simon and Schuster, New York (1937).
(7) Isaac Asimov, *Asimov's Biographical Encyclopedia of Science and Technology*, Pan Books Ltd., London (1975).
(8) O. Neugebauer, *The Exact Sciences in Antiquity*, Harper and Brothers (1962).

Chapter 3

THE HELLENISTIC ERA

Alexander of Macedon

How much influence did Aristotle have on his pupil, Alexander of Macedon? We know that in 327 B.C. Alexander, (who was showing symptoms of megalomania), executed Aristotle's nephew, Callisthenes; so Aristotle's influence cannot have been very complete. On the other hand, we can think of Alexander driving his reluctant army beyond the Caspian Sea to Parthia, beyond Parthia to Bactria, beyond Bactria to the great wall of the Himalayas, and from there south to the Indus, where he turned back only because of the rebellion of his homesick officers. This attempt to reach the uttermost limits of the world seems to have been motivated as much by a lust for knowledge as by a lust for power.

Alexander was not a Greek, but nevertheless he regarded himself as an apostle of Greek culture. As the Athenian orator, Isocrates, remarked, "The word 'Greek' is not so much a term of birth as of mentality, and is applied to a common culture rather than to a common descent."

Although he was cruel and wildly temperamental, Alexander could also display an almost hypnotic charm, and this charm was a large factor in his success. He tried to please the people of the countries through which he passed by adopting some of their customs. He married two barbarian princesses, and, to the dismay of his Macedonian officers, he also adopted the crown and robes of a Persian monarch.

Wherever Alexander went, he founded Greek-style cities, many of which were named Alexandria. In Babylon, In 323 B.C., after a drunken orgy, Alexander caught a fever and died at the age of 33. His loosely-constructed empire immediately fell to pieces. The three largest pieces were seized by three of his generals. The Persian Empire went to Seleucis, and became

Fig. 3.1 *A depiction of Alexander of Macedon. Public domain, Wikimedia Commons*

known as the Seleucid Empire. Antigonius became king of Macedon and protector of the Greek city-states. A third general, Ptolemy, took Egypt.

Although Alexander's dream of a politically united world collapsed immediately after his death, his tour through almost the entire known world had the effect of blending the ancient cultures of Greece, Persia, India and Egypt, and producing a world culture. The era associated with this culture is usually called the Hellenistic Era (323 B.C. - 146 B.C.). Although the

Hellenistic culture was a mixture of all the great cultures of the ancient world, it had a decidedly Greek flavor, and during this period the language of educated people throughout the known world was Greek.

Alexandria

Nowhere was the cosmopolitan character of the Hellenistic Era more apparent than at Alexandria in Egypt. No city in history has ever boasted a greater variety of people. Ideally located at the crossroads of world trading routes, Alexandria became the capital of the world - not the political capital, but the cultural and intellectual capital.

Miletus in its prime had a population of 25,000; Athens in the age of Pericles had about 100,000 people; but Alexandria was the first city in history to reach a population of over a million!

Strangers arriving in Alexandria were impressed by the marvels of the city - machines which sprinkled holy water automatically when a five-drachma coin was inserted, water-driven organs, guns powered by compressed air, and even moving statues, powered by water or steam!

For scholars, the chief marvels of Alexandria were the great library and the Museum established by Ptolemy I. Credit for making Alexandria the intellectual capital of the world must go to Ptolemy I and his successors (all of whom were named Ptolemy except the last of the line, the famous queen, Cleopatra). Realizing the importance of the schools which had been founded by Pythagoras, Plato and Aristotle, Ptolemy I established a school at Alexandria. This school was called the Museum, because it was dedicated to the muses.

Near to the Museum, Ptolemy built a great library for the preservation of important manuscripts. The collection of manuscripts which Aristotle had built up at the Lyceum in Athens became the nucleus of this great library. The library at Alexandria was open to the general public, and at its height it was said to contain 750,000 volumes. Besides preserving important manuscripts, the library became a center for copying and distributing books.

The material which the Alexandrian scribes used for making books was papyrus, which was relatively inexpensive. The Ptolemys were anxious that Egypt should keep its near-monopoly on book production, and they refused to permit the export of papyrus. Pergamum, a rival Hellenistic city in Asia Minor, also boasted a library, second in size only to the great library at Alexandria. The scribes at Pergamum, unable to obtain papyrus from Egypt, tried to improve the preparation of the skins traditionally used for writing in Asia. The resulting material was called *membranum pergamentum*, and in English, this name has become "parchment".

Euclid

One of the first scholars to be called to the newly-established Museum was Euclid. He was born in 325 B.C. and was probably educated at Plato's Academy in Athens. While in Alexandria, Euclid wrote the most successful text-book of all time, the *Elements of Geometry*. The theorems in this splendid book were not, for the most part, originated by Euclid. They were the work of many generations of classical Greek geometers. Euclid's contribution was to take the theorems of the classical period and to arrange them in an order which is so logical and elegant that it almost defies improvement. One of Euclid's great merits is that he reduces the number of axioms to a minimum, and he does not conceal the doubiousness of certain axioms.

Euclid's axiom concerning parallel lines has an interesting history: This axiom states that "Through a given point not on a given line, one and only one line can be drawn parallel to a given line". At first, mathematicians doubted that it was necessary to have such an axiom. They suspected that it could be proved by means of Euclid's other more simple axioms. After much thought, however, they decided that the axiom is indeed one of the necessary foundations of classical geometry. They then began to wonder whether there could be another kind of geometry where the postulate concerning parallels is discarded. These ideas were developed in the 18th and 19th centuries by Lobachevski, Bolyai, Gauss and Riemann, and in the 20th century by Levi-Civita. In 1915, the mathematical theory of non-Euclidian geometry finally became the basis for Einstein's general theory of relativity.

Besides classical geometry, Euclid's book also contains some topics in number theory. For example, he discusses irrational numbers, and he proves that the number of primes is infinite. He also discusses geometrical optics.

Euclid's *Elements* has gone through more than 1,000 editions since the invention of printing - more than any other book, with the exception of the Bible. Its influence has been immense. For more than two thousand years, Euclid's *Elements of Geometry* has served as a model for rational thought.

Fig. 3.2 *Euclid, detail from "The School of Athens", a painting by Raphael. It is not proven that this is Euclid. Some references point this person out as Archimedes. Public domain, Wikimedia Commons*

Eratosthenes

Eratosthenes (276 B.C. - 196 B.C.), the director of the library at Alexandria, was probably the most cultured man of the Hellenistic Era. His interests and abilities were universal. He was an excellent historian, in fact the first historian who ever attempted to set up an accurate chronology of events. He was also a literary critic, and he wrote a treatise on Greek comedy. He made many contributions to mathematics, including a study of prime numbers and a method for generating primes called the "sieve of Eratosthenes".

As a geographer, Eratosthenes made a map of the world which, at that time, was the most accurate that had ever been made. The positions of various places on Eratosthenes' map were calculated from astronomical observations. The latitude was calculated by measuring the angle of the polar star above the horizon, while the longitude probably was calculated from the apparent local time of lunar eclipses.

As an astronomer, Eratosthenes made an extremely accurate measurement of the angle between the axis of the earth and the plane of the sun's apparent motion; and he also prepared a map of the sky which included the positions of 675 stars.

Eratosthenes' greatest achievement however, was an astonishingly precise measurement of the radius of the earth. The value which he gave for the radius was within 50 miles of what we now consider to be the correct value! To make this remarkable measurement, Eratosthenes of course assumed that the earth is spherical, and he also assumed that the sun is so far away from the earth that rays of light from the sun, falling on the earth, are almost parallel. He knew that directly south of Alexandria there was a city called Seyne, where at noon on a midsummer day, the sun stands straight overhead. Given these facts, all he had to do to find the radius of the earth was to measure the distance between Alexandria and Seyne. Then, at noon on a midsummer day, he measured the angle which the sun makes with the vertical at Alexandria. From these two values, he calculated the circumference of the earth to be a little over 25,000 miles. This was so much larger than the size of the known world that Eratosthenes concluded (correctly) that most of the earth's surface must be covered with water; and he stated that "If it were not for the vast extent of the Atlantic, one might sail from Spain to India along the same parallel."

Eratosthenes' friends (one of them was Archimedes) joked with him about his dilettantism. They claimed that he was spreading his talents too thinly, and they gave him the nickname, "Beta", meaning that in all the fields in which he chose to exert himself, Eratosthenes was the second best in the world, rather than the best. This was unjust: In geography, Eratosthenes was unquestionably "Alpha"!

Eratosthenes' brilliant work in geography illustrates a difference between classical Greek science and Hellenistic science. In the classical Greek world, philosophers were far removed from everyday affairs. However, in busy, commercial Alexandria, men like Eratosthenes were in close contact with practical problems, such as the problems of navigation, metallurgy and engineering. This close contact with practical problems gave Hellenistic science a healthy realism which was lacking in the overly-theoretical science of classical Greece.

Aristarchus

The Hellenistic astronomers not only measured the size of the earth - they also measured the sizes of the sun and the moon, and their distances from the earth. Among the astronomers who worked on this problem was Aristarchus (c. 320 B.C. - c. 250 B.C.). Like Pythagoras, he was born on the island of Samos, and he may have studied in Athens under Strato. However, he was soon drawn to Alexandria, where the most exciting scientific work of the time was being done.

Aristarchus calculated the size of the moon by noticing the shape of the shadow of the earth thrown on the face of the moon during a solar eclipse. From the shape of the earth's shadow, he concluded that the diameter of the moon is about a third the diameter of the earth. (This is approximately correct).

From the diameter of the moon and the angle between its opposite edges when it is seen from the earth, Aristarchus could calculate the distance of the moon from the earth. Next he compared the distance from the earth to the moon with the distance from the earth to the sun. To do this, he waited for a moment when the moon was exactly half-illuminated. Then the earth, moon and sun formed a right triangle, with the moon at the corner corresponding to the right angle. Aristarchus, standing on the earth, could measure the angle between the moon and the sun. He already knew the distance from the earth to the moon, so now he knew two angles and one side of the right triangle. This was enough to allow him to calculate the other sides, one of which was the sun-earth distance. His value for this distance was not very accurate, because small errors in measuring the angles were magnified in the calculation.

Aristarchus concluded that the sun is about twenty times as distant from the earth as the moon, whereas in fact it is about four hundred times as distant. Still, even the underestimated distance which Aristarchus found convinced him that the sun is enormous! He calculated that the sun has about seven times the diameter of the earth, and three hundred and fifty times the earth's volume. Actually, the sun's diameter is more than a hundred times the diameter of the earth, and its volume exceeds the earth's volume by a factor of more than a million!

Even his underestimated value for the size of the sun was enough to convince Aristarchus that the sun does not move around the earth. It seemed ridiculous to him to imagine the enormous sun circulating in an orbit around the tiny earth. Therefore he proposed a model of the solar system in which the earth and all the planets move in orbits around the sun, which remains motionless at the center; and he proposed the idea that the earth spins about its axis once every day.

Although it was the tremendous size of the sun which suggested this model to Aristarchus, he soon realized that the heliocentric model had many calculational advantages: For example, it made the occasional retrograde motion of certain planets much easier to explain. Unfortunately, he did not work out detailed table for predicting the positions of the planets. If he had done so, the advantages of the heliocentric model would have been so obvious that it might have been universally adopted almost two thousand years before the time of Copernicus, and the history of science might have been very different.

Aristarchus was not the first person to suggest that the earth moves in an orbit like the other planets. The Pythagorean philosophers, especially Philolaus (c. 480 B.C. - c. 420 B.C.), had also suggested a moving earth. However, the Pythagorean model of the solar system was marred by errors, while the model proposed by Aristarchus was right in every detail.

Aristarchus was completely right, but being right does not always lead to popularity. His views were not accepted by the majority of astronomers, and he was accused of impiety by the philosopher Cleanthes, who urged the authorities to make Aristarchus suffer for his heresy. Fortunately, the age was tolerant and enlightened, and Aristarchus was never brought to trial.

The model of the solar system on which the Hellenistic astronomers finally agreed was not that of Aristarchus but an alternative (and inferior) model developed by Hipparchus (c. 190 B.C. - c. 120 B.C.). Hipparchus made many great contributions to astronomy and mathematics. For example, he was the first person to calculate and publish tables of trigonometric functions. He also invented many instruments for accurate naked-eye observations. He discovered the "precession of equinoxes", introduced a classification of stars according to their apparent brightness, and made a star-map which far outclassed the earlier star-map of Eratosthenes. Finally, he introduced a model of the solar system which allowed fairly accurate calculation of the future positions of the planets, the sun and the moon.

In English, we use the phrase "wheels within wheels" to describe something excessively complicated. This phrase is derived from the model of the solar system introduced by Hipparchus! In his system, each planet has a large wheel which revolves with uniform speed about the earth (or in some cases, about a point near to the earth). Into this large wheel was set a smaller wheel, called the "epicycle", which also revolved with uniform speed. A point on the smaller wheel was then supposed to duplicate the motion of the planet. In some cases, the model of Hipparchus needed still more "wheels within wheels" to duplicate the planet's motion. The velocities and sizes of the wheels were chosen in such a way as to "save the appearances".

The model of Hipparchus was popularized by the famous Egyptian astronomer, Claudius Ptolemy (c. 75 A.D. - c. 135 A.D.), in a book which dominated astronomy up to the time of Copernicus. Ptolemy's book was referred to by its admirers as *Megale Mathematike Syntaxis* (The Great Mathematical Composition). During the dark ages which followed the fall of Rome, Ptolemy's book was preserved and translated into Arabic by the civilized Moslems, and its name was shortened to *Almagest* (The Greatest). It held the field until, in the 15th century, the brilliant heliocentric model of Aristarchus was rescued from oblivion by Copernicus.

Archimedes

Archimedes was the greatest mathematician of the Hellenistic Era. In fact, together with Newton and Gauss, he is considered to be one of the greatest mathematicians of all time.

Archimedes was born in Syracuse in Sicily in 287 B.C.. He was the son of an astronomer, and he was also a close relative of Hieron II, the king of Syracuse. Like most scientists of his time, Archimedes was educated at the Museum in Alexandria, but unlike most, he did not stay in Alexandria. He returned to Syracuse, probably because of his kinship with Hieron II. Being a wealthy aristocrat, Archimedes had no need for the patronage of the Ptolemys.

Many stories are told about Archimedes: For example, he is supposed to have been so absent-minded that he often could not remember whether he had eaten. Another (perhaps apocryphal) story has to do with the discovery of "Archimedes Principle" in hydrostatics. According to the story, Hieron had purchased a golden crown of complex shape, and he had begun to suspect that the goldsmith had cheated him by mixing silver with gold. Since Hieron knew that his bright relative, Archimedes, was an expert in calculating the volumes of complex shapes, he took the crown to Archimedes and asked him to determine whether it was made of pure gold (by calculating its specific gravity). However, the crown was too irregularly shaped, and even Archimedes could not calculate its volume.

While he was sitting in his bath worrying about this problem, Archimedes reflected on the fact that his body seemed less heavy when it was in the water. Suddenly, in a flash of intuition, he saw that the amount by which his weight was reduced was equal to the weight of the displaced water. He leaped out of his bath shouting *"Eureka! Eureka!"* ("I've found it!") and ran stark naked through the streets of Syracuse to the palace of Hieron to tell him of the discovery.

The story of Hieron's crown illustrates the difference between the Hellenistic period and the classical period. In the classical period, geometry was a branch of religion and philosophy. For aesthetic reasons, the tools which a classical geometer was allowed too use were restricted to a compass and a straight-edge. Within these restrictions, many problems are insoluble. For example, within the restrictions of classical geometry, it is impossible to solve the problem of trisecting an angle. In the story of Hieron's crown, Archimedes breaks free from the classical restrictions and shows himself willing to use every conceivable means to achieve his purpose.

One is reminded of Alexander of Macedon who, when confronted with the Gordian Knot, is supposed to have drawn his sword and cut the knot in two! In a book *On Method*, which he sent to his friend Eratosthenes, Archimedes even confesses to cutting out figures from paper and weighing them as a means of obtaining intuition about areas and centers of gravity. Of course, having done this, he then derived the areas and centers of gravity by more rigorous methods.

One of Archimedes' great contributions to mathematics was his development of methods for finding the areas of plane figures bounded by curves, as well as methods for finding the areas and volumes of solid figures bounded by curved surfaces. To do this, he employed the "doctrine of limits". For example, to find the area of a circle, he began by inscribing a square inside the circle. The area of the square was a first approximation to the area of the circle. Next, he inscribed a regular octagon and calculated its area, which was a closer approximation to the area of the circle. This was followed by a figure with 16 sides, and then 32 sides, and so on. Each increase in the number of sides brought him closer to the true area of the circle.

Archimedes also circumscribed polygons about the circle, and thus he obtained an upper limit for the area, as well as a lower limit. The true area was trapped between the two limits. In this way, Archimedes showed that the value of pi lies between 223/71 and 220/70.

Sometimes Archimedes' use of the doctrine of limits led to exact results. For example, he was able to show that the ratio between the volume of a sphere inscribed in a cylinder to the volume of the cylinder is 2/3, and that the area of the sphere is 2/3 the area of the cylinder. He was so pleased with this result that he asked that a sphere and a cylinder be engraved on his tomb, together with the ratio, 2/3.

Another problem which Archimedes was able to solve exactly was the problem of calculating the area of a plane figure bounded by a parabola. In his book *On method*, Archimedes says that it was his habit to begin working on a problem by thinking of a plane figure as being composed of a very large number of narrow strips, or, in the case of a solid, he thought of it as being built up from a very large number of slices. This is exactly the approach which is used in integral calculus.

Archimedes must really be credited with the invention of both differential and integral calculus. He used what amounts to integral calculus to find the volumes and areas not only of spheres, cylinders and cones, but also of spherical segments, spheroids, hyperboloids and paraboloids of revolution; and his method for constructing tangents anticipates differential calculus.

Unfortunately, Archimedes was unable to transmit his invention of the calculus to the other mathematicians of his time. The difficulty was that there was not yet any such thing as algebraic geometry. The Pythagoreans had never recovered from the shock of discovering irrational numbers, and they had therefore abandoned algebra in favor of geometry. The union of algebra and geometry, and the development of a calculus which even non-geniuses could use, had to wait for Descartes, Fermat, Newton and Leibnitz.

Archimedes was the father of statics (as well as of hydrostatics). He calculated the centers of gravity of many kinds of figures, and he made a systematic, quantitative study of the properties of levers. He is supposed to have said: "Give me a place to stand on, and I can move the world!" This brings us to another of the stories about Archimedes: According to the story, Hieron was a bit sceptical, and he challenged Archimedes to prove his statement by moving something rather enormous, although not necessarily as large as the world. Archimedes good-humoredly accepted the challenge, hooked up a system of pulleys to a fully-loaded ship in the harbor, seated himself comfortably, and without excessive effort he singlehandedly pulled the ship out of the water and onto the shore.

Archimedes had a very compact notation for expressing large numbers. Essentially his system was the same as our own exponential notation, and it allowed him to handle very large numbers with great ease. In a curious little book called *The Sand Reckoner*, he used this notation to calculate the number of grains of sand which would be needed to fill the universe. (Of course, he had to make a crude guess about the size of the universe.) Archimedes wrote this little book to clarify the distinction between things which are very large but finite and things which are infinite. He wanted to show that nothing finite - not even the number of grains of sand needed to fill the universe - is too large to be measured and expressed in numbers. *The Sand Reckoner* is important as an historical document, because in it Archimedes incidentally mentions the revolutionary heliocentric model of Aristarchus, which does not occur in the one surviving book by Aristarchus himself.

In addition to his mathematical genius, Archimedes showed a superb mechanical intuition, similar to that of Leonardo da Vinci. Among his inventions are a planetarium and an elegant pump in the form of a helical tube. This type of pump is called the "screw of Archimedes", and it is still in use in Egypt. The helix is held at an angle to the surface of the water, with its lower end half-immersed. When the helical tube is rotated about its long axis, the water is forced to flow uphill!

His humanity and his towering intellect brought Archimedes universal respect, both during his own lifetime and ever since. However, he was not allowed to live out his life in peace; and the story of his death is both dramatic and symbolic:

In c. 212 B.C., Syracuse was attacked by a Roman fleet. The city would have fallen quickly if Archimedes had not put his mind to work to think of ways to defend his countrymen. He devised systems of mirrors which focused the sun's rays on the attacking ships and set them on fire, and cranes which plucked the ships from the water and overturned them.

In the end, the Romans hardly dared to approach the walls of Syracuse. However, after several years of siege, the city fell to a surprise attack. Roman soldiers rushed through the streets, looting, burning and killing. One of them found Archimedes seated calmly in front of diagrams sketched in the sand, working on a mathematical problem. When the soldier ordered him to come along, the great mathematician is supposed to have looked up from his work and replied: "Don't disturb my circles." The soldier immediately killed him.

The death of Archimedes and the destruction of the Hellenistic civilization illustrate the fragility of civilization. It was only a short step from Archimedes to Galileo and Newton; only a short step from Eratosthenes to Colombus, from Aristarchus to Copernicus, from Aristotle to Darwin or from Hippocrites to Pasteur. These steps in the cultural evolution of

Fig. 3.3 *"The death of Archimedes", a painting by Thomas Degeorge. Public domain, Wikimedia Commons*

mankind had to wait nearly two thousand years, because the brilliant Hellenistic civilization was destroyed, and Europe was plunged back into the dark ages.

Roman engineering

During the period between 202 B.C. and 31 B.C., Rome gradually extended its control over the Hellenistic states. By intervening in a dynastic struggle between Cleopatra and her brother Ptolemy, Julius Caesar was able to obtain control of Egypt. He set fire to the Egyptian fleet in the harbour of Alexandria. The fire spread to the city. Soon the great library of Alexandria was in flames, and most of its 750,0000 volumes were destroyed. If these books had survived, our knowledge of the history, science and literature of the ancient world would be incomparably richer. Indeed, if the library had survived, the whole history of the world might have been very different.

The Roman conquest produced 600 years of political stability in the west, and it helped to spread civilization into northern Europe. The Roman genius was for practical organization, and for useful applications of knowledge such as engineering and public health.

Roman roads, bridges and aquaducts, many of them still in use, testify to the superb skill of Roman engineers. The great system of aquaducts which supplied Rome with water brought the city a million cubic meters every day. Under the streets of Rome, a system of sewers (*cloacae*), dating from the 6th century B.C., protected the health of the citizens.

The abacus was used in Rome as an aid to arithmetic. This device was originally a board with a series of groves in which pebbles (*calculi*) were slid up and down. Thus the English word "calculus" is derived from the Latin name for a pebble.

The impressive technical achievements of the Roman Empire were in engineering, public health and applied science, rather than in pure science. In the 5th century A.D., the western part of the Roman Empire was conquered by barbaric tribes from northern Europe, and the west entered a dark age.

Chapter 4

CIVILIZATIONS OF THE EAST

China

After the fall of Rome in the 5th century A.D., Europe became a culturally backward area. However, the great civilizations of Asia and the Middle East continued to flourish, and it was through contact with these civilizations that science was reborn in the west.

During the dark ages of Europe, a particularly high level of civilization existed in China. The art of working in bronze was developed in China during the Shang dynasty (1,500 B.C. - 1,100 B.C.) and it reached a high pitch of excellence in the Chou dynasty (1,100 B.C. - 250 B.C.).

In the Chou period, many of the cultural characteristics which we recognize as particularly Chinese were developed. During this period, the Chinese evolved a code of behaviour based on politeness and ethics. Much of this code of behaviour is derived from the teachings of K'ung Fu-tzu (Confucius), a philosopher and government official who lived between 551 B.C. and 479 B.C.. In his writings about ethics and politics, K'ung Fu-tzu advocated respect for tradition and authority, and the effect of his teaching was to strengthen the conservative tendencies in Chinese civilization. He was not a religious leader, but a moral and political philosopher, like the philosophers of ancient Greece. He is traditionally given credit for the compilation of the Five Classics of Chinese Literature, which include books of history, philosophy and poetry, together with rules for religious ceremonies.

The rational teachings of K'ung Fu-tzu were complemented by the more mystical and intuitive doctrines of Lao-tzu and his followers. Lao-tzu lived at about the same time as K'ung Fu-tzu, and he founded the Taoist religion. The Taoists believed that unity with nature could be achieved by passively blending oneself with the forces of nature.

至聖孔子
名丘字仲尼山東
兗州府曲阜縣人

Fig. 4.1 *A painting of K'ung Fu-tzu (551 B.C.- 479 B.C.). Public domain, Wikimedia Commons*

On the whole, politicians and scholars followed the practical teachings of K'ung Fu-tzu, while poets and artists became Taoists. The intuitive sensitivity to nature inspired by Taoist beliefs allowed these artists and poets to achieve literature and art of unusual vividness and force with great economy of means. The Taoist religion has much in common with Buddhism, and its existence in China paved the way for the spread of Buddhism from India to China and Japan.

From 800 B.C. onwards, the central authority of the Chou dynasty weakened, and China was ruled by local landlords. This period of disunity was ended in 246 B.C. by Shih Huang Ti, a chieftain from the small northern state of Ch'in, who became the first real emperor of China. (In fact, China derives its name from the state of Ch'in).

Shih Huang Ti was an effective but ruthless ruler. It was during his reign (246 B.C. -210 B.C.) that the great wall of China was built. This wall, built to protect China from the savage attacks of the mounted Mongolian hordes, is one of the wonders of the world. It runs 1,400 miles, over all kinds of terrain, marking a rainfall boundary between the rich agricultural land to the south and the arid steppes to the north.

In most places, the great wall is 25 feet high and 15 feet thick. To complete this fantastic building project, Shih Huang Ti carried absolutism to great extremes, uprooting thousands of families and transporting them to the comfortless north to work on the wall. He burned all the copies of the Confucian classics which he could find, since his opponents quoted these classics to show that his absolutism had exceeded proper bounds.

Soon after the death of Shih Huang Ti, there was a popular reaction to the harshness of his government, and Shih's heirs were overthrown. However, Shih Huang Ti's unification of China endured, although the Ch'in dynasty (250 B.C. - 202 B.C.) was replaced by the Han dynasty (202 B.C. -220 A.D.). The Han emperors extended the boundaries of China to the west into Turkistan, and thus a trade route was opened, through which China exported silk to Persia and Rome.

During the Han period, China was quite receptive to foreign ideas, and was much influenced by the civilization of India. For example, the Chinese pagoda was inspired by the Buddhist shrines of India. The Han emperors adopted Confucianism as the official philosophy of China, and they had the Confucian classics recopied in large numbers. The invention of paper at the end of the first century A.D. facilitated this project, and it greatly stimulated scholarship and literature.

The Han emperors honoured scholarship and, in accordance with the political ideas of K'ung Fu-tzu, they made scholarship a means of access to high governmental positions. During the Han dynasty, the imperial government carried through many large-scale irrigation and flood-control projects. These projects were very successful. They increased the food production of China, and gave much prestige to the imperial government.

Like the Roman Empire, the Han dynasty was ended by attacks of barbarians from the north. However, the Huns who overran northern China in 220 A.D. were quicker to adopt civilization than were the tribes which conquered Rome. Also, in the south, the Chinese remained independent; and therefore the dark ages of China were shorter than the European dark ages.

In 581 A.D., China was reunited under the Sui dynasty, whose emperors expelled most of the Huns, built a system of roads and canals, and constructed huge granaries for the prevention of famine. These were worthwhile projects, but in order to accomplish them, the Sui emperors used very harsh methods. The result was that their dynasty was soon overthrown and replaced by the T'ang dynasty (618 A.D. - 906 A.D.).

The T'ang period was a brilliant one for China. Just as Europe was sinking further and further into a mire of superstition, ignorance and bloodshed, China entered a period of peace, creativity and culture. During this period, China included Turkistan, northern Indochina and Korea. The T'ang emperors re-established and strengthened the system of civil-service examinations which had been initiated during the Han dynasty.

Printing

It was during the T'ang period that the Chinese made an invention of immense importance to the cultural evolution of mankind. This was the invention of printing. Together with writing, printing is one of the key inventions which form the basis of human cultural evolution.

Printing was invented in China in the 8th or 9th century A.D., probably by Buddhist monks who were interested in producing many copies of the sacred texts which they had translated from Sanscrit. The act of reproducing prayers was also considered to be meritorious by the Buddhists.

The Chinese had for a long time followed the custom of brushing engraved official seals with ink and using them to stamp documents. The type of ink which they used was made from lamp-black, water and binder. In fact, it was what we now call "India ink". However, in spite of its name, India ink is a Chinese invention, which later spread to India, and from there to Europe.

We mentioned that paper of the type which we now use was invented in China in the first century A.D.. Thus, the Buddhist monks of China had all the elements which they needed to make printing practical: They had good ink, cheap, smooth paper, and the tradition of stamping documents with ink-covered engraved seals. The first block prints which they produced date from the 8th century A.D.. They were made by carving a block of wood the size of a printed page so that raised characters remained, brushing ink onto the block, and pressing this onto a sheet of paper.

The oldest known printed book, the "Diamond Sutra", is dated 868 A.D., and it consists of only six printed pages. In was discovered in 1907 by an English scholar who obtained permission from Buddhist monks in Chinese Turkistan to open some walled-up monastery rooms, which were said to have been sealed for 900 years. The rooms were found to contain a library of about 15,000 manuscripts, among which was the Diamond Sutra.

Block printing spread quickly throughout China, and also reached Japan, where woodblock printing ultimately reached great heights in the work of such artists as Hiroshige and Hokusai. The Chinese made some early experiments with movable type, but movable type never became popular in China, because the Chinese written language contains 10,000 characters. However, printing with movable type was highly successful in Korea as early as the 15th century A.D..

The unsuitability of the Chinese written language for the use of movable type was the greatest tragedy of the Chinese civilization. Writing had been developed at a very early stage in Chinese history, but the system remained a pictographic system, with a different character for each word. A phonetic system of writing was never developed.

The failure to develop a phonetic system of writing had its roots in the Chinese imperial system of government. The Chinese empire formed a vast area in which many different languages were spoken. It was necessary to have a universal language of some kind in order to govern such an empire. The Chinese written language solved this problem admirably.

Suppose that the emperor sent identical letters to two officials in different districts. Reading the letters aloud, the officials might use entirely different words, although the characters in the letters were the same. Thus the Chinese written language was a sort of "Esperanto" which allowed communication between various language groups, and its usefulness as such prevented its replacement by a phonetic system.

The disadvantages of the Chinese system of writing were twofold: First, it was difficult to learn to read and write; and therefore literacy was confined to a small social class whose members could afford a prolonged education. The system of civil-service examinations made participation in the government dependant on a high degree of literacy; and hence the old, established scholar-gentry families maintained a long-term monopoly on power, wealth and education. Social mobility was possible in theory, since the civil service examinations were open to all, but in practice, it was nearly unattainable.

The second great disadvantage of the Chinese system of writing was that it was unsuitable for printing with movable type. An "information explosion" occurred in the west following the introduction of printing with movable type, but this never occurred in China. It is ironical that although both paper and printing were invented by the Chinese, the full effect of these immensely important inventions bypassed China and instead revolutionized the west.

The invention of block printing during the T'ang dynasty had an enormously stimulating effect on literature, and the T'ang period is regarded as the golden age of Chinese lyric poetry. A collection of T'ang poetry, compiled in the 18th century, contains 48,900 poems by more than 2,000 poets.

The technique of producing fine ceramics from porcelain was invented during the T'ang dynasty; and the art of making porcelain reached its highest point in the Sung dynasty (960-1279), which followed the T'ang period. During the Sung dynasty, Chinese landscape painting also reached a high degree of perfection. In this period, the Chinese began to use the magnetic compass for navigation.

The first Chinese text clearly describing the magnetic compass dates from 1088 A.D.. However, the compass is thought to have been invented in China at a very much earlier date. The original Chinese compass was a spoon carved from lodestone, which revolved on a smooth diviner's board. The historian Joseph Needham believes that sometime between the 1st and 6th centuries A.D. it was discovered in China that the directive property of the lodestone could be transferred to small iron needles. These could be placed on bits of wood and floated in water. It is thought that by the beginning of the Sung dynasty, the Chinese were also aware of the deviation of the magnetic north from the true geographical north. By 1190 A.D., knowledge of the compass had spread to the west, where it revolutionized navigation and lead to the great voyages of discovery which characterized the 15th century.

The Sung dynasty was followed by a period during which China was ruled by the Mongols (1279-1328). Among the Mongol emperors was the famous Kublai Khan, grandson of Genghis Khan. He was an intelligent and capable ruler who appreciated Chinese civilization and sponsored many cultural projects. It was during the Mongol period that Chinese drama and fiction were perfected. During this period, the mongols ruled not only China, but also southern Russia and Siberia, central Asia and Persia. They were friendly towards Europeans, and their control of the entire route across Asia opened direct contacts between China and the west.

Fig. 4.2 *A painting of Marco Polo dressed as a Tartar. Public domain, Wikimedia Commons*

Among the first Europeans to take advantage of this newly-opened route were a family of Venetian merchants called Polo. After spending four years crossing central Asia and the terrible Gobi desert, they reached China in 1279. They were warmly welcomed by Kublai Khan, who invited them to his summer palace at Shangtu ("Xanadu"). The Great Khan took special interest in Marco Polo, a young man of the family who had accompanied his uncles Nicolo and Maffeo on the journey. Marco remained in China for seventeen years as a trusted diplomat in the service of Kublai Khan.

Later, after returning to Italy, Marco Polo took part in a war between Venice and Genoa. He was captured by the Genoese, and while in prison he dictated the story of his adventures to a fellow prisoner who happened to be a skilful author of romances. The result was a colorful and readable book which helped to reawaken the west after the middle ages. The era of exploration which followed the middle ages was partly inspired by Marco Polo's book. (Colombus owned a copy and made enthusiastic notes in the margins!) In his book, Marco Polo describes the fabulous wealth of China, as well as Chinese use of paper money, coal and asbestos.

Other Chinese inventions which were transmitted to the west include metallurgical blowing engines operated by water power, the rotary fan and rotary winnowing machine, the piston bellows, the draw-loom, the wheel-barrow, efficient harnesses for draught animals, the cross bow, the kite, the technique of deep drilling, cast iron, the iron-chain suspension bridge, canal lock-gates, the stern-post rudder and gunpowder. Like paper, printing and the magnetic compass, gunpowder and its use in warfare were destined to have an enormous social and political impact.

India

Evidence of a very early river-valley civilization in India has been found at a site called Mohenjo-Daro. However, in about 2,500 B.C., this early civilization was destroyed by some great disaster, perhaps a series of floods; and for the next thousand years, little is known about the history of India. During this dark period between 2,500 B.C. and 1,500 B.C., India was invaded by the Indo-Aryans, who spoke Sanscrit, a language related to Greek. The Indo-Aryians partly drove out and partly enslaved the smaller and darker native Dravidians. However, there was much intermarriage between the groups, and to prevent further intermarriage, the Indo-Aryians introduced a caste system sanctioned by religion.

According to Hindu religious belief, the soul of a person who has died is reborn in another body. If, throughout his life, the person has faithfully performed the duties of his caste, then his or her soul may be reborn into a higher caste. Finally, after existing as a Brahman, the soul may be so purified that it can be released from the cycle of death and rebirth.

In the 6th century B.C., Gautama Buddha founded a new religion in India. Gautama Buddha was convinced that all the troubles of humankind spring from attachment to earthly things. He felt that the only escape from sorrow is through the renunciation of earthly desires. He also urged his disciples to follow a high ethical code, the Eightfold Way. Among the sayings of Buddha are the following:

"Hatred does not cease by hatred at any time; hatred ceases by love."

"Let a man overcome anger by love; let him overcome evil by good."

"All men tremble at punishment. All men love life. Remember that you are like them, and do not cause slaughter."

One of the early converts to Buddhism was the emperor Ashoka Maurya, who reigned in India between 273 B.C. and 232 B.C.. During one of his wars of conquest, Ashoka Maurya became so sickened by the slaughter that he resolved never again to use war as an instrument of policy. He became one of the most humane rulers in history, and he also did much to promote the spread of Buddhism throughout Asia.

Under the Mauryan dynasty (322 B.C. - 184 B.C.), the Gupta dynasty (320 B.C. - 500 A.D.) and also under the rajah Harsha (606 A.D. - 647 A.D.), India had periods of unity, peace and prosperity. At other times, the country was divided and upset by internal wars. The Gupta period especially is regarded as the golden age of India's classical past. During this period, India led the world in such fields as medicine and mathematics.

The Guptas established both universities and hospitals. According to the Chinese Buddhist pilgrim, Fa-Hsien, who visited India in 405 A.D., "The nobles and householders have founded hospitals within the city to which the poor of all countries, the destitute, crippled and diseased may go. They receive every kind of help without payment."

Indian doctors were trained in cleansing wounds, in using ointments and in surgery. They also developed antidotes for poisons and for snakebite, and they knew some techniques for the prevention of disease through vaccination.

When they had completed their training, medical students in India took an oath, which resembled the Hyppocratic oath: "Not for yourself, not for the fulfillment of any earthly desire or gain, but solely for the good of suffering humanity should you treat your patients."

In Indian mathematics, algebra and trigonometry were especially highly developed. For example, the astronomer Brahmagupta (598 A.D. - 660 A.D.) applied algebraic methods to astronomical problems. The notation for zero and the decimal system were invented in India, probably during the 8th or 9th century A.D.. These mathematical techniques were later transmitted to Europe by the Arabs.

Many Indian techniques of manufacture were also transmitted to the west by the Arabs. Textile manufacture in particular was highly developed in India, and the Arabs, who were the middlemen in the trade with the west, learned to duplicate some of the most famous kinds of cloth. One kind of textile which they copied was called "quttan" by the Arabs, a word which in English has become "cotton". Other Indian textiles included cashmere (Kashmir), chintz and calico (from Calcutta, which was once called Calicut). Muslin derives its name from Mosul, an Arab city where it was manufactured, while damask was made in Damascus.

Indian mining and metallurgy were also highly developed. The Europeans of the middle ages prized fine laminated steel from Damascus; but it was not in Damascus that the technique of making steel originated. The Arabs learned steelmaking from the Persians, and Persia learned it from India.

The Nestorians and Islam

After the burning of the great library at Alexandria and the destruction of Hellenistic civilization, most of the books of the classical Greek and Hellenistic philosophers were lost. However, a few of these books survived and were translated from Greek, first into Syriac, then into Arabic and finally from Arabic into Latin. By this roundabout route, fragments from the wreck of the classical Greek and Hellenistic civilizations drifted back into the consciousness of the west.

We mentioned that the Roman empire was ended in the 5th century A.D. by attacks of barbaric Germanic tribes from northern Europe. However, by that time, the Roman empire had split into two halves. The eastern half, with its capital at Byzantium (Constantinople), survived until 1453, when the last emperor was killed vainly defending the walls of his city against the Turks.

The Byzantine empire included many Syriac-speaking subjects; and in fact, beginning in the 3rd century A.D., Syriac replaced Greek as the major language of western Asia. In the 5th century A.D., there was a split in the Christian church of Byzantium;and the Nestorian church, separated from the official Byzantine church. The Nestorians were bitterly persecuted by the Byzantines, and therefore they migrated, first to Mesopotamia, and later to south-west Persia. (Some Nestorians migrated as far as China.)

During the early part of the middle ages, the Nestorian capital at Gondisapur was a great center of intellectual activity. The works of Plato, Aristotle, Hippocrates, Euclid, Archimedes, Ptolemy, Hero and Galen were translated into Syriac by Nestorian scholars, who had brought these books with them from Byzantium.

Among the most distinguished of the Nestorian translators were the members of a family called Bukht-Yishu (meaning "Jesus hath delivered"), which produced seven generations of outstanding scholars. Members of this family were fluent not only in Greek and Syriac, but also in Arabic and Persian.

In the 7th century A.D., the Islamic religion suddenly emerged as a conquering and proselytizing force. Inspired by the teachings of Mohammad (570 A.D. - 632 A.D.), the Arabs and their converts rapidly conquered western Asia, northern Africa, and Spain. During the initial stages of the conquest, the Islamic religion inspired a fanaticism in its followers which was often hostile to learning. However, this initial fanaticism quickly changed to an appreciation of the ancient cultures of the conquered territories; and during the middle ages, the Islamic world reached a very high level of culture and civilization.

Thus, while the century from 750 to 850 was primarily a period of translation from Greek to Syriac, the century from 850 to 950 was a period of translation from Syriac to Arabic. It was during this latter century that Yuhanna Ibn Masawiah (a member of the Bukht-Yishu family, and medical advisor to Caliph Harun al-Rashid) produced many important translations into Arabic.

The skill of the physicians of the Bukht-Yishu family convinced the Caliphs of the value of Greek learning; and in this way the family played an extremely important role in the preservation of the western cultural heritage. Caliph al-Mamun, the son of Harun al-Rashid, established at Baghdad a library and a school for translation, and soon Baghdad replaced Gondisapur as a center of learning.

The English word "chemistry" is derived from the Arabic words *"al-chimia"*, which mean "the changing". The earliest alchemical writer in Arabic was Jabir (760-815), a friend of Harun al-Rashid. Much of his writing deals with the occult, but mixed with this is a certain amount of real chemical knowledge. For example, in his *Book of Properties*, Jabir gives the following recipe for making what we now call lead hydroxycarbonate (white lead), which is used in painting and pottery glazes:

Fig. 4.3 *A painting of Caliph Harun al-Rashid. Public domain, Wikimedia Commons*

"Take a pound of litharge, powder it well and heat it gently with four pounds of vinegar until the latter is reduced to half its original volume. The take a pound of soda and heat it with four pounds of fresh water until the volume of the latter is halved. Filter the two solutions until they are quite clear, and then gradually add the solution of soda to that of the litharge. A white substance is formed, which settles to the bottom. Pour off the supernatant water, and leave the residue to dry. It will become a salt as white as snow."

Another important alchemical writer was Rahzes (c. 860 - c. 950). He was born in the ancient city of Ray, near Teheran, and his name means "the man from Ray". Rhazes studied medicine in Baghdad, and he became chief physician at the hospital there. He wrote the first accurate descriptions of smallpox and measles, and his medical writings include methods for setting broken bones with casts made from plaster of Paris. Rahzes was the first person to classify substances into vegetable, animal and mineral. The word *"al-kali"*, which appears in his writings, means "the calcined" in Arabic. It is the source of our word "alkali", as well as of the symbol K for potassium.

The greatest physician of the middle ages, Avicinna, (Abu-Ali al Hussain Ibn Abdullah Ibn Sina, 980-1037), was also a Persian, like Rahzes. More than a hundred books are attributed to him. They were translated into Latin in the 12th century, and they were among the most important medical books used in Europe until the time of Harvey. Avicinina also wrote on alchemy, and he is important for having denied the possibility of transmutation of elements.

In mathematics, one of the most outstanding Arabic writers was al-Khwarizmi (c. 780 - c. 850). The title of his book, *Ilm al-jabr wa'd muqabalah*, is the source of the English word "algebra". In Arabic *al-jabr* means "the equating". Al-Khwarizmi's name has also become an English word, "algorism", the old word for arithmetic. Al-Khwarizmi drew from both Greek and Hindu sources, and through his writings the decimal system and the use of zero were transmitted to the west.

One of the outstanding Arabic physicists was al-Hazen (965-1038). He made the mistake of claiming to be able to construct a machine which could regulate the flooding of the Nile. This claim won him a position in the service of the Egyptian Caliph, al-Hakim. However, as al-Hazen observed Caliph al-Hakim in action, he began to realize that if he did not construct his machine *immediately*, he was likely to pay with his life! This led al-Hazen to the rather desperate measure of pretending to be insane, a ruse which he kept up for many years. Meanwhile he did excellent work in optics, and in this field he went far beyond anything done by the Greeks.

Al-Hazen studied the reflection of light by the atmosphere, an effect which makes the stars appear displaced from their true positions when they are near the horizon; and he calculated the height of the atmospheric layer above the earth to be about ten miles. He also studied the rainbow, the halo, and the reflection of light from spherical and parabolic mirrors. In his book, *On the Burning Sphere*, he shows a deep understanding of the properties of convex lenses. Al-Hazen also used a dark room with a pin-hole opening to study the image of the sun during an eclipes. This is the first mention of the *camera obscura*, and it is perhaps correct to attribute the invention of the *camera obscura* to al-Hazen.

Another Islamic philosopher who had great influence on western thought was Averröes, who lived in Spain from 1126 to 1198. His writings took the form of thoughtful commentaries on the works of Aristotle. He shocked both his Moslem and his Christian readers by maintaining that the world was not created at a definite instant, but that it instead evolved over a long period of time, and is still evolving.

Fig. 4.4 *In mathematics, one of the most outstanding Arabic writers was al-Khwarizmi (c.780 - c.850), commemorated here on a Russian stamp Public domain, Wikimedia Commons*

Like Aristotle, Averröes seems to have been groping towards the ideas of evolution which were later developed in geology by Steno, Hutton and Lyell and in biology by Darwin and Wallace. Much of the scholastic philosophy which developed at the University of Paris during the 13th century was aimed at refuting the doctrines of Averröes; but nevertheless, his ideas survived and helped to shape the modern picture of the world.

Suggestions for further reading

(1) Joseph Needham, *Science and Civilization in China*, (4 volumes), Cambridge University Press (1954-1971).
(2) Charles Singer, *A Short History of Scientific Ideas to 1900*, Oxford University Press (1959).
(3) Ernst J. Grube, *The World of Islam*, Paul Hamlyn Ltd., London (1966).
(4) Carl Brockelmann, *History of the Islamic Peoples*, Routledge and Kegan Paul (1949).
(5) Marshall Clagett, *The Science of Mechanics in the Middle Ages*, The University of Wisconsin Press, Madison (1959).

Chapter 5

SCIENCE IN THE RENAISSANCE

East-west contacts

Towards the end of the middle ages, Europe began to be influenced by the advanced Islamic civilization. European scholars were anxious to learn, but there was an "iron curtain" of religious intolerance which made travel in the Islamic countries difficult and dangerous for Christians. However, in the 12th century, parts of Spain, including the city of Toledo, were reconquered by the Christians. Toledo had been an Islamic cultural center, and many Moslem scholars, together with their manuscripts, remained in the city when it passed into the hands of the Christians. Thus Toledo became a center for the exchange of ideas between east and west; and it was in this city that many of the books of the classical Greek and Hellenistic philosophers were translated from Arabic into Latin.

In the 12th century, the translation was confined to books of science and philosophy. Classical Greek literature was forbidden by both the Christian and Moslem religions; and the beautiful poems and dramas of Homer, Sophocles and Euripides were not translated into Latin until the time of the Renaissance Humanists.

During the Mongol period (1279-1328), direct contact between Europe and China was possible because the Mongols controlled the entire route across central Asia; and during this period Europe received from China three revolutionary inventions: printing, gunpowder and the magnetic compass.

Another bridge between east and west was established by the crusades. In 1099, taking advantage of political divisions in the Moslem world, the Christians conquered Jerusalem and Palestine, which they held until 1187. This was the first of a series of crusades, the last of which took place in

1270. European armies, returning from the Middle East, brought with them a taste for the luxurious spices, textiles, jewelry, leatherwork and fine steel weapons of the orient; and their control of the Mediterranian sea routes made trade with the east both safe and profitable. Most of the profit from this trade went to a few cities, particularly to Venice and Florence.

At the height of its glory as a trading power, the Venetian Republic maintained six fleets of nationally owned ships, which could be chartered by private enterprise. All the ships of this fleet were of identical construction and rigged with identical components, so that parts could be replaced with ease at depots of the Venetian consular service abroad. The ships of these fleets could either serve as merchant ships, or be converted into warships by the addition of guns. Protected by a guard of such warships, large convoys of Venetian merchant ships could sail without fear of plunder by pirates.

In 1420, at the time of Venice's greatest commercial expansion, the doge, Tommaso Mocenigo, estimated the annual turnover of Venetian commerce to be ten million ducats, of which two million was profit. With this enormous income to spend, the Venetians built a city of splendid palaces, which rose like a shimmering vision above the waters of the lagoon.

The Venetians were passionately fond of pleasure, pagentry and art. The cross-shaped church of Saint Mark rang with the music of great composers, such as Gabrieli and Palestrina; and elegant triumphal music accompanied the doge as he went each year to throw a golden ring into the waters of the lagoon, an act which symbolized the marriage of Venice to the sea.

Like the Athenians after their victory in the Persian war, the Venetians were both rich and confident. Their enormous wealth allowed them to sponsor music, art, literature and science. The painters Titian, Veronese, Giorgione and Tintoretto, the sculptor Verrochio and the architect Palladio all worked in Venice at the height of the city's prosperity.

The self-confidence of the Venetians produced a degree of intellectual freedom which was not found elsewhere in Europe at that time, except in Florence. At the University of Padua, which was supported by Venetian funds, students from all countries were allowed to study regardless of their religious beliefs. It was at Padua that Copernicus studied, and there Andreas Vesalius began the research which led to his great book on anatomy. At one point in his career, Galileo also worked at the University of Padua.

The prosperity of 15th century Florence, like that of Venice, was based on commerce. In the case of Florence, the trade was not by sea, but along the main north-south road of Italy, which crossed the Arno at Florence. In addition to this trade, Florence also had an important textile industry. The

Florentines imported wool from France, Flanders, Holland and England. They wove the wool into cloth and dyed it, using superior techniques, many of which had come to them from India by way of the Islamic civilization. Later, silk weaving (again using eastern techniques) became important. Florentine banking was also highly developed, and our present banking system is derived from Florentine commercial practices.

Humanism

In the 15th and 16th centuries, Florence was ruled by a syndicate of wealthy merchant families, the greatest of whom were the Medicis. Cosimo de' Medici, the unofficial ruler of Florence from 1429 to 1464, was a banker whose personal wealth exceeded that of most contemporary kings. In spite of his great fortune, Cosimo lived in a relatively modest style, not wishing to attract attention or envy; and in general, the Medici influence tended to make life in Florence more modest than life in Venice.

Cosimo de' Medici is important in the history of ideas as one of the greatest patrons of the revival of Greek learning. In 1439, the Greek Patriarch and the Emperor John Palaeologus attended in Florence a council for the reunification of the Greek and Latin churches. The Greek-speaking Byzantine scholars who accompanied the Patriarch brought with them a number of books by Plato which excited the intense interest and admiration of Cosimo de' Medici.

Cosimo immediately set up a Platonic Academy in Florence, and chose a young man named Marsilio Ficino as its director. In one of his letters to Ficino, Cosimo says:

"Yesterday I came to the villa of Careggi, not to cultivate my fields, but my soul. Come to us, Marsilio, as soon as possible. Bring with you our Plato's book *De Summo Bono*. This, I suppose, you have already translated from the Greek language into Latin, as you promised. I desire nothing so much as to know the road to happiness. Farewell, and do not come without the Orphian lyre!"

Cosimo's grandson, Lorenzo the Magnificent, continued his grandfather's policy of reviving classical Greek learning, and he became to the golden age of Florence what Pericles had been too the golden age of Athens. Among the artists whom Lorenzo sponsored were Michelangelo, Botticelli and Donatello. Lorenzo established a system of bursaries and prizes for the support of students. He also gave heavy financial support to the University of Pisa, which became a famous university under Lorenzo's patronage. (It was later to be the university of Galileo and Fermi.)

At Florence, Greek was taught by scholars from Byzantium; and Poliziano, who translated Homer into Latin could say with justice: "Greek learning, long extinct in Greece itself, has come to life and lives again in Florence. There Greek literature is taught and studied, so that Athens, root and branch, has been transported to make her abode - not in Athens in ruins and in the hands of barbarians, but in Athens as she was, with her breathing spirit and her very soul."

Leonardo da Vinci

Against this background, it may seem strange that Lorenzo the Magnificent did not form a closer relationship with Leonardo da Vinci, the most talented student of Verrocchio's school in Florence. One might have expected a close friendship between the two men, since Lorenzo, only four years older than Leonardo, was always quick to recognize exceptional ability.

The explanation probably lies in Leonardo's pride and sensitivity, and in the fact that, while both men were dedicated to knowledge, they represented different points of view. Lorenzo was full of enthusiasm for the revival of classical learning, while Leonardo had already taken the next step: Rejecting all blind obedience to authority, including the authority of the ancients, he relied on his own observations. Lorenzo was fluent in Latin and Greek, and was widely educated in Greek philosophy, while Leonardo was ignorant of both languages and was largely self-taught in philosophy and science (although he had studied mathematics at the school of Benedetto d'Abacco).

While he did not form a close friendship with Lorenzo the Magnificent, Leonardo was lucky in becoming the friend and protegé of the distinguished Florentine mathematician, physician, geographer and astronomer, Paolo Toscanelli, who was also the friend and advisor of Colombus. (Toscanelli furnished Colombus with maps of the world and encouraged him in his project of trying to reach India and China by sailing westward. Toscanelli's maps mistakenly showed the Atlantic Ocean with Europe on one side, and Asia on the other!)

Gradually, under Toscanelli's influence, young Leonardo's powerful and original mind was drawn away from the purely representational aspects of art, and he became more and more involved in trying to understand the underlying structure and mechanism of the things which he observed in nature - the bodies of men and animals, the flight of birds, the flow of fluids and the features of the earth.

Both in painting and in science, Leonardo looked directly to nature for guidence, rather than to previous masters. He wrote:

"The painter will produce pictures of small merit if he takes as his standard the pictures of others; but if he will study from natural objects, he will produce good fruits. And I would say about these mathematical studies, that those who study the authorities and not the works of nature are descendents but not sons of nature."

In another place, Leonardo wrote:

"But first I will test with experiment before I proceed further, because my intention is to consult experience first, and then with reasoning to show why such experience is bound to operate in such a way. And that is the true rule by which those who analyse the effects of nature must proceed; and although nature begins with the cause and ends with the experience, we must follow the opposite course, namely (as I said before) begin with the experience and by means of it investigate the cause."

Lorenzo the Magnificent finally did help Leonardo in a backhanded way: In 1481, when Leonardo was 29 years old, Lorenzo sent him as an emissary with a gift to the Duke of Milan, Ludovico Sforza. Although Milan was far less culturally developed than Florence, Leonardo stayed there for eighteen years under the patronage of Sforza. He seemed to work better in isolation, without the competition and criticism of the Florentine intellectuals.

In Milan, Leonardo began a series of anatomical studies which he developed into a book, intended for publication. Leonardo's anatomical drawings make previous work in this field seem like the work of children, and in many respects his studies were not surpassed for hundreds of years. Some of his anatomical drawings were published in a book by Fra Pacioli, and they were very influential; but most of the thousands of pages of notes which Leonardo wrote have only been published in recent years.

The notebooks of Leonardo da Vinci cover an astonishing range of topics: mathematics, physics, astronomy, optics, engineering, architecture, city planning, geology, hydrodynamics and aerodynamics, anatomy, painting and perspective, in addition to purely literary works. He was particularly interested in the problem of flight, and he made many studies of the flight of birds and bats in order to design a flying machine. Among his notes are designs for a helicopter and a parachute, as well as for a propellor-driven flying machine.

In astronomy, Leonardo knew that the earth rotates about its axis once every day, and he understood the physical law of inertia which makes this motion imperceptible to us except through the apparent motion of the stars. In one of his notebooks, Leonardo wrote: "The sun does not move." However, he did not publish his ideas concerning astronomy. Leonardo was always planning to organize and publish his notes, but he was so busy with his many projects that he never finished the task. At one point, he wrote what sounds like a cry of despair: "Tell me, tell me if anything ever was finished!"

Leonardo ended his life in the court of the king of France, Francis I. The king gave him a charming chateau in which to live, and treated him with great respect. Francis I visited Leonardo frequently in order to discuss philosophy, science and art; and when Leonardo died, the king is said to have wept openly.

Fig. 5.1 *Leonardo da Vinci as an old man. Self-portrait in red chalk. Public domain, Wikimedia Commons*

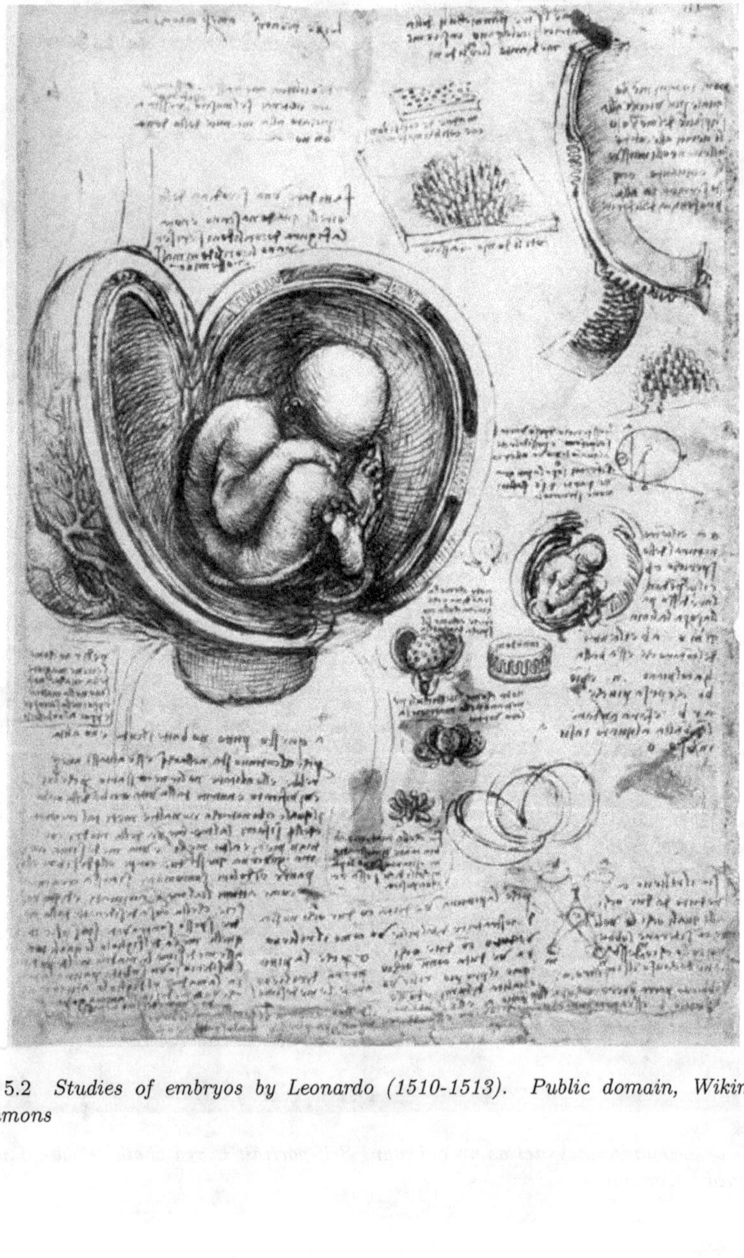

Fig. 5.2 *Studies of embryos by Leonardo (1510-1513). Public domain, Wikimedia Commons*

Copernicus

The career of Leonardo da Vinci illustrates the first phase of the "information explosion" which has produced the modern world: Inexpensive paper was being manufactured in Europe, and it formed the medium for Leonardo's thousands of pages of notes. His notes and sketches would never have been possible if he had been forced to use expensive parchment as a medium. On the other hand, the full force of Leonardo's genius and diligence was never felt because his notes were not printed. Copernicus, who was a younger contemporary of Leonardo, had a much greater effect on the history of ideas, because his work was published. Thus, while paper alone made a large contribution to the information explosion, it was printing combined with paper which had an absolutely decisive and revolutionary impact: The modern scientific era began with the introduction of printing.

Nicolas Copernicus (1473-1543) was orphaned at the age of ten, but fortunately for science he was adopted by his uncle, Lucas Watzelrode, the Prince-Bishop of Ermland (a small semi-independent state which is now part of Poland). Through his uncle's influence, Copernicus was made a Canon of the Cathedral of Frauenberg in Ermland at the age of twenty-three. He had already spent four years at the University of Krakow, but his first act as Canon was to apply for leave of absence to study in Italy.

At that time, Italy was very much the center of European intellectual activity. Copernicus stayed there for ten years, drawing a comfortable salary from his cathedral, and wandering from one Italian University to another. He studied medicine and church law at Padua and Bologna, and was made a Doctor of Law at the University of Ferrara. Thus, thanks to the influence of his uncle, Copernicus had an education which few men of his time could match. He spent altogether fourteen years as a student at various universities, and he experienced the bracing intellectual atmosphere of Italy at the height of the Renaissance.

In 1506, Bishop Lucas recalled Copernicus to Ermland, where the young Canon spent the next six years as his uncle's personal physician and administrative assistant. After his uncle's death, Copernicus finally took up his duties as Canon at the cathedral-fortress of Frauenberg on the Baltic coast of Ermland; and he remained there for the rest of his life, administering the estates of the cathedral, acting as a physician to the people of Ermland, and working in secret on his sun-centered cosmology.

Even as a student in Krakow, Copernicus had thought about the problem of removing the defects in the Ptolomeic system. In Italy, where the books of the ancient philosophers had just become available in the original Greek, Copernicus was able to search among their writings for alternative proposals. In Ptolemy's system, not all the "wheels within wheels" turn with a uniform velocity, although it is possible to find a point of observation called the *"punctum equans"* from which the motion seems to be uniform. Concerning this, Copernicus wrote:

"A system of this sort seems neither sufficiently absolute, nor sufficiently pleasing to the mind... Having become aware of these defects, I often considered whether there could be found a more reasonable arrangement of circles, in which everything would move uniformly about its proper center, as the rule of absolute motion requires.."

While trying to remove what he regarded as a defect in the Ptolemeic system by rearranging the wheels, Copernicus rediscovered the sun-centered cosmology of Aristarchus. However, he took a crucial step which went beyond Aristarchus: What Copernicus did during the thirty-one years which he spent in his isolated outpost on the Baltic was to develop the heliocentric model into a complete system, from which he calculated tables of planetary positions.

The accuracy of Copernicus' tables was a great improvement on those calculated from the Ptolemeic system, and the motions of the planets followed in a much more natural way. The inner planets, Mercury and Venus, stayed close to the sun because of the smallness of their orbits, while the occasional apparently retrograde motion of the outer planets could be explained in a very natural way by the fact that the more rapidly-moving earth sometimes overtook and passed one of the outer planets. Furthermore, the speed of the planets diminished in a perfectly regular way according to their distances from the sun.

In spite of these successes, Copernicus hesitated to publish the book which he had written outlining his theory. He feared ridicule, and he feared that his position in the church hierarchy would be endangered if he put forward such unorthodox and possibly heretical ideas. In his youth, he had participated in the Italian Renaissance, and he had even translated a book of Greek poems into Latin, thus declaring himself to be on the side of the humanists in the controversy over whether pagan Greek literature ought to be revived. However, old age and isolation in medieval Ermland had turned him into a thoroughly conservative churchman.

The intellectual freedom of the early 15th century had begun to disappear because of the increasingly bitter controversy between Martin Luther and the established church. As a result of the attacks of Luther, the Roman church had become more strict. Following the edict of his bishop, Copernicus was forced to send away his housekeeper of many years, a woman who was probably his unofficial wife.

Against the background of this atmosphere of intolerance, it is easy to understand why Copernicus hesitated to publish his unorthodox theory. Probably he would never have done so had it not been for the arrival at Frauenberg of an ardent young disciple, Georg Joachim Rheticus, a professor of mathematics and astronomy from the University of Wittenberg.

Rheticus had heard rumors about the sun-centered cosmology of Copernicus, and he arrived at Frauenberg "at the extreme outskirts of the earth" full of enthusiasm and hero worship, determined to learn from Copernicus the details of his heliocentric system. He brought with him as gifts the first printed editions of Euclid and Ptolemy in the original Greek.

Copernicus could not resist the flattering admiration and enthusiasm of Rheticus, but he was much embarrassed to have a visitor from Wittenberg, the very center of the Lutheran heresy. Therefore he hastily packed Rheticus off to Loebau Castle in Kulm. Tiedimann Geise, the closest friend of Copernicus, had been made Bishop of Kulm, and Loebau Castle was his official residence.

Rheticus and Bishop Geise worked together at Loebau Castle, trying in every way they could think of to persuade Copernicus to publish his great book, *De Revolutionibus Orbium Coelestium*; but the cautious old Canon resisted all their arguments. Finally they hammered out a compromise: Rheticus was to take a short course on the sun-centered system from Copernicus. Then he would write a little book which would be a *Preliminary Account* of Copernicus' great work; and the name of Copernicus would not be mentioned in the *Preliminary Account* except in a very oblique way.

In other words, Rheticus agreed to stick his neck out, and if it was not chopped off, then Copernicus might possibly agree to publish his book. This was done, and Rheticus seemed to survive the publication of his little book. In fact the *Preliminary Account* was quite well received.

Copernicus could no longer resist the combined forces of Rheticus and Geise. He handed over his precious manuscript to Rheticus, who left triumphantly for Nürnberg to have it printed. (At that time, printing was most advanced in the Protestant parts of Germany. Like the Buddhist monks of China, the Lutherans had strong religious motives for promoting the development of printing. They believed that the Bible ought to be read by ordinary people. Also, Luther's battle against the established church was being fought by means of printed pamphlets.)

His great *Revolutionibus* was finally being printed, but in 1512, Copernicus himself fell mortally ill with a cerebral hemmorhage. His faithful friend, Bishop Geise, recorded that "For many days he had been deprived of his memory and his mental vigour; he only saw his completed book at the last moment, on the day that he died."

The publication of the *Revolutionibus* did not cause an immediate stir; nor was Copernicus himself the sort of person who might have been expected to overthrow the established patterns of human thought. He was an extremely learned man, but his outlook was distinctly conservative. Nevertheless, hidden in the Copernican cosmology, there were implications which caused an intellectual revolution once they were understood. The earth was dethroned from its position as the center of the universe. Also, if Copernicus was right, the universe had to be almost unimaginably enormous.

According the the Copernican cosmology, the earth moves around the sun in an orbit whose radius is ninety-three million miles. As the earth moves in its enormous orbit, it is sometimes closer to a particular star, and sometimes farther away. Therefore the observed positions of the stars relative to each other ought to change as the earth moves around its orbit. This effect, called "stellar parallax", could not be observed with the instruments which were available in the 16th century.

The explanation which Copernicus gave for the absence of stellar parallax was that "Compared to the distance of the fixed stars, the earth's distance from the sun is neglegably small!" If this is true for the nearest stars, then what about the distance to the farthest stars?

Vast and frightening chasms of infinity seemed to open under the feet of those who understood the implications of the Copernican cosmology. Humans were no longer rulers of a small, tidy universe especially created for themselves. They were suddenly "lost in the stars", drifting on a tiny speck of earth through unimaginably vast depths of space. Hence the cry of Blaise Pascal: *"Le silence eternal de ce éspaces infinis m'effraie!"*, "The eternal silence of these infinite spaces terrifies me!"

Fig. 5.3 *Woodcut portrait of Copernicus by Nicolaus Reusner, ca. 1578. Public domain, Wikimedia Commons*

Tycho Brahe

The next step in the Copernican revolution was taken by two men who presented a striking contrast to one another. Tycho Brahe (1546-1601) was a wealthy and autocratic Danish nobleman, while Johannes Kepler (1571-1630) was a neurotic and poverty-stricken teacher in a provincial German school. Nevertheless, in spite of these differences, the two men collaborated for a time, and Johannes Kepler completed the work of Tycho Brahe.

At the time when Tycho was born, Denmark included southern Sweden; and ships sailing to and from the Baltic had to pay a toll as they passed through the narrow sound between Helsingør (Elsinore) in Denmark, and Helsingborg in what is now Sweden. On each side of the sound was a castle, with guns to control the sea passage. Tycho Brahe's father, a Danish nobleman, was Governor of Helsingborg Castle.

Tycho's uncle was also a military man, a Vice-Admiral in the navy of the Danish king, Frederik II. This uncle was childless, and Tycho's father promised that the Vice-Admiral could adopt one of his own children. By a fortunate coincidence, twins were born to the Governor's wife. However, when one of the twins died, Tycho's father was unwilling to part with the survivor (Tycho). The result was that, in the typically high-handed style of the Brahe family, the Vice-Admiral kidnapped Tycho. The Governor at first threatened murder, but soon calmed down and accepted the situation with good grace.

The adoption of Tycho Brahe by his uncle was as fortunate for science as the adoption of Copernicus by Bishop Watzelrode, because the Vice-Admiral soon met his death in an heroic manner which won the particular gratitude of the Danish Royal Family:

Admiral Brahe, returning from a battle against the Swedes, was crossing a bridge in the company of King Frederik II. As the king rode across the bridge, his horse reared suddenly, throwing him into the icy water below. The king would have drowned if Admiral Brahe had not leaped into the water and saved him. However, the Admiral saved the king's life at the cost of his own. He caught pneumonia and died from it. The king's gratitude to Admiral Brahe was expressed in the form of special favor shown to his adopted son, Tycho, who had in the meantime become an astronomer (against the wishes of his family).

As a boy of fourteen, Tycho Brahe had witnessed a partial eclipse of the sun, which had been predicted in advance. It struck him as "something divine that men could know the motions of the stars so accurately that they were able a long time beforehand to predict their places and relative positions". Nothing that his family could say would dissuade him from studying astronomy, and he did so not only at the University of Copenhagen, but also at Leipzig, Wittenberg, Rostock, Basel and Augsberg.

During this period of study, Tycho began collecting astronomical instruments. His lifelong quest for precision in astronomical observation dated from his seventeenth year, when he observed a conjunction of Saturn and Jupiter. He found that the best tables available were a month in error in predicting this event. Tycho had been greatly struck by the fact that (at least as far as the celestial bodies were concerned), it was possible to predict the future; but here the prediction was in error by a full month! He resolved to do better.

Tycho first became famous among astronomers through his observations on a new star, which suddenly appeared in the sky in 1572. He used the splendid instruments in his collection to show that the new star was very distant from the earth - certainly beyond the sphere of the moon - and that it definitely did not move with respect to the fixed stars. This was, at the time, a very revolutionary conclusion. According to Aristotle, (who was still regarded as the greatest authority on matters of natural philosophy), all generation and decay should be confined to the region beneath the sphere of the moon. Tycho's result meant that Aristotle could be wrong!

Tycho thought of moving to Basel. He was attracted by the beauty of the town, and he wanted to be nearer to the southern centers of culture. However, in 1576 he was summoned to appear before Frederik II. Partly in recognition of Tycho's growing fame as an astronomer, and partly to repay the debt of gratitude which he owed to Admiral Brahe, the king made Tycho the ruler of Hven, an island in the sound between Helsingborg and Helsingør. Furthermore, Frederik granted Tycho generous funds from his treasury to construct an observatory on Hven.

With these copious funds, Tycho Brahe constructed a fantastic castle-observatory which he called Uraniborg. It was equipped not only with the most precise astronomical instruments the world had ever seen, but also with a chemical laboratory, a paper mill, a printing press and a dungeon for imprisoning unruly tenants.

Tycho moved in with a retinue of scientific assistants and servants. The only thing which he lacked was his pet elk. This beast had been transported from the Brahe estate at Knudstrup to Landskrona Castle on the Sound, and it was due to be brought on a boat to the island of Hven. However, during the night, the elk wandered up a stairway in Landskrona Castle and found a large bowl of beer in an unoccupied room. Like its master, the elk was excessively fond of beer, and it drank so much that, returning down the stairway, it fell, broke its leg, and had to be shot.

Fig. 5.4 *Tycho Brahe. Public domain, Wikimedia Commons*

Tycho ruled his island in a thoroughly autocratic and grandiose style, the effect of which was heightened by his remarkable nose. In his younger days, Tycho had fought a duel with another student over the question of who was the better mathematician. During the duel, the bridge of Tycho's nose had been sliced off. He had replaced the missing piece by an artificial bridge which he had made of gold and silver alloy, and this was held in place by means of a sticky ointment which he always carried with him in a snuff box.

Tycho entertained in the grandest possible manner the stream of scholars who came to Hven to see the wonders of Uraniborg. Among his visitors were King James VI of Scotland (who later ascended the English throne as James I), and the young prince who later became Christian IV of Denmark.

With the help of his numerous assistants, Tycho observed and recorded the positions of the sun, moon, planets and stars with an accuracy entirely unprecedented in the history of astronomy. He corrected both for atmospheric refraction and for instrumental errors, with the result that his observations were accurate to within two minutes of arc. This corresponds to the absolute limit of what can be achieved without the help of a telescope.

Not only were Tycho's observations made with unprecedented accuracy - they were also made *continuously* over a period of 35 years. Before Tycho's time, astronomers had haphazardly recorded an observation every now and then, but no one had thought of making systematic daily records of the positions of each of the celestial bodies. Tycho was able to make a "motion picture" record of the positions of the planets because he could divide the work among his numerous assistants.

In 1577, a spectacular comet appeared in the sky. Tycho treated it in the same way that he treated the planets, making scrupulously careful and continuous records of its position. He showed by parallax studies that the comet had to be farther away from the earth than the orbit of the moon. Again Aristotle was shown to be wrong! Aristotle had recognized that comets violated the rules which he had set down for celestial motion, but he believed comets to be atmospheric phenomena.

In a book which he wrote about the comet in 1577, Tycho proposed his own cosmology. It was halfway between Ptolemy and Copernicus, and was designed to eliminate the shocking idea of a moving earth. In Tycho's system, Mercury, Venus, Mars, Jupiter and Saturn all moved in orbits around the sun, but the sun moved in an orbit around the earth, which remained stationary at the center of the universe.

Tycho believed his system to be true because, even though he tried very hard with his superb instruments, he could not observe the stellar parallax which must exist if the earth really moves in an orbit around the sun. The parallax does in fact exist, but because the distance to the nearest stars is so immense, it cannot be observed without the use of a large telescope. It was finally observed in the 19th century by the German astronomer, F.W. Bessel (the inventor of Bessel functions).

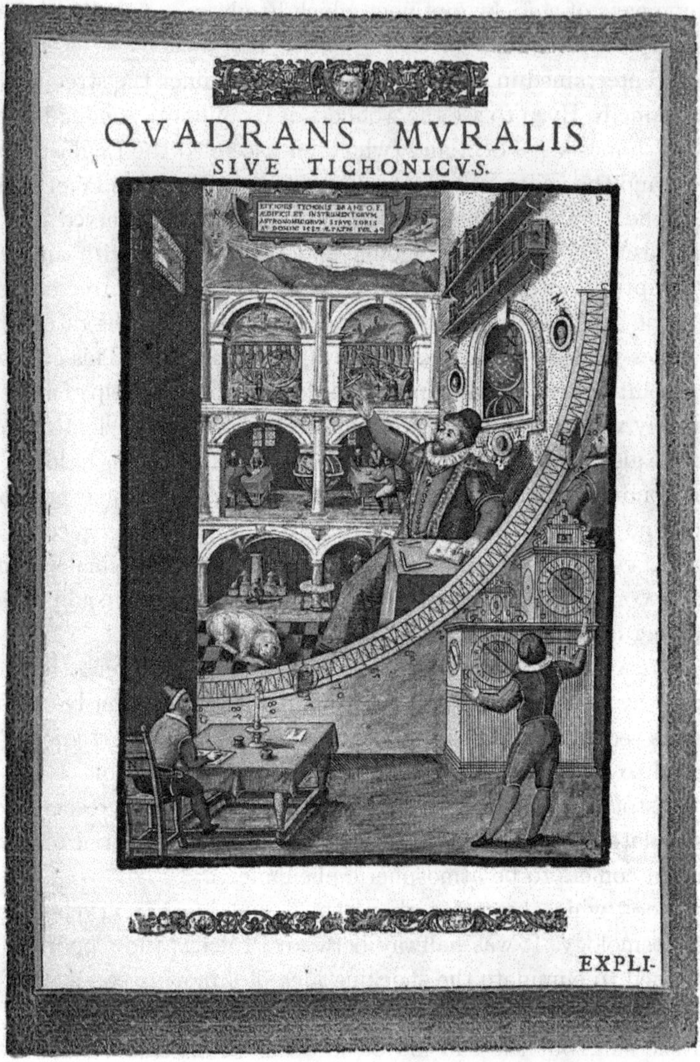

Fig. 5.5 *Tycho's Wall Quadrant. An engraving of Tycho Brahe in his Uraniborg observatory on the island of Hven, probably from the 1598 printing of his Astronomiae instauratae mechanica, hand coloured. Public domain, Wikimedia Commons*

All went well with Tycho on the island of Hven for twelve years. Then, in 1588, Frederik II died (of alcoholism), and his son ascended the throne as Christian IV. Frederik II had been especially grateful to Admiral Brahe for saving his life, and he treated the Admiral's adopted son, Tycho, with great indulgence. However, Christian IV was unwilling to overlook the increasingly scandalous and despotic way in which Tycho was ruling Hven; and he reduced the subsidies which Tycho Brahe had been receiving from the royal treasury. The result was that Tycho, feeling greatly insulted, dismantled his instruments and moved them to Prague, together with his retinue of family, scientific assistants, servants and jester.

In Prague, Tycho became the Imperial Mathematician of the Holy Roman Emperor, Rudolph II. (We should mention in passing that royal patrons such as Rudolph were more interested in astrology than in astronomy: The chief duty of the Imperial Mathematician was to cast horoscopes for the court!) After the move to Prague, one of Tycho's senior scientific assistants became dissatisfied and left. To replace him, Tycho recruited a young German mathematician named Johannes Kepler.

Johannes Kepler

Two thousand years before the time of Kepler, Pythagoras had dreamed of finding mathematical harmony in the motions of the planets. Kepler and Newton were destined to fulfil his dream. Kepler was also a true follower of Pythagoras in another sense: Through his devotion to philosophy, he transcended the personal sufferings of a tortured childhood and adolescence. He came from a family of misfits whose neurotic quarrelsomeness was such that Kepler's father narrowly escaped being hanged, and his mother was accused of witchcraft by her neighbors. She was imprisoned, and came close to being burned.

At the age of 4, Kepler almost died of smallpox, and his hands were badly crippled. Concerning his adolescence, Kepler wrote: "I suffered continually from skin ailments, often severe sores, often from the scabs of chronic putrid wounds in my feet, which healed badly and kept breaking out again. On the middle finger of my right hand, I had a worm, and on the left, a huge sore."

Kepler's mental strength compensated for his bodily weakness. His brilliance as a student was quickly recognized, and he was given a scholarship to study theology at the University of Tübingen. He was agonizingly lonely and unpopular among his classmates.

Kepler distinguished himself as a student at Tübingen, and shortly before his graduation, he was offered a post as a teacher of mathematics and astronomy at the Protestant School in Graz. With the post went the title of "Mathematician of the Provence of Styria". (Gratz was the capital of Styria, a province of Austria).

Johannes Kepler was already an ardent follower of Copernicus; and during the summer of his first year in Graz, he began to wonder why the speed of the planets decreased in a regular way according to their distances from the sun, and why the planetary orbits had the particular sizes which Copernicus assigned to them.

On July 9, 1595, in the middle of a lecture which he was giving to his class, Kepler was electrified by an idea which changed the entire course of his life. In fact, the idea was totally wrong, but it struck Kepler with such force that he thought he had solved the riddle of the universe with a single stroke!

Kepler had drawn for his class an equilateral triangle with a circle circumscribed about it, so that the circle passed through all three corners of the triangle. Inside, another circle was inscribed, so that it touched each side of the triangle. It suddenly struck Kepler that the ratio between the sizes of the two circles resembled the ratio between the orbits of Jupiter and Saturn. His mercurial mind immediately leaped from the two-dimensional figure which he had drawn to the five regular solids of Pythagoras and Plato.

In three dimensions, only five different completely symmetrical many-sided figures are possible: the tetrahedron, cube, octahedron, icosohedron and the dodecahedron. There the list stops. As Euclid proved, it is a peculiarity of three-dimensional space that there are only five possible regular polyhedra. These five had been discovered by Pythagoras, and they had been popularized by Plato, the most famous of the Pythagorean philosophers. Because Plato made so much of the five regular solids in his dialogue *Timaeus*, they became known as the "Platonic solids".

In a flash of (completely false) intuition, Kepler saw why there had to be exactly six planets: The six spheres of the planetary orbits were separated by the five Platonic solids! This explained the sizes of the orbits too: Each sphere except the innermost and the outermost was inscribed in one solid and circumscribed about another!

Kepler, who was then twenty-three years old, was carried away with enthusiasm. He immediately wrote a book about his discovery and called it *Mysterium Cosmigraphicum*, "The Celestial Mystery". The book begins with an introduction strongly supporting the Copernican cosmology. After that comes the revelation of Kepler's marvelous (and false) solution to the cosmic mystery by means of the five Platonic solids. Kepler was unable

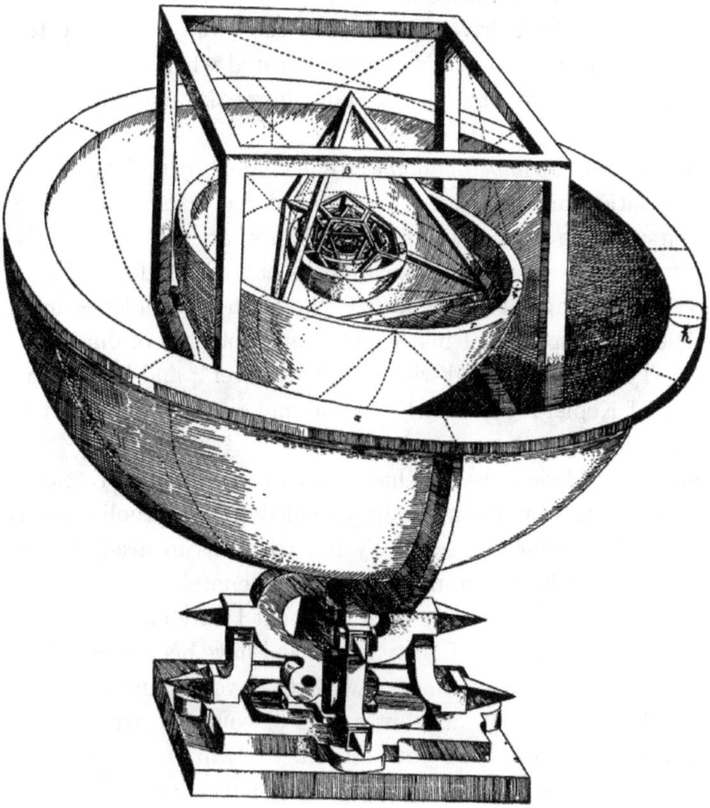

Fig. 5.6 *Kepler's (entirely wrong) model of the solar system based on the Platonic solids. Public domain, Wikimedia Commons*

to make the orbit of Jupiter fit his model, but he explains naively that "nobody will wonder at it, considering the great distance". The figures for the other planets did not quite fit either, but Kepler believed that the distances given by Copernicus were inaccurate.

Finally, after the mistaken ideas of the book, comes another idea, which comes close to the true picture of gravitation. Kepler tries to solve the problem of why the outer planets move more slowly than the inner ones, and he says:

"If we want to get closer to the truth and establish some correspondence in the proportions, then we must choose between these two assumptions: Either the souls of the planets are less active the farther they are from the sun, or there exists only one moving soul in the center of the orbits, that is the sun, which drives the planets the more vigorously the closer the planet is, but whose force is quasi-exhausted when acting on the outer planets, because of the long distance and the weakening of the force which it entails."

In *Mysterium Cosmographicum*, Kepler tried to find an exact mathematical relationship between the speeds of the planets and the sizes of their orbits; but he did not succeed in this first attempt. He finally solved this problem many years later, towards the end of his life.

Kepler sent a copy of his book to Tycho Brahe with a letter in which he called Tycho "the prince of mathematicians, not only of our time, but of all time". Tycho was pleased with this "fan letter"; and he recognized the originality of Kepler's book, although he had reservations about its main thesis.

Meanwhile, religious hatred had been deepening and Kepler, like all other Protestants, was about to be expelled from Catholic Austria. He appealed to Tycho for help, and Tycho, who was in need of a scientific assistant, wrote to Kepler from the castle of Benatek near Prague:

"You have no doubt already been told that I have most graciously been called here by his Imperial Majesty and that I have been received in a most friendly and benevolent manner. I wish that you would come here, not forced by the adversity of fate, but rather of your own will and desire for common study. But whatever your reason, you will find in me your friend, who will not deny you his advice and help in adversity"

To say that Kepler was glad for this opportunity to work with Tycho Brahe is to put the matter very mildly. The figures of Copernicus did not really fit Kepler's model, and his great hope was that Tycho's more accurate observations would give a better fit. In his less manic moments, Kepler also recognized that his model might not be correct after all, but he hoped that Tycho's data would allow him to find the true solution.

Kepler longed to get his hands on Tycho's treasure of accurate data, and concerning these he wrote:

"Tycho possesses the best observations, and thus so-to-speak the material for building the new edifice. He also has collaborators, and everything else he could wish for. He only lacks the architect who would put all this to use according to his own design. For although he has a happy disposition and real architectural skill, he is nevertheless obstructed in his progress by the multitude of the phenomena, and by the fact that the truth is deeply hidden in them. Now old age is creeping upon him, enfeebling his spirit and his forces"

Fig. 5.7 *Portrait of Johannes Kepler. Public domain, Wikimedia Commons*

In fact, Tycho had only a short time to live. Kepler arrived in Prague in 1600, and in 1601 he wrote:

"On October 13, Tycho Brahe, in the company of Master Minkowitz, had dinner at the illustrious Rosenborg's table, and held back his water beyond the demands of courtesy. When he drank more, he felt the tension in his bladder increase, but he put politeness before health. When he got home, he was scarcely able to urinate.. After five sleepless nights, he could still only pass water with the greatest pain, and even so the passage was impeded. The insomnia continued, with internal fever gradually leading to delirium; and the food which he ate, from which he could not be kept, exacerbated the evil... On his last night, he repeated over and over again, like someone composing a poem: 'Let me not seem to have lived in vain'."

A few days after Tycho's death, Kepler was appointed to succeed him as Imperial Mathematician of the Holy Roman Empire. Kepler states that the problem of analysing Tycho's data took such a hold on him that he nearly went out of his mind. With a fanatic diligence rarely equalled in the history of science, he covered thousands of pages with calculations. Finally, after many years of struggle and many false starts, he wrung from Tycho's data three precise laws of planetary motion:

1) The orbits of the planets are ellipses, with the sun at one focal point.

2) A line drawn from the sun to any one of the planets sweeps out equal areas in equal intervals of time.

3) The square of the period of a planet is proportional to the cube of the mean radius of its orbit.

Thanks to Kepler's struggles, Tycho certainly had not lived in vain. Kepler's three laws were to become the basis for Newton's great universal laws of motion and gravitation. Kepler himself imagined a universal gravitational force holding the planets in their orbits around the sun, and he wrote:

"If two stones were placed anywhere in space, near to each other, and outside the reach of force of any other material body, then they would come together after the manner of magnetic bodies, at an intermediate point, each approaching the other in proportion to the other's mass... "

"If the earth ceased to attract the waters of the sea, the seas would rise up and flow to the moon... If the attractive force of the moon reaches down to the earth, it follows that the attractive force of the earth, all the more, extends to the moon, and even farther... "

"Nothing made of earthly substance is absolutely light; but matter which is less dense, either by nature or through heat, is relatively lighter... Out of the definition of lightness follows its motion; for one should not believe that when lifted up it escapes to the periphery of the world, or that it is not attracted to the earth. It is merely less attracted than heavier matter, and is therefore displaced by heavier matter."

Kepler also understood the correct explanation of the tides. He explained them as being produced primarily by the gravitational attraction of the moon, while being influenced to a lesser extent by the gravitational field of the sun.

Unfortunately, when Kepler published these revolutionary ideas, he hid them in a tangled jungle of verbiage and fantasy which repelled the most important of his readers, Galileo Galilei. In fact, the English were the first to appreciate Kepler. King James I (whom Tycho entertained on Hven) invited Kepler to move to England, but he declined the invitation. Although the skies of Europe were darkened by the Thirty Years War, Kepler could not bring himself to leave the German cultural background where he had been brought up and where he felt at home. Meanwhile, his contemporary, Galileo Galilei, who should have profited greatly from Kepler's insights, ignored Kepler and broke off correspondence with him.

Suggestions for further reading

(1) Irma A. Richter (editor), *Selections from the Notebooks of Leonardo da Vinci*, Oxford University Press (1977).
(2) Lorna Lewis, *Leonardo the Inventor*, Heinemann Educational Books, London (1974).
(3) Iris Noble, *Leonardo da Vinci*, Blackie, London (1968).
(4) C.H. Monk, *Leonardo da Vinci*, Hamlyn, London (1975).
(5) Thomas S. Kuhn, *The Copernican Revolution*, Harvard University Press (1957).
(6) Angus Armitage, *The World of Copernicus*, The New American Library, New York (1951).
(7) Arthur Koestler, *The Watershed*, Heinemann, London (1961).
(8) D.W. Singer, *Gordiano Bruno: His Life and Thought*, Greenwood Press, New York (1968).
(9) Edward A. Gosselin and Lawrence S. Lerner, Galileo and the Long Shadow of Bruno, Archives Internationales d'Histoire des Sciences, *25*, 223 (1975).
(10) Martin Olsson, *Uraniborg och Stjarneborg*, Almquist and Wiksell, Stockholm (1968).
(11) Galileo Galilei, *Dialogues Concerning Two New Sciences*, Dover, New York (1954).
(12) I. Bernard Cohen, *The Birth of a New Physics*, Heinemann, London (1960).
(13) D.L. Hurd and J.J. Kipling (editors), *The Origins and Growth of Physical Science*, Penguin Books Ltd. (1964).

Chapter 6

GALILEO

Experimental physics

Galileo Galilei was born in Pisa in 1564. He was the son of Vincenzo Galilei, an intellectual Florentine nobleman whose fortune was as small as his culture was great. Vincenzo Galilei was a mathematician, composer and music critic, and from him Galileo must have learned independence of thought, since in one of his books Vincenzo wrote: "It appears to me that those who try to prove a assertion by relying simply on the weight of authority act very absurdly." This was to be Galileo's credo throughout his life. He was destined to demolish the decayed structure of Aristotelian physics with sledgehammer blows of experiment.

Vincenzo Galilei, who knew what it was like to be poor, at first tried to make his son into a wool merchant. However, when Galileo began to show unmistakable signs of genius, Vincenzo decided to send him to the University of Pisa, even though this put a great strain on the family's financial resources.

At the university and at home, Galileo was deliberately kept away from mathematics. Following the wishes of his father, he studied medicine, which was much better paid than mathematics. However, he happened to hear a lecture on Euclid given by Ostilio Ricci, a friend of his father who was Mathematician at the court of the Grand Duke Ferdinand de' Medici.

Galileo was so struck by the logical beauty and soundness of the lecture that he begged Ricci to lend him some of the works of Euclid. These he devoured in one gulp, and they were followed by the works of Archimedes. Galileo greatly admired Archimedes' scientific method, and he modeled his own scientific method after it.

After three years at the University of Pisa, Galileo was forced to return home without having obtained a degree. His father had no more money with which to support him, and Galileo was unable to obtain a scholarship, probably because his irreverent questioning of every kind of dogma had made him unpopular with the authorities. However, by this time he had already made his first scientific discovery.

According to tradition, Galileo is supposed to have made this discovery while attending a service at the Cathedral of Pisa. His attention was attracted to a lamp hung from the vault, which the verger had lighted and left swinging. As the swings became smaller, he noticed that they still seemed to take the same amount of time. He checked this by timing the frequency against his pulse. Going home, he continued to experiment with pendula. He found that the frequency of the oscillations is independent of their amplitude, provided that the amplitude is small; and he found that the frequency depends only on the length of the pendulum.

Having timed the swings of a pendulum against his pulse, Galileo reversed the procedure and invented an instrument which physicians could use for timing the pulse of a patient. This instrument consisted of a pendulum whose length could be adjusted until the swings matched the pulse of the patient. The doctor then read the pulse rate from the calibrated length of the pendulum. Galileo's pulse meter was quickly adopted by physicians throughout Europe. Later, the famous Dutch physicist, Christian Huygens (1629-1695), developed Galileo's discovery into the pendulum clock as we know it today.

While he was living at home after leaving the University of Pisa, Galileo invented a balance for measuring specific gravity, based on Archimedes' Principle in hydrostatics.

Through his writings and inventions, particularly through his treatise on the hydrostatic balance, Galileo was becoming well known, and at the age of 26 he was appointed Professor of Mathematics at the University of Pisa. However, neither age nor the dignity of his new title had mellowed him. As a professor, he challenged authority even more fiercely than he had done as a student. He began systematically checking all the dogmas of Aristotle against the results of experiment.

Aristotle had asserted that the speed of a falling object increased according to its weight: Thus, according to Aristotle, an object ten times as heavy as another would fall ten times as fast. This idea was based on the common experience of a stone falling faster than a feather.

Galileo realized that the issue was being complicated by air resistance. There were really two questions to be answered: 1) How would a body fall in the absence of air? and 2) What is the effect of air resistance? Galileo considered the first question to be the more fundamental of the two, and in order to answer it, he experimented with falling weights made of dense materials, such as iron and lead, for which the effect of air resistance was reduced to a minimum.

According to Galileo's student and biographer, Viviani, Galileo, wishing to refute Aristotle, climbed the Leaning Tower of Pisa in the presence of all the other teachers and philosophers and of all the students, and "by repeated experiments proved that the velocity of falling bodies of the same composition, unequal in weight, does not attain the proportion of their weight as Aristotle assigned it to them, but rather that they move with equal velocity." (Some historians doubt Viviani's account of this event, since no mention of it appears in other contemporary sources.)

Galileo maintained that, in a vacuum, a feather would fall to the ground like a stone. This experiment was not possible in Galileo's time, but later it was tried, and Galileo's prediction was found to be true.

Galileo realized that falling bodies gain in speed as they fall, and he wished to find a quantitative law describing this acceleration. However, he had no good method for measuring very small intervals of time. Therefore he decided to study a similar process which was slow enough to measure: He began to study the way in which a ball, rolling down an inclined plane, increases in speed.

Describing these experiments, Galileo wrote:

"..Having placed the board in a sloping position... we rolled the ball along the channel, noting , in a manner presently to be described, the time required to make the descent. We repeated the experiment more than once, in order to measure the time with an accuracy such that the deviation between two observations never exceeded one-tenth of a pulse beat"

"Having performed this operation, and having assured ourselves of its reliability, we now rolled the ball only one quarter of the length of the channel, and having measured the time of its descent, we found it precisely one-half the former. Next we tried other distances, comparing the time for the whole length with that for the half, or with that for two-thirds or three-fourths, or indeed any fraction. In such experiments, repeated a full hundred times, we always found that the spaces traversed were to each other as the squares of the times..."

"For the measurement of time, we employed a large vessel of water placed in an elevated position. To the bottom of this vessel was soldered a pipe of small diameter giving a thin jet of water, which we collected in a small glass during the time of each descent... The water thus collected was weighed after each descent on a very accurate balance. The differences and ratios of these weights gave us the differences and ratios of the times, and with such an accuracy that although the operation was repeated many, many times, there was no appreciable discrepancy in the results"

These experiments pointed to a law of motion for falling bodies which Galileo had already guessed: The acceleration of a falling body is constant; the velocity increases in linear proportion to the time of fall; and the distance traveled increases in proportion to the square of the time.

Extending these ideas and experiments, Galileo found that a projectile has two types of motion superimposed: the uniformly accelerated falling motion just discussed, and, at the same time, a horizontal motion with uniform velocity. He showed that, neglecting air resistance, these two types of motion combine to give the projectile a parabolic trajectory.

Galileo also formulated the principle of inertia, a law of mechanics which states that in the absence of any applied force, a body will continue at rest, or if in motion, it will continue indefinitely in uniform motion. Closely related to this principle of inertia is the principle of relativity formulated by Galileo and later extended by Einstein: Inside a closed room, it is impossible to perform any experiment to determine whether the room is at rest, or whether it is in a state of uniform motion.

For example, an observer inside a railway train can tell whether the train is in motion by looking out of the window, or by the vibrations of the car; but if the windows were covered and the tracks perfectly smooth, there would be no way to tell. An object dropped in a uniformly-moving railway car strikes the floor directly below the point from which it was dropped, just as it would do if the car were standing still.

The Galilean principle of relativity removed one of the objections which had been raised against the Copernican system. The opponents of Copernicus argued that if the earth really were in motion, then a cannon ball, shot straight up in the air, would not fall back on the cannon but would land somewhere else. They also said that the birds and the clouds would be left behind by the motion of the earth.

In 1597, Kepler sent Galileo a copy of his *Mysterium Cosmographicum.* Galileo read the introduction to the book, which was the first printed support of Copernicus from a professional astronomer, and he replied in a letter to Kepler:

"...I shall read your book to the end, sure of finding much that is excellent in it. I shall do so with the more pleasure because I have for many years been an adherent of the Copernican system, and it explains to me the causes of many of the phenomena of nature which are quite unintelligible on the commonly accepted hypothesis."

"I have collected many arguments in support of the Copernican system and refuting the opposite view, which I have so far not ventured to make public for fear of sharing the fate of Copernicus himself, who, though he acquired immortal fame with some, is yet to an infinite multitude of others (for such is the number of fools) an object of ridicule and derision. I would certainly publish my reflections at once if more people like you existed; as they don't, I shall refrain from publishing."

Kepler replied urging Galileo to publish his arguments in favor of the Copernican system:

"...Have faith, Galileo, and come forward! If my guess is right, there are but few among the prominent mathematicians of Europe who would wish to secede from us, for such is the force of truth." However, Galileo left Kepler's letter unanswered, and he remained silent concerning the Copernican system.

By this time, Galileo was 33 years old, and he had become Professor of Mathematics at the University of Padua. His Aristotelian enemies at the University of Pisa had succeeded in driving him out, but by the time they did so, his fame had become so great that he was immediately offered a position at three times the salary at Padua.

The move was a very fortunate one for Galileo. Padua was part of the free Venetian Republic, outside the power of the Inquisition, and Galileo spent fifteen happy and productive years there. He kept a large house with a master mechanic and skilled craftsmen to produce his inventions (among which was the thermometer). His lectures were attended by enthusiastic audiences, sometimes as large as two thousand; and he had two daughters and a son with a Venetian girl.

The telescope

In 1609, news reached Galileo that a Dutch optician had combined two spectacle lenses in such a way as to make distant objects seem near. Concerning this event, Galileo wrote:

"A report reached my ears that a certain Fleming had constructed a spyglass by means of which visible objects, though very distant from the eye of the observer, were distinctly seen as if nearby. Of this truly remarkable effect, several experiences were related, to which some persons gave credence while others denied them."

"A few days later the report was confirmed to me in a letter from (a former pupil) at Paris; which caused me to apply myself wholeheartedly to inquire into the means by which I might arrive at the invention of a similar instrument. This I did shortly afterward through deep study of the theory of refraction."

"First I prepared a tube of lead at the ends of which I fitted two glass lenses, both plane on one side, while on the other side one was spherically convex and the other concave. Then, placing my eye near the concave lens, I perceived objects satisfactorally large and near, for they appeared three times closer and nine times larger than when seen with the naked eye alone."

"Next I constructed another more accurate instrument, which represented objects as enlarged more than sixty times. Finally, sparing neither labor nor expense, I succeeded in constructing for myself an instrument so excellent that objects seen through it appeared nearly one thousand times larger and over thirty times closer than when regarded with our natural vision."

Galileo showed one of his early telescopes to his patrons, the Signoria of Venice. Writing of this, Galileo says:

"Many noblemen and senators, though of advanced age, mounted to the top of one of the highest towers to watch the ships, which were visible through my glass two hours before they were seen entering the harbor; for it makes a thing fifty miles off as near and clear as if it were only five."

The senate asked Galileo whether he would give the city a similar instrument to aid in its defense against attack by sea. When he did this, they immediately doubled his salary, and they confirmed him in his position for life.

After perfecting the telescope as much as he could, Galileo turned it towards the moon, the planets and the stars. He made a series of revolutionary discoveries which he announced in a short booklet called *Siderius Nuncius*, (The Siderial Messenger). The impact of this booklet was enormous, as can be judged by the report of Sir Henry Wotton, the British Ambassador to Venice:

Fig. 6.1 *Galileo Galilei in a portrait by Domenico Tintoretto. Public domain, Wikimedia Commons*

"Now touching the occurents of the present", Sir Henry wrote, "I send herewith to His Majesty the strangest piece of news (as I may justly call it) that he has ever yet received from any part of the world; which is the annexed book (come abroad this very day) of the Mathematical Professor at Padua, who by the help of an optical instrument (which both enlargeth and approximateth the object) invented first in Flanders and bettered by himself, hath discovered four new planets rolling around the sphere of Jupiter, besides many other unknown fixed stars; likewise the true cause of the *Via Lactae* (Milky Way), so long searched; and lastly that the moon is not spherical but endued with many prominences, and, which is strangest of all, illuminated with the solar light by reflection from the body of the earth, as he seemeth to say. So as upon the whole subject, he hath overthrown all former astronomy.."

"These things I have been so bold to discourse unto your Lordship, whereof here all corners are full. And the author runneth a fortune to be either exceeding famous or exceeding ridiculous. By the next ship your Lordship shall receive from me one of the above instruments, as it is bettered by this man."

Wherever Galileo turned his powerful telescope, he saw myriads of new stars, so utterly outnumbering the previously known stars that mankind's presumption to know anything at all about the universe suddenly seemed pitiful. The Milky Way now appeared as a sea of stars so numerous that Galileo despaired of describing them in detail. The vastness of the universe as postulated by Nicolas Copernicus and Gordiano Bruno (one ridiculed and the other burned alive) was now brought directly to Galileo's senses. In fact, everywhere he looked he saw evidence supporting the Copernican system and refuting Aristotle and Ptolemy.

The four moons of Jupiter, which Galileo had discovered, followed the planet in its motion, thus refuting the argument that if the earth revolved around the sun, the moon would not be able to revolve around the earth. Also, Jupiter with its moons formed a sort of Copernican system in miniature, with the massive planet in the center and the four small moons circling it, the speed of the moons decreasing according to their distance from Jupiter.

Galileo discovered that the planet Venus has phase changes like the moon, and that these phase changes are accompanied by changes in the apparent size of the planet. Copernicus had predicted that if the power of human vision could be improved, exactly these changes in the appearance of Venus would be observed. Galileo's observations proved that Venus moves in an orbit around the sun: When it is on the opposite side of the sun from the earth, it appears small and full; when it lies between the earth and the sun, it is large and crescent.

Galileo also observed mountains on the moon. He measured their height by observing the way in which sunlight touches their peaks just before the lunar dawn, and he found some of the peaks to be several miles high. This disproved the Aristotelian doctrine that the moon is a perfect sphere, and it established a point of similarity between the moon and the earth.

Galileo observed that the dark portion of the moon is faintly illuminated, and he asserted that this is due to light reflected from the earth, another point of similarity between the two bodies. Generally speaking, the impression which Galileo gained from his study of the moon is that it is a body more or less like the earth, and that probably the same laws of physics apply on the moon as on the earth.

All these observations strongly supported the Copernican system, although the final rivet in the argument, the observation of stellar parallax, remained missing until the 19th century. Although he did not possess this absolutely decisive piece of evidence, Galileo thought that he had a strong enough basis to begin to be more open in teaching the Copernican system. His booklet, *Siderius Nuncius* had lifted him to an entirely new order of fame. He had seen what no man had ever seen before, and had discovered new worlds. His name was on everyone's lips, and he was often compared to Colombus.

Still it moves!

In 1610, Galileo left Padua to take up a new post as Mathematician to the court of the Medicis in Florence; and in the spring of 1611, he made a triumphant visit to Rome. Describing this visit, Cardinal del Monte wrote: "If we were living under the ancient Republic of Rome, I really believe that there would have been a column on the Capital erected in Galileo's honor!" The Pope received Galileo in a friendly audience, and Prince Cesi made him a member of the Adademia dei Lincei.

The Jesuit astronomers were particularly friendly to Galileo. They verified his observations and also improved some of them. However, Galileo made many enemies, especially among the entrenched Aristotelian professors in the universities. He enjoyed controversy (and publicity), and he could not resist making fools of his opponents in such a way that they often became bitter personal enemies.

Not only did Galileo's law describing the acceleration of falling bodies contradict Aristotle, but his principle of inertia contradicted the Aristotelian dogma, *omne quod movetur ab alio movetur* - whatever moves must be moved by something else. (The Aristotelians believed that each planet is moved by an angel.) Galileo also denied Aristotle's teaching that generation and decay are confined to the sphere beneath the orbit of the moon.

Although Galileo was at first befriended and honored by the Jesuit astronomers, he soon made enemies of the members of that order through a controversy over priority in the discovery of sunspots. In spite of this controversy, Galileo's pamphlet on sunspots won great acclaim; and Cardinal Maffeo Barberini (who later became Pope Urban VIII) wrote to Galileo warmly praising the booklet.

In 1613, the Medicis gave a dinner party and invited Professor Castelli, one of Galileo's students who had become Professor of Mathematics at Pisa. After dinner, the conversation turned to Galileo's discoveries, and the Grand Duchess Christina, mother of Duke Cosimo de' Medici, asked Castelli his opinion about whether the motion of the earth contradicted the Bible.

When this conversation was reported to Galileo, his response was to publish a pamphlet entitled *Letter to Castelli*, which was later expanded into a larger pamphlet called *Letter to the Grand Duchess Christina.* These pamphlets, which were very widely circulated, contain the following passage:

"...Let us grant, then, that Theology is conversant with the loftiest divine contemplation, and occupies the regal throne among the sciences by this dignity. By acquiring the highest authority in this way, if she does not descend to the lower and humbler speculations of the subordinate sciences, and has no regard for them because they are not concerned with blessedness, then her professors should not arrogate to themselves the authority to decide on controversies in professions which they have neither studied nor practiced. Why this would be as if an absolute despot, being neither a physician nor an architect, but knowing himself free to command, should undertake to administer medicines and erect buildings according to his whim, at the grave peril of his poor patients' lives, and the speedy collapse of his edifices..."

Galileo's purpose in publishing these pamphlets was to overcome the theological objections to the Copernican system. The effect was exactly the opposite. The *Letter to Castelli* was brought to the attention of the Inquisition, and in 1616 the Inquisition prohibited everyone, especially Galileo, from holding or defending the view that the earth turns on its axis and moves in an orbit around the sun.

Galileo was silenced, at least for the moment. For the next eighteen years he lived unmolested, pursuing his scientific research. For example, continuing his work in optics, he constructed a compound microscope.

In 1623, marvelous news arrived: Cardinal Maffio Barberini had been elected Pope. He was a great intellectual, and also Galileo's close friend. Galileo went to Rome to pay his respects to the new Pope, and he was received with much warmth. He had six long audiences with the Pope, who showered him with praise and gifts. The new Pope refused to revoke the Inquisition's decree of 1616, but Galileo left Rome with the impression that he was free to discuss the Copernican system, provided he stayed away from theological arguments.

Galileo judged that the time was right to bring forward his evidence for the Copernican cosmology; and he began working on a book which was to be written in the form of a Platonic dialogue. The characters in the conversation are Salivati, a Copernican philosopher, Sagredo, a neutral but intelligent layman, and Simplicio, a slightly stupid Aristotelian, who always ends by losing the arguments.

The book, which Galileo called *Dialogue on the Two Chief World Systems*, is a strong and only very thinly veiled argument in favor of the Copernican system. When it was published in 1632, the reaction was dramatic. Galileo's book was banned almost immediately, and the censor who had allowed it to be printed was banished in disgrace. When the agents of the Inquisition arrived at the bookstores to confiscate copies of the *Dialogue*, they found that the edition had been completely sold out.

The Pope was furious. He felt that he had been betrayed. Galileo's enemies had apparently convinced the Pope that the character called Simplicio in the book was a caricature of the Pope himself! Galileo, who was seventy years old and seriously ill, was dragged to Rome and threatened with torture. His daughter, Maria Celeste, imposed severe penances and fasting on herself, thinking that these would help her prayers for her father. However, her health was weak, and she became ill.

Meanwhile, Galileo, under threat of torture, had renounced his advocacy of the motion of the earth. According to tradition, as he rose from his knees after the recantation he muttered *"Eppur si muove!"*, ("Still it moves!"). It is unlikely that he muttered anything of the kind, since it would have been fatally dangerous to do so, and since at that moment, Galileo was a broken man. Nevertheless, the retort which posterity has imagined him to make remains unanswerable. As Galileo said, before his spirit was broken by the Inquisition, "...It is not in the power of any creature to make (these ideas) true or false or otherwise than of their own nature and in fact they are."

Galileo was allowed to visit the bedside of his daughter, Maria Celeste, but in her weak condition, the anxiety of Galileo's ordeal had been too much for her. Soon afterward, she died. Galileo was now a prisoner of the Inquisition. He used his time to write a book on his lifelong work on dynamics and on the strength of material structures. The manuscript of this book, entitled *Two New Sciences*, was smuggled out of Italy and published in Holland.

When Galileo became blind, the Inquisition relaxed the rules of his imprisonment, and he was allowed to have visitors. Many people came to see him, including John Milton, who was then 29 years old. One wonders whether Milton, meeting Galileo, had any premonition of his own fate. Galileo was already blind, while Milton was destined to become so. The two men had another point in common: their eloquent use of language. Galileo was a many-sided person, an accomplished musician and artist as well as a great scientist. The impact of his ideas was enhanced by his eloquence as a speaker and a writer. This can be seen from the following passage, taken from Galileo's *Dialogue*, where Sagredo comments on the Platonic dualism between heavenly perfection and earthly corruption:

"...I cannot without great wonder, nay more, disbelief, hear it being attributed to natural bodies as a great honor and perfection that they are impassable, immutable, inalterable, etc.; as, conversely, I hear it esteemed a great imperfection to be alterable, generable and mutable. It is my opinion that the earth is very noble and admirable by reason of the many different alterations, mutations and generations which incessantly occur in it. And if, without being subject to any alteration, it had been one vast heap of sand, or a mass of jade, or if, since the time of the deluge, the waters freezing that covered it, it had continued an immense globe of crystal, whereon nothing had ever grown, altered or changed, I should have esteemed it a wretched lump of no benefit to the Universe, a mass of idleness, and in a word, superfluous, exactly as if it had never been in Nature. The difference for me would be the same as between a living and a dead creature."

"I say the same concerning the moon, Jupiter and all the other globes of the Universe. The more I delve into the consideration of the vanity of popular discourses, the more empty and simple I find them. What greater folly can be imagined than to call gems, silver and gold noble, and earth and dirt base? For do not these persons consider that if there were as great a scarcity of earth as there is of jewels and precious metals, there would be no king who would not gladly give a heap of diamonds and rubies and many ingots of gold to purchase only so much earth as would suffice to plant a jasmine in a little pot or to set a tangerine in it, that he might see it sprout, grow up, and bring forth such goodly leaves, fragrant flowers and delicate fruit?"

The trial of Galileo cast a chill over the intellectual atmosphere of southern Europe, and it marked the end of the Italian Renaissance. However, the Renaissance had been moving northward, and had produced such figures as Dürer and Gutenberg in Germany, Erasmus and Rembrandt in Holland, and Shakespeare in England. In 1642, the same year during which Galileo died in Italy, Isaac Newton was born in England.

Suggestions for further reading

(1) Joseph C. Pitt, *Galileo, Human Knowledge and the Book of Nature; Method Replaces Metaphysics*, Kluwer, Dordrecht, (1992).
(2) Michael Segre, *In the Wake of Galileo*, Rutgers University Press, New Brunswick, N.J., (1991).
(3) Stillman Drake, *Galileo, Pioneer Scientist*, Toronnto University Press, (1990).
(4) Silvio A. Bedini, *The Pulse of Time; Galileo Galilei, the Determination of Longitude and the Pendulum Clock*, Olschki, Fierenze, (1991).
(5) Stillman Drake et al., *Nature, Experiment and the Sciences; Essays on Galileo and the History of Science*, Kluwer, Dordrecht, (1990).
(6) Pietro Redondi, *Galileo Heretic*, Princeton University Press, (1987).
(7) William A. Wallace, *Galileo and his Sources; The Heritage of the Collegio Romani in Galileo's Science*, Princeton University Press, (1984).
(8) William A. Wallace, *Prelude to Galileo*, Reidel, Dordrecht, (1981).
(9) Stillman Drake, *Telescopes, Tides nad Tactics; a Galilean Dialogue about the Starry Messinger and Systems of the World*, University of Chicago Press, (1980).
(10) Stillman Drake, *Galileo*, Oxford University Press, (1980).
(11) K.J.J. Hintikka et al. editors, *Conference on the History and Philosophy of Science*, Reidel, Dordrecht, (1981).

Chapter 7

THE AGE OF REASON

Descartes

Until the night of November 10, 1619, algebra and geometry were separate disciplines. On that autumn evening, the troops of the Elector of Bavaria were celebrating the Feast of Saint Martin at the village of Neuberg in Bohemia. With them was a young Frenchman named René Descartes (1596-1659), who had enlisted in the army of the Elector in order to escape from Parisian society. During that night, Descartes had a series of dreams which, as he said later, filled him with enthusiasm, converted him to a life of philosophy, and put him in possession of a wonderful key with which to unlock the secrets of nature.

The program of natural philosophy on which Descartes embarked as a result of his dreams led him to the discovery of analytic geometry, the combination of algebra and geometry. Essentially, Descartes' method amounted to labeling each point in a plane with two numbers, x and y. These numbers represented the distance between the point and two perpendicular fixed lines, (the coordinate axes). Then every algebraic equation relating x and y generated a curve in the plane.

Descartes realized the power of using algebra to generate and study geometrical figures; and he developed his method in an important book, which was among the books that Newton studied at Cambridge. Descartes' pioneering work in analytic geometry paved the way for the invention of differential and integral calculus by Fermat, Newton and Leibnitz. (Besides taking some steps towards the invention of calculus, the great French mathematician, Pierre de Fermat (1601-1665), also discovered analytic geometry independently, but he did not publish this work.)

Fig. 7.1 *Portrait of René Descartes, after Frans Hals. Public domain, Wikimedia Commons*

Analytic geometry made it possible to treat with ease the elliptical orbits which Kepler had introduced into astronomy, as well as the parabolic trajectories which Galileo had calculated for projectiles.

Descartes also worked on a theory which explained planetary motion by means of "vortices"; but this theory was by no means so successful as his analytic geometry, and eventually it had to be abandoned.

Descartes did important work in optics, physiology and philosophy. In philosophy, he is the author of the famous phrase *"Cogito, ergo sum"*,

"I think; therefore I exist", which is the starting point for his theory of knowledge. He resolved to doubt everything which it was possible to doubt; and finally he was reduced to knowledge of his own existence as the only real certainty.

René Descartes died tragically through the combination of two evils which he had always tried to avoid: cold weather and early rising. Even as a student, he spent a large portion of his time in bed. He was able to indulge in this taste for a womblike existence because his father had left him some estates in Brittany. Descartes sold these estates and invested the money, from which he obtained an ample income. He never married, and he succeeded in avoiding responsibilities of every kind.

Descartes might have been able to live happily in this way to a ripe old age if only he had been able to resist a flattering invitation sent to him by Queen Christina of Sweden. Christina, the intellectual and strong-willed daughter of King Gustav Adolf, was determined to bring culture to Sweden, much to the disgust of the Swedish noblemen, who considered that money from the royal treasury ought to be spent exclusively on guns and fortifications. Unfortunately for Descartes, he had become so famous that Queen Christina wished to take lessons in philosophy from him; and she sent a warship to fetch him from Holland, where he was staying. Descartes, unable to resist this flattering attention from a royal patron, left his sanctuary in Holland and sailed to the frozen north.

The only time Christina could spare for her lessons was at five o'clock in the morning, three times a week. Poor Descartes was forced to get up in the utter darkness of the bitterly cold Swedish winter nights to give Christina her lessons in a draughty castle library; but his strength was by no means equal to that of the queen, and before the winter was over he had died of pneumonia.

Newton

On Christmas day in 1642 (the year in which Galileo died), a recently widowed woman named Hannah Newton gave birth to a premature baby at the manor house of Woolsthorpe, a small village in Lincolnshire, England. Her baby was so small that, as she said later, "he could have been put into a quart mug", and he was not expected to live. He did live, however, and lived to achieve a great scientific synthesis, uniting the work of Copernicus, Brahe, Kepler, Galileo and Descartes.

When Isaac Newton was four years old, his mother married again and went to live with her new husband, leaving the boy to be cared for by his grandmother. This may have caused Newton to become more solemn and introverted than he might otherwise have been. One of his childhood friends remembered him as "a sober, silent, thinking lad, scarce known to play with the other boys at their silly amusements".

As a boy, Newton was fond of making mechanical models, but at first he showed no special brilliance as a scholar. He showed even less interest in running the family farm, however; and a relative (who was a fellow of Trinity College) recommended that he be sent to grammar school to prepare for Cambridge University.

When Newton arrived at Cambridge, he found a substitute father in the famous mathematician Isaac Barrow, who was his tutor. Under Barrow's guidence, and while still a student, Newton showed his mathematical genius by inventing the binomial theorem.

In 1665, Cambridge University was closed because of an outbreak of the plague, and Newton returned for two years to the family farm at Woolsthorpe. He was then twenty-three years old. During the two years of isolation, Newton developed his binomial theorem into the beginnings of differential calculus.

Newton's famous experiments in optics also date from these years. The sensational experiments of Galileo were very much discussed at the time, and Newton began to think about ways to improve the telescope. Writing about his experiments in optics, Newton says:

"In the year 1666 (at which time I applied myself to the grinding of optic glasses of other figures than spherical), I procured me a triangular prism, to try therewith the celebrated phenomena of colours. And in order thereto having darkened my chamber, and made a small hole in the window shuts to let in a convenient quantity of the sun's light, I placed my prism at its entrance, that it might thereby be refracted to the opposite wall."

"It was at first a very pleasing divertisment to view the vivid and intense colours produced thereby; but after a while, applying myself to consider them more circumspectly, I became surprised to see them in an oblong form, which, according to the received laws of refraction I expected should have been circular."

Newton then describes his crucial experiment. In this experiment, the beam of sunlight from the hole in the window shutters was refracted by two prisms in succession. The first prism spread the light into a rainbow-like band of colors. From this spectrum, he selected a beam of a single color, and allowed the beam to pass through a second prism; but when light of a single color passed through the second prism, the color did not change, nor was the image spread out into a band. No matter what Newton did to it, red light always remained red, once it had been completely separated from the other colors; yellow light remained yellow, green remained green, and blue remained blue.

Newton then measured the amounts by which the beams of various colors were bent by the second prism; and he discovered that red light was bent the least. Next in sequence came orange, yellow, green, blue and finally violet, which was deflected the most. Newton recombined the separated colors, and he found that together, they once again produced white light.

Concluding the description of his experiments, Newton wrote:

"...and so the true cause of the length of the image (formed by the first prism) was detected to be no other than that light is not similar or homogenial, but consists of *deform rays, some of which are more refrangible than others.*"

"As rays of light differ in their degrees of refrangibility, so they also differ in their disposition to exhibit this or that particular colour... To the same degree of refrangibility ever belongs the same colour, and to the same colour ever belongs the same degree of refrangibility."

"...The species of colour and the degree of refrangibility belonging to any particular sort of rays is not mutable by refraction, nor by reflection from natural bodies, nor by any other cause that I could yet observe. When any one sort of rays hath been well parted from those of other kinds, it hath afterwards obstinately retained its colour, notwithstanding my utmost endeavours to change it."

During the plague years of 1665 and 1666, Newton also began the work which led to his great laws of motion and universal gravitation. Referring to the year 1666, he wrote:

"I began to think of gravity extending to the orb of the moon; and having found out how to estimate the force with which a globe revolving within a sphere presses the surface of the sphere, from Kepler's rule of the periodical times of the planets being in a sesquialternate proportion of their distances from the centres of their orbs, I deduced that the forces which keep the planets in their orbs must be reciprocally as the squares of the distances from the centres about which they revolve; and thereby compared the force requisite to keep the moon in her orb with the force of gravity at the surface of the earth, and found them to answer pretty nearly."

"All this was in the plague years of 1665 and 1666, for in those days I was in the prime of my age for invention, and minded mathematics and philosophy more than at any time since."

Galileo had studied the motion of projectiles, and Newton was able to build on this work by thinking of the moon as a sort of projectile, dropping towards the earth, but at the same time moving rapidly to the side. The combination of these two motions gives the moon its nearly-circular path.

From Kepler's third law, Newton had deduced that the force with which the sun attracts a planet must fall off as the square of the distance between the planet and the sun. With great boldness, he guessed that this force is *universal*, and that every object in the universe attracts every other object with a gravitational force which is directly proportional to the product of the two masses, and inversely proportional to the square of the distance between them.

Newton also guessed correctly that in attracting an object outside its surface, the earth acts as though its mass were concentrated at its center. However, he could not construct the proof of this theorem, since it depended on integral calculus, which did not exist in 1666. (Newton himself invented integral calculus later in his life.)

In spite of the missing proof, Newton continued and "...compared the force requisite to keep the moon in her orb with the force of gravity at the earth's surface, and found them to answer pretty nearly". He was not satisfied with this incomplete triumph, and he did not show his calculations to anyone. He not only kept his ideas on gravitation to himself, (probably because of the missing proof), but he also refrained for many years from publishing his work on the calculus. By the time Newton published, the calculus had been invented independently by the great German mathematician and philosopher, Gottfried Wilhelm Leibniz (1646-1716); and the result was a bitter quarrel over priority. However, Newton did publish his experiments in optics, and these alone were enough to make him famous.

In 1669, Newton's teacher, Isaac Barrow, generously resigned his post as Lucasian Professor of Mathematics so that Newton could have it. Thus, at the age of 27, Newton became the head of the mathematics department at Cambridge. He was required to give eight lectures a year, but the rest of his time was free for research.

Newton's prism experiments had led him to believe that the only possible way to avoid blurring of colors in the image formed by a telescope was to avoid refraction entirely. Therefore he designed and constructed the first reflecting telescope. In 1672, he presented a reflecting telescope to the newly-formed Royal Society, which then elected him to membership.

Meanwhile, the problems of gravitation and planetary motion were increasingly discussed by the members of the Royal Society. In January, 1684, three members of the Society were gathered in a London coffee house. One of them was Robert Hooke (1635-1703), author of *Micrographia* and Professor of Geometry at Gresham College, a brilliant but irritable man. He had begun his career as Robert Boyle's assistant, and had gone on to do important work in many fields of science. Hooke claimed that he could calculate the motion of the planets by assuming that they were attracted to the sun by a force which diminished as the square of the distance.

Listening to Hooke were Sir Christopher Wren (1632-1723), the designer of St. Paul's Cathedral, and the young astronomer, Edmund Halley (1656-1742). Wren challenged Hooke to produce his calculations; and he offered to present Hooke with a book worth 40 shillings if he could prove his inverse square force law by means of rigorous mathematics. Hooke tried for several months, but he was unable to win Wren's reward.

Meanwhile, in August, 1684, Halley made a journey to Cambridge to talk with Newton, who was rumored to know very much more about the motions of the planets than he had revealed in his published papers. According to an almost-contemporary account, what happened then was the following:

"Without mentioning his own speculations, or those of Hooke and Wren, he (Halley) at once indicated the object of his visit by asking Newton what would be the curve described by the planets on the supposition that gravity diminished as the square of the distance. Newton immediately answered: an Ellipse. Struck with joy and amazement, Halley asked how he knew it? 'Why', replied he, 'I have calculated it'; and being asked for the calculation, he could not find it, but promised to send it to him."

Newton soon reconstructed the calculation and sent it to Halley; and Halley, filled with enthusiasm and admiration, urged Newton to write out in detail all of his work on motion and gravitation. Spurred on by Halley's encouragement and enthusiasm, Newton began to put his research in order. He returned to the problems which had occupied him during the plague years, and now his progress was rapid because he had invented integral calculus. This allowed him to prove rigorously that terrestrial gravitation acts as though all the earth's mass were concentrated at its center. Newton also had available an improved value for the radius of the earth, measured by the French astronomer Jean Picard (1620-1682). This time, when he approached the problem of gravitation, everything fell into place.

By the autumn of 1684, Newton was ready to give a series of lectures on dynamics, and he sent the notes for these lectures to Halley in the form of a small booklet entitled *On the Motion of Bodies.* Halley persuaded Newton to develop these notes into a larger book, and with great tact and patience he struggled to keep a controversy from developing between Newton, who was neurotically sensitive, and Hooke, who was claiming his share of recognition in very loud tones, hinting that Newton was guilty of plagiarism.

Although Newton was undoubtedly the greatest physicist of all time, he had his shortcomings as a human being; and he reacted by striking out from his book every single reference to Robert Hooke. The Royal Society at first offered to pay for the publication costs of Newton's book, but because a fight between Newton and Hooke seemed possible, the Society discretely backed out. Halley then generously offered to pay the publication costs himself, and in 1686 Newton's great book was printed. It is entitled *Philosophae Naturalis Principia Mathematica*, (The Mathematical Principles of Natural Philosophy), and it is divided into three sections.

The first book sets down the general principles of mechanics. In it, Newton states his three laws of motion, and he also discusses differential and integral calculus (both invented by himself).

In the second book, Newton applies these methods to systems of particles and to hydrodynamics. For example, he calculates the velocity of sound in air from the compressibility and density of air; and he treats a great variety of other problems, such as the problem of calculating how a body moves when its motion is slowed by a resisting medium, such as air or water.

The third book is entitled *The System of the World.* In this book, Newton sets out to derive the entire behavior of the solar system from his three laws of motion and from his law of universal gravitation. From these, he not only derives all three of Kepler's laws, but he also calculates the periods of the planets and the periods of their moons; and he explains such details as the flattened, non-spherical shape of the earth, and the slow precession of its axis about a fixed axis in space. Newton also calculated the irregular motion of the moon resulting from the combined attractions of the earth and the sun; and he determined the mass of the moon from the behavior of the tides.

Newton's *Principia* is generally considered to be the greatest scientific work of all time. To present a unified theory explaining such a wide variety of phenomena with so few assumptions was a magnificent and unprecedented achievement; and Newton's contemporaries immediately recognized the importance of what he had done.

The great Dutch physicist, Christian Huygens (1629-1695), inventor of the pendulum clock and the wave theory of light, travelled to England with the express purpose of meeting Newton. Voltaire, who for reasons of personal safety was forced to spend three years in England, used the time to study Newton's *Principia*; and when he returned to France, he persuaded his mistress, Madame du Chatelet, to translate the *Principia* into French; and Alexander Pope, expressing the general opinion of his contemporaries, wrote a famous couplet, which he hoped would be carved on Newton's tombstone:

"Nature and Nature's law lay hid in night.
God said: 'Let Newton be!', and all was light!"

The Newtonian synthesis was the first great achievement of a new epoch in human thought, an epoch which came to be known as the "Age of Reason" or the "Enlightenment". We might ask just what it was in Newton's work that so much impressed the intellectuals of the 18th century. The answer is that in the Newtonian system of the world, the entire evolution of the solar system is determined by the laws of motion and by the positions and velocities of the planets and their moons at a given instant of time. Knowing these, it is possible to predict all of the future and to deduce all of the past.

The Newtonian system of the world is like an enormous clock which has to run on in a predictable way once it is started. In this picture of the world, comets and eclipses are no longer objects of fear and superstition. They too are part of the majestic clockwork of the universe. The Newtonian laws are simple and mathematical in form; they have complete generality; and they are unalterable. In this picture, although there are no miracles or exceptions to natural law, nature itself, in its beautiful works, can be regarded as miraculous.

Newton's contemporaries knew that there were other laws of nature to be discovered besides those of motion and gravitation; but they had no doubt that, given time, all of the laws of nature would be discovered. The climate of intellectual optimism was such that many people thought that these discoveries would be made in a few generations, or at most in a few centuries.

In 1704, Newton published a book entitled *Opticks*, expanded editions of which appeared in 1717 and 1721. Among the many phenomena discussed in this book are the colors produced by thin films. For example, Newton discovered that when he pressed two convex lenses together, the thin film of air trapped between the lenses gave rise to rings of colors ("Newton's rings"). The same phenomenon can be seen in the in the colors of soap bubbles or in films of oil on water.

In order to explain these rings, Newton postulated that "..every ray of light in its passage through any refracting surface is put into a transient constitution or state, which in the progress of the ray returns at equal intervals, and disposes the ray at every return to be easily transmitted through the next refracting surface and between the returns to be easily reflected from it."

Newton's rings were later understood on the basis of the wave theory of light advocated by Huygens and Hooke. Each color has a characteristic wavelength, and is easily reflected when the ratio of the wavelength to the film thickness is such that the wave reflected from the bottom surface of the film interferes constructively with the wave reflected from the top surface. However, although he ascribed periodic "fits of easy reflection" and "fits of easy transmission" to light, and although he suggested that a particular wavelength is associated with each color, Newton rejected the wave theory of light, and believed instead that light consists of corpuscles emitted from luminous bodies.

Newton believed in his corpuscular theory of light because he could not understand on the basis of Huygens' wave theory how light casts sharp shadows. This is strange, because in his *Opticks* he includes the following passage:

"Grimaldo has inform'd us that if a beam of the sun's light be let into a dark room through a very small hole, the shadows of things in this light will be larger than they ought to be if the rays went on by the bodies in straight lines, and that these shadows have three parallel fringes, bands or ranks of colour'd light adjacent to them. But if the hole be enlarg'd, the fringes grow broad and run into one another, so that they cannot be distinguish'd"

After this mention of the discovery of diffraction by the Italian physicist, Francesco Maria Grimaldi (1618-1663), Newton discusses his own studies of diffraction. Thus, Newton must have been aware of the fact that light from a very small source does not cast completely sharp shadows!

Newton felt that his work on optics was incomplete, and at the end of his book he included a list of "Queries", which he would have liked to have investigated. He hoped that this list would help the research of others. In general, although his contemporaries were extravagant in praising him,

Newton's own evaluation of his work was modest. "I do not know how I may appear to the world", he wrote, "but to myself I seem to have been only like a boy playing on the seashore and diverting myself in now and then finding a smoother pebble or a prettier shell than ordinary, whilst the great ocean of truth lay all undiscovered before me."

Fig. 7.2 *Portrait of Isaac Newton (1642-1727) by Sir Godfrey Kneller. Public domain, Wikimedia Commons*

Huygens and Leibniz

Meanwhile, on the continent, mathematics and physics had been developing rapidly, stimulated by the writings of René Descartes. One of the most distinguished followers of Descartes was the Dutch physicist, Christian Huygens (1629-1695).

Huygens was the son of an important official in the Dutch government. After studying mathematics at the University of Leiden, he published the first formal book ever written about probability. However, he soon was diverted from pure mathematics by a growing interest in physics.

In 1655, while working on improvements to the telescope together with his brother and the Dutch philosopher Benedict Spinoza, Huygens invented an improved method for grinding lenses. He used his new method to construct a twenty-three foot telescope, and with this instrument he made a number of astronomical discoveries, including a satellite of Saturn, the rings of Saturn, the markings on the surface of Mars and the Orion Nebula.

Huygens was the first person to estimate numerically the distance to a star. By assuming the star Sirius to be exactly as luminous as the sun, he calculated the distance to Sirius, and found it to be 2.5 trillion miles. In fact, Sirius is more luminous than the sun, and its true distance is twenty times Huygens' estimate.

Another of Huygens' important inventions is the pendulum clock. Improving on Galileo's studies, he showed that for a pendulum swinging in a circular arc, the period is not precisely independent of the amplitude of the swing. Huygens then invented a pendulum with a modified arc, not quite circular, for which the swing was exactly isochronous. He used this improved pendulum to regulate the turning of cog wheels, driven by a falling weight; and thus he invented the pendulum clock, almost exactly as we know it today.

In discussing Newton's contributions to optics, we mentioned that Huygens opposed Newton's corpuscular theory of light, and instead advocated a wave theory. Huygens believed that the rapid motion of particles in a hot body, such as a candle flame, produces a wave-like disturbance in the surrounding medium; and he believed that this wavelike disturbance of the "ether" produces the sensation of vision by acting on the nerves at the back of our eyes.

In 1678, while he was working in France under the patronage of Louis XIV, Huygens composed a book entitled *Traité de la Lumiere*, (Treatise on Light), in which he says:

Fig. 7.3 *Christian Huygens. Public domain, Wikimedia Commons*

"...It is inconceivable to doubt that light consists of the motion of some sort of matter. For if one considers its production, one sees that here upon the earth it is chiefly engendered by fire and flame, which undoubtedly contain bodies in rapid motion, since they dissolve and melt many other bodies, even the most solid; or if one considers its effects, one sees that when light is collected, as by concave mirrors, it has the property of burning as fire does, that is to say, it disunites the particles of bodies. This is assuredly the mark of motion, at least in the true philosophy in which one conceives the causes of all natural effects in terms of mechanical motions..."

"Further, when one considers the extreme speed with which light spreads on every side, and how, when it comes from different regions, even from those directly opposite, the rays traverse one another without hindrance, one may well understand that when we see a luminous object, it cannot be by any transport of matter coming to us from the object, in the way in which a shot or an arrow traverses the air; for assuredly that would too greatly impugn these two properties of light, especially the second of them. It is in some other way that light spreads; and that which can lead us to comprehend it is the knowledge which we have of the spreading of sound in the air."

Huygens knew the velocity of light rather accurately from the work of the Danish astronomer, Ole Rømer (1644-1710), who observed the moons of Jupiter from the near and far sides of the earth's orbit. By comparing the calculated and observed times for the moons to reach a certain configuration, Rømer was able to calculate the time needed for light to propagate across the diameter of the earth's orbit. In this way, Rømer calculated the velocity of light to be 227,000 kilometers per second. Considering the early date of this first successful measurement of the velocity of light, it is remarkably close to the accepted modern value of 299,792 kilometers per second. Thus Huygens knew that although the speed of light is enormous, it is not infinite.

Huygens considered the propagation of a light wave to be analogous to the spreading of sound, or the widening of the ripple produced when a pebble is thrown into still water. He developed a mathematical principle for calculating the position of a light wave after a short interval of time if the initial surface describing the wave front is known. Huygens considered each point on the initial wave front to be the source of spherical wavelets, moving outward with the speed of light in the medium. The surface marking the boundary between the region outside all of the wavelets and the region inside some of them forms the new wave front.

If one uses Huygens' Principle to calculate the wave fronts and rays for light from a point source propagating past a knife edge, one finds that a part of the wave enters the shadow region. This is, in fact, precisely the effect which was observed by both Grimaldi and Newton, and which was given the name "diffraction" by Grimaldi. In the hands of Thomas Young (1773-1829) and Augustin Jean Fresnel (1788-1827), diffraction effects later became a strong argument in favor of Huygens' wave theory of light.

(You can observe diffraction effects yourself by looking at a point source of light, such as a distant street lamp, through a piece of cloth, or through a small slit or hole. Another type of diffraction can be seen by looking at light reflected at a grazing angle from a phonograph record. The light will appear to be colored. This effect is caused by the fact that each groove is a source of wavelets, in accordance with Huygens' Principle. At certain angles, the wavelets will interfere constructively, the angles for constructive interference being different for each color.)

Interestingly, modern quantum theory (sometimes called wave mechanics) has shown that *both* Huygens' wave theory of light and Newton's corpuscular theory contain aspects of the truth! Light has both wave-like and particle-like properties. Furthermore, quantum theory has shown that small particles of matter, such as electrons, also have wave-like properties! For example, electrons can be diffracted by the atoms of a crystal in a manner exactly analogous to the diffraction of light by the grooves of a phonograph record. Thus the difference of opinion between Huygens and Newton concerning the nature of light is especially interesting, since it foreshadows the wave-particle duality of modern physics.

Among the friends of Christian Huygens was the German philosopher and mathematician Gottfried Wilhelm Leibniz (1646-1716). Leibniz was a man of universal and spectacular ability. In addition to being a mathematician and philosopher, he was also a lawyer, historian and diplomat. He invented the doctrine of balance of power, attempted to unify the Catholic and Protestant churches, founded academies of science in Berlin and St. Petersburg, invented combinatorial analysis, introduced determinants into mathematics, independently invented the calculus, invented a calculating machine which could multiply and divide as well as adding and subtracting, acted as advisor to Peter the Great and originated the theory that "this is the best of all possible worlds" (later mercilessly satirized by Voltaire in *Candide*).

Leibniz learned mathematics from Christian Huygens, whom he met while travelling as an emissary of the Elector of Mainz. Since Huygens too was a man of very wide interests, he found the versatile Leibniz congenial, and gladly agreed to give him lessons. Leibniz continued to correspond with Huygens and to receive encouragement from him until the end of the older man's life.

In 1673, Leibniz visited England, where he was elected to membership by the Royal Society. During the same year, he began his work on cal-

Fig. 7.4 *Portrait of Gottfried Wilhelm Leibniz by J.F. Wentzel. Public domain, Wikimedia Commons*

culus, which he completed and published in 1684. Newton's invention of differential and integral calculus had been made much earlier than the independent work of Leibniz, but Newton did not publish his discoveries until 1687. This set the stage for a bitter quarrel over priority between the admirers of Newton and those of Leibniz. The quarrel was unfortunate for everyone concerned, especially for Leibniz himself. He had taken a position in the service of the Elector of Hanover, which he held for forty years. However, in 1714, the Elector was called to the throne of England as George I. Leibniz wanted to accompany the Elector to England, but was left behind, mainly because of the quarrel with the followers of Newton. Leibniz died two years later, neglected and forgotten, with only his secretary attending the funeral.

The Bernoullis and Euler

Among the followers of Leibniz was an extrordinary family of mathematicians called Bernoulli. They were descended from a wealthy merchant family in Basle, Switzerland. The head of the family, Nicolas Bernoulli the Elder, tried to force his three sons, James (1654-1705), Nicolas II (1662-1716) and John (1667-1748) to follow him in carrying on the family business. However, the eldest son, James, had taught himself the Leibnizian form of calculus, and instead became Professor of Mathematics at the University of Basle. His motto was *"Invicto patre sidera verso"* ("Against my father's will, I study the stars").

Nicolas II and John soon caught their brother's enthusiasm, and they learned calculus from him. John became Professor of Mathematics in Gröningen and Nicolas II joined the faculty of the newly-formed Academy of St. Petersburg. John Bernoulli had three sons, Nicolas III (1695-1726), Daniel (1700-1782) and John II (1710-1790), all of whom made notable contributions to mathematics and physics. In fact, the family of Nicolas Bernoulli the Elder produced a total of nine famous mathematicians in three generations!

Daniel Bernoulli's brilliance made him stand out even among the other members of his gifted family. He became professor of mathematics at the Academy of Sciences in St. Petersburg when he was twenty-five. After eight Russian winters however, he returned to his native Basle. Since the chair in mathematics was already occupied by his father, he was given a vacant chair, first in anatomy, then in botany, and finally in physics. In spite of the variety of his titles, however, Daniel's main work was in applied mathematics, and he has been called the father of mathematical physics.

One of the good friends of Daniel Bernoulli and his brothers was a young man named Leonhard Euler (1707-1783). He came to their house once a week to take private lessons from their father, John Bernoulli. Euler was destined to become the most prolific mathematician in history, and the Bernoullis were quick to recognize his great ability. They persuaded Euler's father not to force him into a theological career, but instead to allow him to go with Nicolas III and Daniel to work at the Academy in St. Petersburg.

Euler married the daughter of a Swiss painter and settled down to a life of quiet work, producing a large family and an unparalleled output of papers. A recent edition of Euler's works contains 70 quatro volumes of published research and 14 volumes of manuscripts and letters. His books and papers are mainly devoted to algebra, the theory of numbers, analysis, mechanics, optics, the calculus of variations (invented by Euler), geometry, trigonometry and astronomy; but they also include contributions to shipbuilding science, architecture, philosophy and musical theory!

Euler achieved this enormous output by means of a calm and happy disposition, an extraordinary memory and remarkable powers of concentration, which allowed him to work even in the midst of the noise of his large family. His friend Thiébault described Euler as sitting "..with a cat on his shoulder and a child on his knee - that was how he wrote his immortal works".

In 1771, Euler became totally blind. Nevertheless, aided by his sons and his devoted scientific assistants, he continued to produce work of fundamental importance. It was his habit to make calculations with chalk on a board for the benefit of his assistants, although he himself could not see what he was writing. Appropriately, Euler was making such computations on the day of his death. On September 18, 1783, Euler gave a mathematics lesson to one of his grandchildren, and made some calculations on the motions of balloons. He then spent the afternoon discussing the newly-discovered planet Uranus with two of his assistants. At five o'clock, he suffered a cerebral hemorrhage, lost consciousness, and died soon afterwards. As one of his biographers put it, "The chalk fell from his hand; Euler ceased to calculate, and to live".

In the eighteenth century it was customary for the French Academy of Sciences to propose a mathematical topic each year, and to award a prize for the best paper dealing with the problem. Léonard Euler and Daniel Bernoulli each won the Paris prize more than ten times, and they share the distinction of being the only men ever to do so. John Bernoulli is said to have thrown his son out of the house for winning the Paris prize in a year when he himself had competed for it.

Euler and the Bernoullis did more than anyone else to develop the Leibnizian form of calculus into a workable tool and to spread it throughout Europe. They applied it to a great variety of problems, from the shape of ships' sails to the kinetic theory of gasses. An example of the sort of problem which they considered is the vibrating string.

In 1727, John Bernoulli in Basle, corresponding with his son Daniel in St. Petersburg, developed an approximate set of equations for the motion of a vibrating string by considering it to be a row of point masses, joined together by weightless springs. Then Daniel boldly passed over to the continuum limit, where the masses became infinitely numerous and small.

The result was Daniel Bernoulli's famous wave equation, which is what we would now call a partial differential equation. He showed that the wave equation has sinusoidal solutions, and that the sum of any two solutions is also a solution. This last result, his superposition principle, is a mathematical proof of a property of wave motion noticed by Huygens. The fact that many waves can propagate simultaneously through the same medium without interacting was one of the reasons for Huygens' belief that light is wavelike, since he knew that many rays of light from various directions can cross a given space simultaneously without interacting. Because of their work with partial differential equations, Daniel Bernoulli and Léonard Euler are considered to be the founders of modern theoretical physics.

Political philosophy of the Enlightenment

The 16th, 17th and 18th centuries have been called the "Age of Discovery", and the "Age of Reason", but they might equally well be called the "Age of Observation". On every side, new worlds were opening up to the human mind. The great voyages of discovery had revealed new continents, whose peoples demonstrated alternative ways of life. The telescopic exploration of the heavens revealed enormous depths of space, containing myriads of previously unknown stars; and explorations with the microscope revealed a new and marvelously intricate world of the infinitesimally small.

In the science of this period, the emphasis was on careful observation. This same emphasis on observation can be seen in the Dutch and English painters of the period. The great Dutch masters, such as Jan Vermeer (1632-1675), Frans Hals (1580-1666), Pieter de Hooch (1629-1678) and Rembrandt van Rijn (1606-1669), achieved a careful realism in their paintings and drawings which was the artistic counterpart of the observations of the pioneers of microscopy, Anton van Leeuwenhoek and Robert Hooke. These artists were supported by the patronage of the middle class, which had become prominent and powerful both in England and in the Netherlands because of the extensive world trade in which these two nations were engaged.

Members of the commercial middle class needed a clear and realistic view of the world in order to succeed with their enterprises. (An aristocrat of the period, on the other hand, might have been more comfortable with a somewhat romanticized and out-of-focus vision, which would allow him to overlook the suffering and injustice upon which his privilages were based.) The rise of the commercial middle class, with its virtues of industriousness, common sense and realism, went hand in hand with the rise of experimental science, which required the same virtues for its success.

In England, the House of Commons (which reflected the interests of the middle class), had achieved political power, and had demonstrated (in the Puritan Rebellion of 1640 and the Glorious Revolution of 1688) that Parliament could execute or depose any monarch who tried to rule without its consent. In France, however, the situation was very different.

After passing through a period of disorder and civil war, the French tried to achieve order and stability by making their monarchy more absolute. The movement towards absolute monarchy in France culminated in the long reign of Louis XIV, who became king in 1643 and who ruled until he died in 1715.

The historical scene which we have just sketched was the background against which the news of Newton's scientific triumph was received. The news was received by a Europe which was tired of religious wars; and in France, it was received by a middle class which was searching for an ideology in its struggle against the *ancien régime*.

To the intellectuals of the 18th century, the orderly Newtonian cosmos, with its planets circling the sun in obedience to natural law, became an imaginative symbol representing rationality. In their search for a society more in accordance with human nature, 18th century Europeans were greatly encouraged by the triumphs of science. Reason had shown itself to be an adequate guide in natural philosophy. Could not reason and natural law also be made the basis of moral and political philosophy? In attempting to carry out this program, the philosophers of the Enlightenment laid the foundations of psychology, anthropology, social science, political science and economics.

One of the earliest and most influential of these philosophers was John Locke (1632-1705), a contemporary and friend of Newton. In his *Second Treatise on Government*, published in 1690, John Locke's aim was to refute the doctrine that kings rule by divine right, and to replace that doctrine by an alternative theory of government, derived by reason from the laws of nature. According to Locke's theory, men originally lived together without formal government:

"Men living together according to reason," he wrote, "without a common superior on earth with authority to judge between them, is properly the state of nature... A state also of equality, wherein all the power and jurisdiction is reciprocal, no one having more than another; there being nothing more evident than that creatures of the same species, promiscuously born to all the same advantages of nature and the use of the same facilities, should also be equal amongst one another without subordination or subjection..."

"But though this be a state of liberty, yet it is not a state of licence... The state of nature has a law to govern it, which obliges every one; and reason, which is that law, teaches all mankind who will but consult it, that being equal and independent, no one ought to harm another in his life, health, liberty or possessions."

In Locke's view, a government is set up by means of a social contract. The government is given its powers by the consent of the citizens in return for the services which it renders to them, such as the protection of their lives and property. If a government fails to render these services, or if it becomes tyrannical, then the contract has been broken, and the citizens must set up a new government.

Locke's influence on 18th century thought was very great. His influence can be seen, for example, in the wording of the American Declaration of Independence. In England, Locke's political philosophy was accepted by almost everyone. In fact, he was only codifying ideas which were already in wide circulation and justifying a revolution which had already occurred. In France, on the other hand, Locke's writings had a revolutionary impact.

Credit for bringing the ideas of both Newton and Locke to France, and making them fashionable, belongs to Francois Marie Arouet (1694-1778), better known as "Voltaire". Besides persuading his mistress, Madame de Chatelet, to translate Newton's *Principia* into French, Voltaire wrote an extremely readable commentary on the book; and as a result, Newton's ideas became highly fashionable among French intellectuals. Voltaire lived with Madame du Chatalet until she died, producing the books which established him as the leading writer of Europe, a prophet of the Age of Reason, and an enemy of injustice, feudalism and superstition.

The Enlightenment in France is considered to have begun with Voltaire's return from England in 1729; and it reached its high point with the publication of of the *Encyclopedia* between 1751 and 1780. Many authors contributed to the *Encyclopedia*, which was an enormous work, designed to sum up the state of human knowledge.

Fig. 7.5 *Portrait of John Locke, by Sir Godfrey Kneller. Public domain, Wikimedia Commons*

Turgot and Montesquieu wrote on politics and history; Rousseau wrote on music, and Buffon on natural history; Quesnay contributed articles on agriculture, while the Baron d'Holbach discussed chemistry. Other articles were contributed by Condorcet, Voltaire and d'Alembert. The whole enterprise was directed and inspired by the passionate faith of Denis Diderot (1713-1784). The men who took part in this movement called themselves *"philosophes"*. Their creed was a faith in reason, and an optimistic belief in the perfectability of human nature and society by means of education, political reforms, and the scientific method.

The *philosophes* of the Enlightenment visualized history as a long progression towards the discovery of the scientific method. Once discovered, this method could never be lost; and it would lead inevitably (they believed) to both the material and moral improvement of society. The *philosophes* believed that science, reason, and education, together with the principles of political liberty and equality, would inevitably lead humanity forward to a new era of happiness. These ideas were the faith of the Enlightenment; they influenced the French and American revolutions; and they are still the basis of liberal political belief.

Suggestions for further reading

(1) Phillip Bricker and R.I.G. Hughs, *Philosophical Perspectives on Newtonian Science*, M.I.T. Press, Cambridge, Mass., (1990).
(2) Zev Bechler, *Newton's Physics and the Conceptual Structure of the Scientific Revolution*, Kluwer, Dordrecht, (1991).
(3) Zev Bechler, *Contemporary Newtonian Research*, Reidel, Dordrecht, (1982).
(4) I. Bernard Cohen, *The Newtonian Revolution*, Cambridge University Press, (1980).
(5) B.J.T. Dobbs, *The Janus Face of Genius; The Role of Alchemy in Newton's Thought*, Cambridge University Press, (1991).
(6) Paul B. Scheurer and G. Debrock, *Newton's Scientific and Philosophical Legacy*, Kluwer, Dordrecht, (1988).
(7) A. Rupert Hall, *Isaac Newton, Adventurer in Thought*, Blackwell, Oxford, (1992).
(8) Frank Durham and Robert D. Purrington, *Some Truer Method; Reflections on the Heritage of Newton*, Columbia University Press, New York, (1990).
(9) John Fauvel, *Let Newton Be*, Oxford University Press, (1989).
(10) René Taton and Curtis Wilson, *Planetary Astronomy from the Renaissance to the Rise of Astrophysics*, Cambridge University Press, (1989).
(11) Brian Vickers, *English Science, Bacon to Newton*, Cambridge University Press, (1989).

(12) John G. Burke, *The Uses of Science in the Age of Newton*, University of California Press, (1983).
(13) A.I. Sabra, *Theories of Light from Descartes to Newton*, Cambridge University Press, (1991).
(14) E.N. da Costa Andrade, *Isaac Newton*, Folcroft Library Editions, (1979).
(15) Gideon Freudenthal, *Atom and Individual in the Age of Newton*, Reidel, Dordrecht, (1986).
(16) Henry Guerlac, *Newton on the Continent*, Cornell University Press, (1981).
(17) A.R. Hall, *Philosophers at War; the Quarrel Between Newton and Leibnitz*, Cambridge University Press, (1980).
(18) Gale E. Christianson, *In the Presence of the Creator; Isaac Newton and his Times*, Free Press, New York, (1984).
(19) Lesley Murdin, *Under Newton's Shadow; Astronomical Practices in the Seventeenth Century*, Hilger, Bristol, (1985).
(20) H.D. Anthony, *Sir Isaac Newton*, Collier, New York (1961).
(21) Sir Oliver Lodge, *Pioneers of Science*, Dover, New York (1960).

Chapter 8

THE INDUSTRIAL REVOLUTION

Technical change

We have just seen how the development of printing in Europe produced a brilliant, chainlike series of scientific discoveries. During the 17th century, the rate of scientific progress gathered momentum, and in the 18th and 19th centuries, the practical applications of scientific knowledge revolutionized the methods of production in agriculture and industry.

The changes produced by the industrial revolution at first resulted in social chaos - enormous wealth in some classes of society, and great suffering in other classes; but later, after the appropriate social and political adjustments had been made, the improved methods of production benefited all parts of society in a more even way.

There is, in fact, a general pattern which we can notice in the social impact of technology: Technical changes usually occur rapidly, while social and political adjustments take more time. The result is an initial period of social disruption following a technical change, which continues until the structure of society has had time to adjust. Thus, for example, the introduction of a money-based economy into a society which has previously been based on a pattern of traditional social duties always creates an initial period of painful disruption.

In the case of the Industrial Revolution, feudal society, with its patterns of village life and its traditional social obligations, was suddenly replaced by an industrial society whose rules were purely economic, and in which labor was regarded as a commodity. At first, the change produced severe social disruption and suffering; but now, after two centuries of social and political adjustment, the industrialized countries are generally considered to have benefited from the change.

Cullen, Black and Watt

The two driving forces behind the Industrial Revolution were world trade and scientific discovery. During the 18th century, both these forces were especially strongly felt in Scotland and in the north-western part of England. The distilling industry in Scotland grew enormously because of world trade; and the resulting interest in what happens when liquids are vaporized and condensed produced one of the major scientific and technical developments of the Industrial Revolution.

The first step in this development was taken by William Cullen, a professor of medicine at the universities of Glasgow and Edinburgh. In a paper entitled *Of the Cold Produced by Evaporation* (1749), Cullen wrote that he had noticed that "...water and some other liquids, in evaporating, produce some degree of cold".

Cullen therefore began to make experiments in which he dipped a thermometer in and out of a liquid and observed the drop in temperature. He noticed that the effect was increased by "...moving the thermometer very nimbly to and fro in the air; or if, while the ball was wet with spirit of wine, it was blown upon with a pair of bellows". In this way, Cullen achieved a temperature 44 degrees below the freezing point of water. He next tried producing vacuums above various liquids with the help of an air pump:

"We set the vessel containing the ether", Cullen wrote, "In another a little larger, containing water. Upon exhausting the receiver and the vessel's remaining a few minutes *in vacuo*, we found the most part of the water frozen, and the vessel containing the ether surrounded with a thick crust of ice."

One of Cullen's favorite students at Edinburgh was Joseph Black (1728-1799). He became Cullen's scientific assistant, and later, in 1756, he was elected to the Chair of Medicine at Glasgow University. Continuing Cullen's work on the cold produced by evaporating liquids, Black discovered and studied quantitatively the phenomenon of latent heats, e.g., the very large quantities of heat which are necessary to convert ice into water, or to convert water into steam.

Black was led to his discovery of latent heats not only by Cullen's work, but also by his own observations on Scottish weather. Writing of the discovery, one of Black's friends at Glasgow recorded that "...since a fine winter day of sunshine did not at once clear the hills of snow, nor a frosty night suddenly cover the ponds with ice, Dr. Black was already convinced that much heat was absorbed and fixed in the water which slowly trickled from the wreaths of snow; and on the other hand, that much heat emerged from it while it was slowly changing into ice. For, during a thaw, a thermometer will always sink when removed from the air into melting snow; and during a severe frost it will rise when plunged into freezing water. Therefore in the first case, the snow is receiving heat, and in the last, the water is allowing it to emerge again."

Fig. 8.1 *James Watt. Public domain, Wikimedia Commons*

At Glasgow University, where Joseph Black was Professor of Medicine, there was a shop where scientific instruments were made and sold. The owner of the shop was a young man named James Watt (1736-1819), who came from a family of ship builders and teachers of mathematics and navigation. Besides being an extremely competent instrument maker, Watt was a self-taught scientist of great ability, and his shop became a meeting place for scientifically inclined students. Dr. Black was also a frequent visitor to Watt's shop, and a strong friendship formed between the professor and the highly intelligent young instrument maker.

In 1763, Glasgow University asked James Watt to repair a model of a Newcomen steam engine. This type of steam engine had been used for several years to pump water out of mines. It had a single cylinder which filled with steam so that the piston was driven to one end. Then water was sprayed into the cylinder, condensing the steam; and the vacuum drew the piston back to the other end of the cylinder, thus completing the cycle.

James Watt tried to repair the university's small-scale model of the Newcomen engine, but he failed to make it work well. He could see that it was extraordinarily inefficient in its use of fuel, and he began making experiments to find out why it was so wasteful. Because of James Watt's friendship with Joseph Black, he quickly found the answer in the phenomena of latent heats and specific heats: The engine was inefficient because of the large amounts of energy needed to convert water into steam and to heat the iron cylinder.

In 1765, Watt designed an improved engine with a separate condenser. The working cylinder could then be kept continuously hot, and the condensing steam could be returned through the boiler, so that its latent heat could be used to preheat the incoming water. To have an idea for a new, energy-saving engine was one thing, however, and to make the machine practical was another. James Watt had experience as instrument maker, but no experience in large-scale engineering.

In 1767, Watt was engaged to make a survey for a canal which was to join the Forth and the Clyde through Loch Lomond. Because of this work, he had to make a trip to London to explain the canal project to a parliamentary committee; and on the return trip he met Dr. Erasmus Darwin in Birmingham. Darwin, who was interested in steam engines, quickly recognized Watt's talent and the merit of his idea.

Erasmus Darwin (1731-1802) was the most famous physician of the period, but his interests were by no means confined to medicine. He anticipated his grandson, Charles Darwin, by developing the first reasonably well thought-out theory of evolution; and, at the time when he met James Watt he was enthusiastically trying to design a steam locomotive. His collaborators in this project were Benjamin Franklin and the pioneering Birmingham industrialist, Matthew Boulton.

Fig. 8.2 *Erasmus Darwin (grandfather of Charles Darwin). Public domain, Wikimedia Commons*

In August, 1767, Erasmus Darwin wrote to Watt: "The plan of your steam improvements I have religiously kept secret, but begin to see myself some difficulties in your execution, which did not strike me when you were here. I have got another and another hobby horse since I saw you. I wish that the Lord would send you to pass a week with me, and Mrs. Watt with you; - a week, a month, a year!"

Dr. Darwin introduced James Watt to Matthew Boulton, and a famous partnership was formed. The partnership of Boulton and Watt was destined to make the steam engine practical, and thus to create a new age - an age in which humans would would rely for power neither on their own muscles nor on the muscles of slaves, but would instead control almost unlimited power through their engines.

James Watt was lucky to meet Erasmus Darwin and to be introduced to Matthew Boulton, since Boulton was the most talented and progressive manufacturer in England - the best possible man to understand the significance of Watt's great invention and to help in its development.

Boulton

Matthew Boulton was the son of a Birmingham manufacturer, and at the age of seventeen, he had invented a type of metal buckle inlaid with glass, which proved to be extremely popular and profitable. By the time that he was twenty-one, his father had made him manager of the business. At twenty-eight, Matthew Boulton married an heiress, receiving a very large dowry. When his wife died four years later, Boulton married her younger sister, and he was given a second large fortune.

Instead of retiring from manufacturing and becoming a country gentleman, as most of his contemporaries would have done, Boulton used his wealth to try out new ideas. He tried especially to improve the quality of the goods manufactures in Birmingham. Since he was already an extremely rich man, he was more interested in applying art and science to manufacturing than he was in simply making money.

Boulton's idea was to bring together under one roof the various parts of the manufacturing process which had been scattered among many small workshops by the introduction of division of labor. He believed that improved working conditions would result in an improved quality of products.

With these ideas in mind, Matthew Boulton built a large mansion-like house on his property at Soho, outside Birmingham, and installed in it all the machinery necessary for the complete production of a variety of small steel products. Because of his personal charm, and because of the comfortable working conditions at the Soho Manufactory, Boulton was able to attract the best and most skillful craftsmen in the region; and by 1765, the number of the staff at Soho had reached 600.

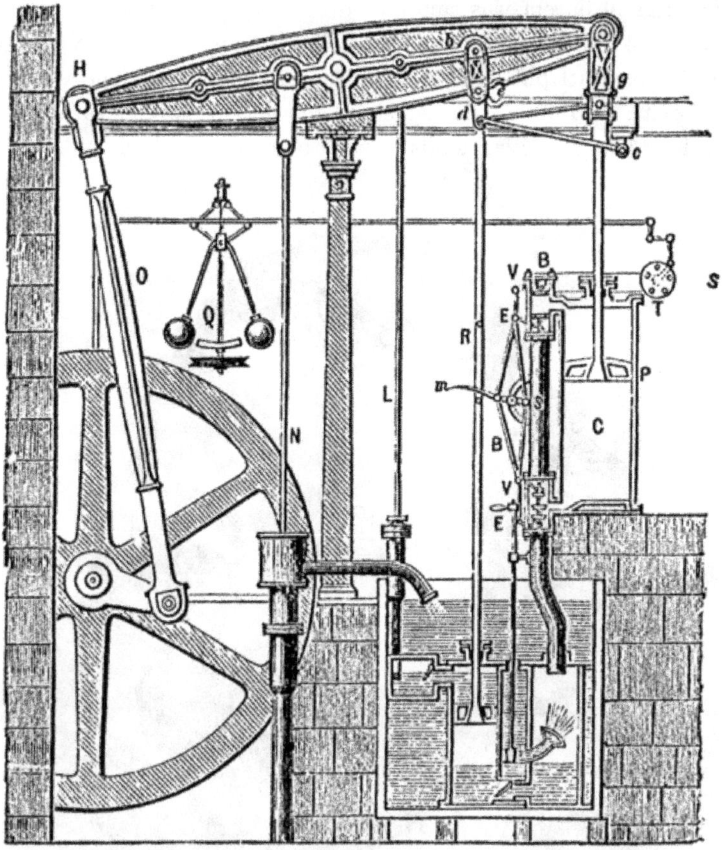

Fig. 8.3 *A diagram of the Boulton-Watt steam engine of 1784. Steam enters through a pipe (S,T) from a boiler on the right, not shown in the diagram. Public domain, Wikimedia Commons*

Boulton continued to manufacture utilitarian goods, on which he made a profit, but he also introduced a line of goods of high artistic merit on which he gained prestige but lost money. He made fine gilt brass candelabra for both George III and Catherine the Great; and he was friendly with George III, who consulted him on technical questions.

At this point, Erasmus Darwin introduced James Watt to Matthew Boulton, and they formed a partnership for the development of the steam engine. The high quality of craftsmanship and engineering skill which Matthew Boulton was able to put at Watt's disposal allowed the young inventor to turn his great idea into a reality. However, progress was slow, and the original patent was running out.

Boulton skillfully lobbied in Parliament for an extension of the patent and, as James Watt put it, "Mr. Boulton's amiable and friendly character, together with his fame as an ingenious and active manufacturer procured me many and very active friends in both houses of Parliament".

Fig. 8.4 *Joseph Priestley. Public domain, Wikimedia Commons*

In 1775, the firm of Boulton and Watt was granted an extension of the master steam engine patent until 1800. From a legal and financial standpoint, the way was now clear for the development of the engine; and a major technical difficulty was overcome when the Birmingham ironmaster and cannon-maker, John Wilkinson, invented a method for boring large cylinders accurately by fixing the cutting tool to a very heavy and stable boring shaft.

By 1780, Boulton and Watt had erected 40 engines, about half of which pumped water from the deep Cornish tin mines. Even their early models were at least four times as efficient as the Newcomen engine, and Watt continually improved the design. At Boulton's urging, James Watt designed rotary engines, which could be used for driving mills; and he also invented a governor to regulate the speed of his engines, thus becoming a pioneer of automation. By the time its patent of the separate condenser had run out in 1800, the firm of Boulton and Watt had made 500 engines. After 1800, the rate of production of steam engines became exponential, and when James Watt died in 1819, his inventions had given employment, directly or indirectly, to an estimated two million people.

The Soho manufactory became an almost obligatory stop on any distinguished person's tour of England. Samuel Johnson, for example, wrote that he was received at Soho with great civility; and Boswell, who visited Soho on another occasion, was impressed by "the vastness and contrivance" of the machinery. He wrote that he would never forget Matthew Boulton's words to him as they walked together through the manufactory: "I sell here, Sir, what all the world desires to have - Power!"

The Lunar Society

Matthew Boulton loved to entertain; and he began to invite his friends in science and industry to regular dinners at his home. At these dinners, it was understood by all the guests that science and philosophy were to be the topics of the conversation. This group of friends began to call themselves the "Lunar Society", because of their habit of meeting on nights when the moon was full so that they could find their way home easily afterwards.

During the early stages of the Industrial Revolution, the Lunar Society of Birmingham played a role in the development of scientific ideas which was almost as important as the role played by the Royal Society of London at the time of Isaac Newton. Among the members of this group of friends, besides Erasmus Darwin and James Watt, were the inventive and artistic pottery manufacturer, Josiah Wedgwood (the other grandfather of Charles

Darwin), and the author, chemist and Unitarian minister, Joseph Priestley (1733-1804).

Joseph Priestley's interests were typical of the period: The center of scientific attention had shifted from astronomy to the newly-discovered phenomena of electricity, heat and chemistry, and to the relationship between them. Priestly, who was a prolific and popular author of books on many topics, decided to write a *History of Electricity.* He not only collected all the results of previous workers in an organized form, but also, while repeating their experiments, he made a number of original discoveries. For example, Joseph Priestley was the first to discover the inverse square law of attraction and repulsion between electrical charges, a law which was later verified by the precise experiments of Henry Cavendish (1731-1810) and Charles Coulomb (1736-1806).

The chemistry of gases was also very much in vogue during this period. Joseph Black's medical thesis at Edinburgh University had opened the field with an elegant quantitative treatment of chemical reactions involving carbon dioxide. Black had shown that when chalk (calcium carbonate) is heated, it is changed into a caustic residue (calcium oxide) and a gas (carbon dioxide).

Black had carefully measured the weight lost by the solid residue when the gas was driven off, and he had shown that precisely the same weight was regained by the caustic residue when it was exposed to the atmosphere and reconverted to chalk. His work suggested not only that weight is conserved in chemical reactions, but also that carbon dioxide is present in the atmosphere. Black's work had initiated the use of precise weighing in chemistry, a technique which later was brought to perfection by the great French chemist, Anton Lavoisier (1743-1794).

Joseph Priestley, (who had been supplied with a large burning-glass by his brother-in-law, the wealthy ironmaster, John Wilkinson), carried out an experiment similar to Black's. He used the glass to focus the rays of the sun on a sample of what we now call red oxide of mercury. He collected the gas which was driven off, and tested its properties, recording that "...what surprized me more than I can well express was that a candle burned in this air with a remarkably vigorous flame". He also found that a mouse could live much longer in the new gas than in ordinary air.

On a trip to France, Priestley communicated these results to Anton Lavoisier, who named the gas "oxygen" and established fully its connection with combustion and respiration. At almost the same time, the Swedish chemist, Karl Wilhelm Scheele (1742-1786), discovered oxygen independently.

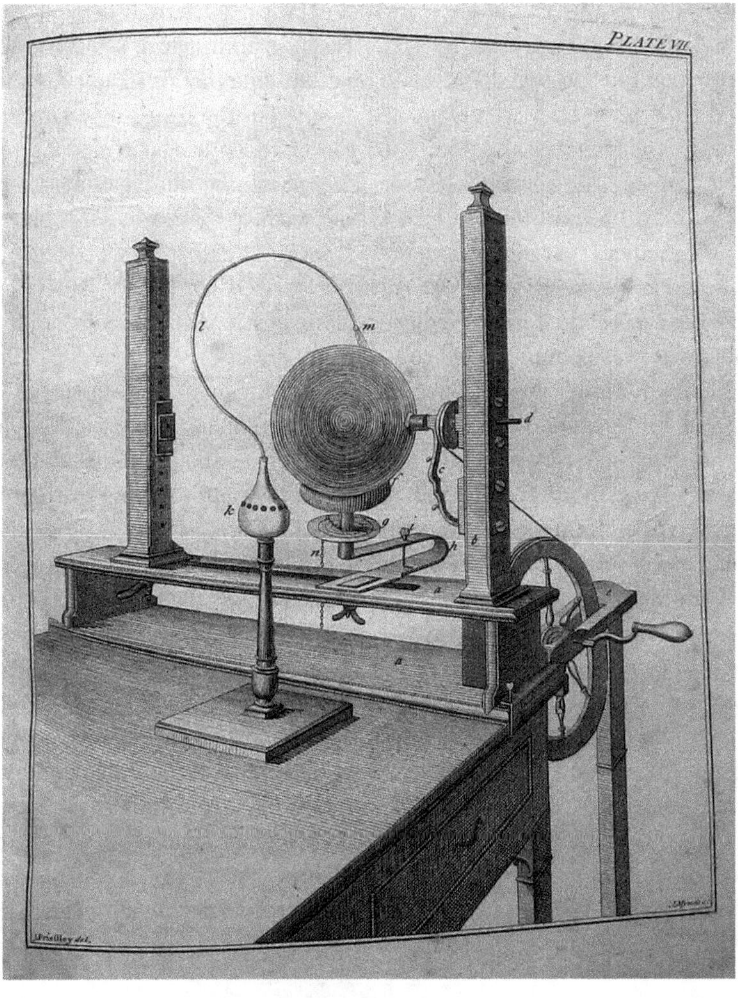

Fig. 8.5 *The 7th plate from the 1st edition (1768) of Joseph Priestley's "A Familiar Introduction to the Study of Electricity", depicting an electrical machine designed by Priestley. Public domain, Wikimedia Commons*

Joseph Priestley isolated and studied nine other new gases; and he invented the technique of collecting gases over mercury. This was much better than collecting them over water, since the gases did not dissolve in mercury. He extended Joseph Black's studies of carbon dioxide, and he invented a method for dissolving carbon dioxide in beverages under pressure, thus becoming the father of the modern soft drink industry!

The tremendous vogue for gas chemistry in the late 18th century can also be seen in the work of the eccentric multimillionaire scientist, Henry Cavendish, who discovered hydrogen by dissolving metals in acids, and then showed that when hydrogen is burned in oxygen, the resulting compound is pure water. Cavendish also combined the nitrogen in the atmosphere with oxygen by means of electrical sparks. The remaining bubble of atmospheric gas, which stubbornly refused to combine with oxygen, was later shown to be a new element - argon.

The great interest in gas chemistry shown by intelligent people of the period can be seen in Josiah Wedgwood's suggestions to the painter, George Stubbs, who was commissioned to make a portrait of Wedgwood's children:

"The two family pieces I have hinted at, I mean to contain the children only, and grouped perhaps in some such manner as this - Sukey playing upon her harpsichord with Kitty singing to her, as she often does, and Sally and Mary Ann upon the carpet in some employment suitable to their ages. This to be one picture. The pendant to be Jack standing at a table making fixable air with the glass apparatus etc., and his two brothers accompanying him, Tom jumping up and clapping his hands in joy, and surprized at seeing the stream of bubbles rise up just as Jack has put a little chalk to the acid. Jos with the chemical dictionary before him in a thoughtful mood; which actions will be exactly descriptive of their respective characters."

The force of feudal traditions was still so strong, however, that in spite of Josiah Wedgwood's suggestions, George Stubbs painted the children on horseback, looking precisely like the children of a traditional landlord. The "fixable air" which Wedgwood mentions was the contemporary word for carbon dioxide. Josiah Wedgwood's daughter, Sukey (Susannah), was destined to become the mother of the greatest biologist of all time, Charles Darwin.

Adam Smith

One of Joseph Black's best friends at Glasgow University was the Professor of Moral Philosophy, Adam Smith. In 1759, Smith published a book entitled *The Theory of Moral Sentiments*, which was subtitled: *An Essay towards an Analysis of the Principles by which Men naturally judge concerning the Conduct and Character, first of their Neighbors, and afterwards of themselves.*

In this book, Adam Smith pointed out that people can easily judge the conduct of their neighbors. They certainly know when their neighbors are treating them well, or badly. Having learned to judge their neighbors, they can, by analogy, judge their own conduct. They can tell when they are mistreating their neighbor or being kind by asking themselves: "Would I want him to do this to me?" As Adam Smith put it:

"Our continual observations upon the conduct of others insensibly lead us to form to ourselves certain general rules concerning what is fit and proper to be done or avoided... It is thus the general rules of morality are formed."

When we are kind to our neighbors, they maintain friendly relations with us; and to secure the benefits of their friendship, we are anxious to behave well towards other people. Thus, according to Adam Smith, enlightened self-interest leads men and women to moral behaviour.

In 1776, Adam Smith published another equally optimistic book, with a similar theme: *The Wealth of Nations.* In this book, he examined the reasons why some nations are more prosperous than others. Adam Smith concluded that the two main factors in prosperity are division of labor and economic freedom.

As an example of the benefits of division of labor, he cited the example of a pin factory, where ten men, each a specialist in a particular manufacturing operation, could produce 48,000 pins per day. One man drew the wire, another straightened it, a third pointed the pins, a fourth put on the heads, and so on. If each man had worked separately, doing all the operations himself, the total output would be far less. The more complicated the manufacturing process (Smith maintained), the more it could be helped by division of labor. In the most complex civilizations, division of labor has the greatest utility.

Adam Smith believed that the second factor in economic prosperity is economic freedom, and in particular, freedom from mercantilist government regulations. He believed that natural economic forces tend to produce an optimum situation, in which each locality specializes in the economic operation for which it is best suited.

Smith believed that when each individual aims at his own personal prosperity, the result is the prosperity of the community. A baker does not consciously set out to serve society by baking bread - he only intends to make money for himself; but natural economic forces lead him to perform a public service, since if he were not doing something useful, people would not pay him for it. Adam Smith expressed this idea in the following way:

"As every individual, therefore, endeavours as much as he can, both to employ his capital in support of domestic industry, and so to direct that industry that its produce may be of greatest value, each individual necessarily labours to render the annual revenue of the Society as great as he can."

"He generally, indeed, neither intends to promote the public interest, nor knows how much he is promoting it. By preferring the support of domestic to that of foreign industry, he intends only his own security; and by directing that industry in such a manner as its produce may be of the greatest value, he intends only his own gain; and he is in this, as in many other cases, led by an invisible hand to promote an end which was no part of his intention. Nor is it always the worse for Society that it was no part of it. By pursuing his own interest, he frequently promotes that of society more effectively than when he really intends to promote it."

In Adam Smith's optimistic view, an "invisible hand" guides individuals to promote the public good, while they consciously seek only their own gain. This vision was enthusiastically adopted adopted by the vigorously growing industrial nations of the west. It is the basis of much of modern history; but there proved to be shortcomings in Smith's theory. A collection of individuals, almost entirely free from governmental regulation, each guided only by his or her desire for personal gain - this proved to be a formula for maximum economic growth; but certain modifications were needed before it could lead to widely shared happiness and social justice.

Fig. 8.6 *Adam Smith. Public domain, Wikimedia Commons*

The dark, Satanic mills

Both Matthew Boulton and Josiah Wedgwood were model employers as well as pioneers of the factory system. Matthew Boulton had a pension scheme for his men, and he made every effort to insure that they worked under comfortable conditions. However, when he died in 1809, the firm of Boulton and Watt was taken over by his son, Matthew Robbinson Boulton, in partnership with James Watt Jr.. The two sons did not have their fathers' sense of social responsibility; and although they ran the firm very efficiently, they seemed to be more interested in profit-making than in the welfare of their workers.

A still worse employer was Richard Arkwright (1732-1792), who held patents on a series of machines for carding, drawing and spinning silk, cotton, flax and wool. He was a rough, uneducated man, who rose from humble origins to become a multimillionaire by driving himself almost as hard as he drove his workers. Arkwright perfected machines (invented by others) which could make extremely cheap and strong cotton thread; and as a result, a huge cotton manufacturing industry grew up within the space of a few years. The growth of the cotton industry was especially rapid after Arkwright's patent expired in 1785.

Crowds of workers, thrown off the land by the Enclosure Acts, flocked to the towns, seeking work in the new factories. Wages fell to a near-starvation level, hours of work increased, and working conditions deteriorated. Dr. Peter Gaskell, writing in 1833, described the condition of the English mill workers as follows:

"The vast deterioration in personal form which has been brought about in the manufacturing population during the last thirty years... is singularly impressive, and fills the mind with contemplations of a very painful character... Their complexion is sallow and pallid, with a peculiar flatness of feature caused by the want of a proper quantity of adipose substance to cushion out the cheeks. Their stature is low - the average height of men being five feet, six inches... Great numbers of the girls and women walk lamely or awkwardly... Many of the men have but little beard, and that in patches of a few hairs... (They have) a spiritless and dejected air, a sprawling and wide action of the legs..."

"Rising at or before daybreak, between four and five o'clock the year round, they swallow a hasty meal or hurry to the mill without taking any food whatever... At twelve o'clock the engine stops, and an hour is given for dinner... Again they are closely immured from one o'clock till eight or nine, with the exception of twenty minutes, this being allowed for tea. During the whole of this long period, they are actively and unremittingly engaged in a crowded room at an elevated temperature."

Dr. Gaskell described the housing of the workers as follows:

"One of the circumstances in which they are especially defective is that of drainage and water-closets. Whole ranges of these houses are either totally undrained, or very partially... The whole of the washings and filth from these consequently are thrown into the front or back street, which, often being unpaved and cut into deep ruts, allows them to collect into stinking and stagnant pools; while fifty, or even more than that number, having only a single convenience common to them all, it is in a very short time choked with excrementous matter. No alternative is left to the inhabitants but adding this to the already defiled street."

"It frequently happens that one tenement is held by several families... The demoralizing effects of this utter absence of domestic privacy must be seen before they can be thoroughly appreciated. By laying bare all the wants and actions of the sexes, it strips them of outward regard for decency - modesty is annihilated - the father and the mother, the brother and the sister, the male and female lodger, do not scruple to commit acts in front of each other which even the savage keeps hid from his fellows."

"Most of these houses have cellers beneath them, occupied - if it is possible to find a lower class - by a still lower class than those living above them."

The abuse of child labor was one of the worst features of early industrialism in England. Sometimes small children, starting at the age of six or seven, were forced to work, because wages were so low that the family would otherwise starve; and sometimes the children were orphans, taken from parish workhouses. The following extract from John Fielden's book, *The Curse of the Factory System* (1836), describes the condition of young children working in the cotton industry:

"It is well known that Arkwright's (so called at least) inventions took manufactures out of the cottages and farmhouses of England... and assembled them in the counties of Derbyshire, Nottinghamshire and more particularly, in Lancashire, where the newly-invented machinery was used in large factories built on the side of streams capable of turning the water wheel. Thousands of hands were suddenly required in these places, remote from towns."

"The small and nimble fingers of children being by far the most in request, the custom instantly sprang up of procuring 'apprentices' from the different parish workhouses of London, Birmingham and elsewhere... Overseers were appointed to see to the works, whose interest it was to work the children to the utmost, because their pay was in proportion to the quantity of work which they could exact."

"Cruelty was, of course, the consequence; and there is abundant evidence on record to show that in many of the manufacturing districts, the most heart-rending cruelties were practiced on the unoffending and friendless creatures... that they were flogged, fettered and tortured in the most exquisite refinement of cruelty, that they were, in many cases, starved to the bone while flogged to their work, and that even in some instances they were driven to commit suicide... The profits of manufacture were enormous; but this only whetted the appetite it should have satisfied."

One of the arguments which was used to justify the abuse of labor was that the alternative was starvation. The population of Europe had begun to grow rapidly for a variety of reasons: - because of the application of scientific knowledge to the prevention of disease; because the potato had been introduced into the diet of the poor; and because bubonic plague had become less frequent after the black rat had been replaced by the brown rat, accidentally imported from Asia.

It was argued that the excess population could not be supported unless workers were employed in the mills and factories to produce manufactured goods, which could be exchanged for imported food. In order for the manufactured goods to be competitive, the labor which produced them had to be cheap: hence the abuses. (At least, this is what was argued).

Fig. 8.7 *A girl pulling a tub full of coal through a narrow seam.*

Overpopulation

When the facts about the abuse of industrial workers in England became known, there were various attempts to explain what had gone wrong with the optimistic expectations of the Enlightenment. Among the writers who discussed this problem was the economist David Ricardo (1772-1823). In his book, *The Principles of Political Economy and Taxation* (1817), Ricardo proposed his "iron law of wages".

According to Ricardo, labor is a commodity, and wages are determined by the law of supply and demand: When wages fall below the starvation level, the workers' children die. Labor then becomes a scarce commodity, and the wages rise. On the other hand, when wages rise above the starvation level, the working population multiplies rapidly, labor becomes a plentiful commodity, and wages fall again. Thus, according to Ricardo, there is an "iron law" which holds wages at the minimum level at which life can be supported.

Ricardo's reasoning assumes industrialists to be completely without social conscience or governmental regulation; it fails to anticipate the development of trade unionism; and it assumes that the working population will multiply without restraint as soon as their wages rise above the starvation level. This was an accurate description of what was happening in England during Ricardo's lifetime, but it obviously does not hold for all times and all places.

A more general and complete description of the situation was given by Thomas Robert Malthus (1766-1834). Malthus came from an intellectual family: His father, Daniel Malthus, was a friend of Rousseau, Hume and Goodwin. The famous book on population by the younger Malthus grew out of his conversations with his father.

Daniel Malthus was an enthusiastic believer in the optimistic philosophy of the Enlightenment. Like Goodwin, Condorcet and Voltaire, he believed that the application of scientific progress to agriculture and industry would inevitably lead humanity forward to a golden age. His son, Robert, was more pessimistic. He pointed out that the benefits of scientific progress would probably be eaten up by a growing population.

At his father's urging, Robert Malthus developed his ideas into a book, *An Essay on the Principle of Population*, which he published anonymously in 1798. In this famous book, Malthus pointed out that under optimum conditions, every biological population, including that of humans, is capable of increasing exponentially. For humans under optimum conditions, the population can double every twenty-five years, quadruple every fifty years and increase by a factor of 8 every seventy-five years. It can grow by a factor of 16 every century, and by a factor of 256 every two centuries, and so on.

Obviously, human populations cannot increase at this rate for very long, since if they did, the earth would be completely choked with people in a very few centuries. Therefore, Malthus pointed out, various forces must be operating to hold the population in check. Malthus listed first the "positive checks" to population growth - disease, famine and war - which we now call the "Malthusian forces". In addition, he listed checks of another kind - birth control (which he called "Vice"), late marriage, and "Moral Restraint". Being a clergyman, Malthus naturally favored moral restraint.

According to Malthus, a population need not outrun its food supply, provided that late marriage, birth control or moral restraint are practiced; but without these less painful checks, the population will quickly grow to the point where the grim Malthusian forces - famine, disease and war - begin to act.

Curiously, it was France, a Catholic country, which led the way in the development of birth control. Robert Owen (who was an enlightened English industrialist, and the founder of the cooperative movement), wished to advise his workers about birth control; and so he went to France to learn about the techniques practiced there. In 1825, an article (by Richard Carlile) appeared in *The Republican.* The article described the importation of birth control from France to England as follows:

"...It was suggested to Mr. Owen that, in his new establishments, the healthy state of the inhabitants would tend to breed an excess of children. The matter was illustrated and explained to him, so that he felt the force of it. He was told that on the Continent, the women used some means of preventing conception which were uniformly successful. Mr. Owen set out for Paris to discover the process. He consulted the most eminent physicians, and assured himself of what was the common practice among their women."

"...A piece of soft sponge is tied by a bobbin or penny ribbon, and inserted before sexual intercourse takes place, and is withdrawn again as soon as it has taken place... If the sponge be large enough, that is, as large as a green walnut or a small apple, it will prevent conception, without diminishing the pleasures of married life."

Carlile goes on to say:

"...When the number of working people in any trade or manufacture has for some years been too great, wages are reduced very low, and the working people become little better than slaves... By limiting the number of children, the wages of both children and grown persons will rise; and the hours of working will be no more than they ought to be."

Birth control and late marriage have (until now) kept the grim predictions of Ricardo and Malthus from being fulfilled in the developed industrial nations of the modern world. Most of these nations have gone through a process known as the "demographic transition" - the shift from an equilibrium where population growth is held in check by the Malthusian forces of disease, starvation and war, to one where it is held in check by birth control and late marriage.

The transition begins with a fall in the death rate, caused by various factors, among which the most important is the application of scientific knowledge to the prevention of disease. Cultural patterns require some time to adjust to the lowered death rate, and so the birth rate continues to be high. Families continue to have six or seven children, just as they did when most of the children died before having children of their own. Therefore, at the start of the demographic transition, the population increases sharply. After a certain amount of time, however, cultural patterns usually adjust to the lowered death rate, and a new equilibrium is established, where both the birth rate and the death rate are low.

In Europe, this period of adjustment required about two hundred years. In 1750, the death rate began to fall sharply: By 1800, it had been cut in half, from 35 deaths per thousand people in 1750 to 18 in 1800; and it continued to fall. Meanwhile, the birth rate did not fall, but even increased to 40 births per thousand per year in 1800. Thus the number of children born every year was more than twice the number needed to compensate for the deaths!

By 1800, the population was increasing by more than two percent every year. In 1750, the population of Europe was 150 million; by 1800, it was roughly 220 million; by 1950 it had exceeded 540 million, and in 1970 it was 646 million.

Meanwhile the achievements of medical science and the reduction of the effects of famine and warfare had been affecting the rest of the world: In 1750, the non-European population of the world was only 585 million. By 1850 it had reached 877 million. During the century between 1850 and 1950, the population of Asia, Africa and Latin America more than doubled, reaching 1.8 billion in 1950. In the twenty years between 1950 and 1970, the population of Asia, Africa and Latin America increased still more sharply, and in 1970, this segment of the world's population reached 2.6 billion, bringing the world total to 3.6 billion. The fastest increase was in Latin America, where population almost doubled during the twenty years between 1950 and 1970.

The latest figures show that the population explosion is leveling off in Europe, Russia, North America and Japan, where the demographic transition is almost complete. However, the population of the rest of the world is still increasing at a breakneck speed; and it cannot continue to expand at this rate for very much longer without producing widespread famine.

Colonialism

In the 18th and 19th centuries, the continually accelerating development of science and science-based industry began to affect the whole world. As the factories of Europe poured out cheap manufactured goods, a change took place in the patterns of world trade: Before the Industrial Revolution, trade routes to Asia had brought Asian spices, textiles and luxury goods to Europe. For example, cotton cloth and fine textiles, woven in India, were imported to England. With the invention of spinning and weaving machines, the trade was reversed. Cheap cotton cloth, manufactured in England, began to be sold in India, and the Indian textile industry withered.

The rapid development of technology in the west also opened an enormous gap in military strength between the industrialized nations and the rest of the world. Taking advantage of their superior weaponry, the advanced industrial nations rapidly carved the remainder of the world into colonies, which acted as sources of raw materials and food, and as markets for manufactured goods.

In North America, the native Indian population had proved vulnerable to European diseases, such as smallpox, and large numbers of them had died. The remaining Indians were driven westward by streams of immigrants arriving from Europe. In Central and South America, European diseases proved equally fatal to the Indians.

Often the industrialized nations made their will felt by means of naval bombardements: In 1854, Commodore Perry and an American fleet forced Japan to accept foreign traders by threatening to bombard Tokyo. In 1856, British warships bombarded Canton in China to punish acts of violence against Europeans living in the city. In 1864, a force of European and American warships bombarded Choshu in Japan, causing a revolution. In 1882, Alexandria was bombarded, and in 1896, Zanzibar.

Between 1800 and 1875, the percentage of the earth's surface under European rule increased from 35 percent to 67 percent. In the period between 1875 and 1914, there was a new wave of colonial expansion, and the fraction of the earth's surface under the domination of colonial powers (Europe, the United States and Japan) increased to 85 percent, if former colonies are included.

During the period between 1880 and 1914, English industrial and colonial dominance began to be challenged. Industrialism had spread from England to Belgium, Germany and the United States, and, to a lesser extent, to France, Italy, Russia and Japan. By 1914, Germany was producing twice as much steel as Britain, and the United States was producing four times as much.

New techniques in weaponry were introduced, and a naval armaments race began among the major industrial powers. The English found that their old navy was obsolete, and they had to rebuild. Thus, the period of colonial expansion between 1880 and 1914 was filled with tensions, as the industrial powers raced to arm themselves in competition with each other, and raced to seize as much as possible of the rest of the world.

Much that was beautiful and valuable was lost, as mature traditional cultures collapsed, overcome by the power and temptations of modern industrial civilization. For the Europeans and Americans of the late 19th century and early 20th century, progress was a religion, and imperialism was its crusade. The cruelties of the crusade were justified, in the eyes of the westerners, by their mission to "civilize" and Christianize the rest of the world. To a certain extent, the industrial countries were right in feeling that they had something of value to offer to the rest of the world; and among the people whom they sent out were educators and medical workers who often accepted lives of extreme discomfort and danger in order to be of service.

At the beginning of the 19th century, the world was divided into parts: China was a world in itself; India was a separate world; Africa south of the Sahara was another enclosed world; and the Islamic world was also self-contained, as was the west. By 1900, there was only one world, bound together by constantly-growing ties of trade and communication.

Suggestions for further reading

(1) Marie Boaz, *Robert Boyle and Seventeenth Century Chemistry*, Cambridge University Press (1958).

(2) J.G. Crowther, *Scientists of the Industrial Revolution*, The Cresset Press, London (1962).

(3) R.E. Schofield, *The Lunar Society of Birmingham*, Oxford University Press (1963).

(4) L.T.C. Rolt, *Isambard Kingdom Brunel*, Arrow Books, London (1961).

(5) J.D. Bernal, *Science in History*, Penguin Books Ltd. (1969).

(6) Bertrand Russell, *The Impact of Science on Society*, Unwin Books, London (1952).

(7) Wilbert E. Moore, *The Impact of Industry*, Prentice Hall (1965).

(8) Charles Morazé, *The Nineteenth Century*, George Allen and Unwin Ltd., London (1976).

(9) Carlo M. Cipolla (editor), *The Fontana Economic History of Europe*, Fontana/Collins, Glasgow (1977).

(10) Richard Storry, *A History of Modern Japan*, Penguin Books Ltd. (1960).

(11) Martin Gerhard Geisbrecht, *The Evolution of Economic Society*, W.H. Freeman and Co. (1972).

Chapter 9

EVOLUTION

Linnaeus, Lamarck and E. Darwin

During the 17th and 18th centuries, naturalists had been gathering information on thousands of species of plants and animals. This huge, undigested heap of information was put into some order by the great Swedish naturalist, Carl von Linné (1707-1778), who is usually called by his Latin name, Carolus Linnaeus.

Linnaeus reclassified all living things, and he introduced a binomial nomenclature, so that each plant or animal became known by two names - the name of its genus, and the name of its species. In the classification of Linnaeus, the species within a given genus resemble each other very closely. Linnaeus also grouped related genera into classes, and related classes into orders. Later, the French anatomist, Cuvier (1769-1832), grouped related orders into phyla.

In France, the Chevalier J.B. de Lamarck (1744-1829), was struck by the close relationships between various animal species; and in 1809 he published a book entitled *Philosophie Zoologique*, in which he tried to explain this interrelatedness in terms of a theory of evolution. Lamarck explained the close similarity of the species within a genus by supposing these species to have evolved from a common ancestor. However, the mechanism of evolution which he postulated was seriously wrong, since he believed that acquired characteristics could be inherited.

Lamarck believed, for example, that giraffes stretched their necks slightly by reaching upward to eat the leaves of high trees. He believed that these slightly-stretched necks could be inherited; and in this way, Lamarck thought, the necks of giraffes have gradually become longer over many generations. Although his belief in the inheritability of acquired characteristics was a serious mistake, Lamarck deserves much credit for correctly maintaining that the close similarity between the species of a genus is due to their descent from a common ancestral species.

Meanwhile, in England, the brilliant physician-poet, Erasmus Darwin (1731-1802), who was considered by Coleridge to have "...a greater range of knowledge than any other man in Europe", had published *The Botanic Garden* and *Zoonomia* (1794). Darwin's first book, *The Botanic Garden*, was written in verse, and in the preface he stated that his purpose was "...to inlist imagination under the banner of science.." and to call the reader's attention to "the immortal works of the celebrated Swedish naturalist, Linnaeus". This book was immensely popular during Darwin's lifetime, but modern readers might find themselves wishing that he had used prose instead of poetry.

Darwin's second book, *Zoonomia*, is more interesting, since it contains a clear statement of the theory of evolution:

"...When we think over the great changes introduced into various animals", Darwin wrote, "as in horses, which we have exercised for different purposes of strength and swiftness, carrying burthens or in running races; or in dogs, which have been cultivated for strength and courage, as the bull-dog; or for acuteness of his sense of smell, as in the hound and spaniel; or for the swiftness of his feet, as the greyhound; or for his swimming in the water, or for drawing snow-sledges, as the rough-haired dogs of the north... and add to these the great change of shape and colour which we daily see produced in smaller animals from our domestication of them, as rabbits or pigeons;... when we revolve in our minds the great similarity of structure which obtains in all the warm-blooded animals, as well as quadrupeds, birds and anphibious animals, as in mankind, from the mouse and the bat to the elephant and whale; we are led to conclude that they have alike been produced from a similar living filament."

Erasmus Darwin's son, Robert, married Suzannah Wedgwood, the pretty and talented daughter of the famous potter, Josiah Wedgwood; and in 1809, (the same year in which Lamarck published his *Philosophie Zoologique*), she became the mother of Charles Darwin.

Charles Darwin

As a boy, Charles Darwin was fond of collecting and hunting, but he showed no special ability in school. His father, disappointed by his mediocre performance, once said to him: "You care for nothing but shooting, dogs and rat-catching; and you will be a disgrace to yourself, and to all your family."

Robert Darwin was determined that his son should not turn into an idle, sporting man, as he seemed to be doing, and when Charles was sixteen, he was sent to the University of Edinburgh to study medicine. However, Charles Darwin had such a sensitive and gentle disposition that he could not stand to see operations (performed, in those days, without chloroform). Besides, he had found out that his father planned to leave him enough money to live on comfortably; and consequently he didn't take his medical studies very seriously. However, some of his friends were scientists,and through them, Darwin became interested in geology and zoology.

Robert Darwin realized that his son did not want to become a physician, and, as an alternative, he sent Charles to Cambridge to prepare for the clergy. At Cambridge, Charles Darwin was very popular because of his cheerful, kind and honest character; but he was not a very serious student. Among his many friends, however, there were a few scientists, and they had a strong influence on him. The most important of Darwin's scientific friends were John Stevens Henslow, the Professor of Botany at Cambridge, and Adam Sedgwick, the Professor of Geology.

Remembering the things which influenced him at that time, Darwin wrote:

"During my last year at Cambridge, I read with care and profound interest Humboldt's *Personal Narritive of Travels to the Equinoctal Regions of America.* This work, and Sir J. Hirschel's *Introduction to the Study of Natural Philosophy*, stirred up in me a burning desire to add even the most humble contribution to the noble structure of Natural Science. No one of a dozen books influenced me nearly so much as these. I copied out from Humboldt long passages about Teneriffe, and read them aloud to Henslow, Ramsay and Dawes... and some of the party declared that they would endeavour to go there; but I think they were only half in earnest. I was, however, quite in earnest, and got an introduction to a merchant in London to enquire about ships."

During the summer of 1831, Charles Darwin went to Wales to help Professor Sedgwick, who was studying the extremely ancient rock formations found there. When he returned to his father's house after this geological expedition, he found a letter from Henslow. This letter offered Darwin the post of unpaid naturalist on the *Beagle*, a small brig which was being sent by the British government to survey the coast of South America and to carry a chain of chronological measurements around the world.

Fig. 9.1 *Charles Darwin as a young man. Public domain, Wikimedia Commons*

Darwin was delighted and thrilled by this offer. He had a burning desire both to visit the glorious, almost-unknown regions described by his hero, Alexander von Humboldt, and to "add even the most humble contribution to the noble structure of Natural Science". His hopes and plans were blocked, however, by the opposition of his father, who felt that Charles was once again changing his vocation and drifting towards a life of sport and idleness. "If you can find any man of common sense who advises you to go", Robert Darwin told his son, "I will give my consent".

Deeply depressed by his father's words, Charles Darwin went to visit the estate of his uncle, Josiah Wedgwood, at Maer, where he always felt more comfortable than he did at home. In Darwin's words what happened next was the following:

"...My uncle sent for me, offering to drive me over to Shrewsbury and talk with my father, as my uncle thought that it would be wise in me to accept the offer. My father always maintained that my uncle was one of the most sensible men in the world, and he at once consented in the kindest possible manner. I had been rather extravagant while at Cambridge, and to console my father, I said that 'I should be deuced clever to spend more than my allowance whilst on board the *Beagle*', but he answered with a smile, 'But they tell me you are very clever!'."

Thus, on December 27, 1831, Charles Darwin started on a five-year voyage around the world. Not only was this voyage destined to change Darwin's life, but also, more importantly, it was destined to change man's view of his place in nature.

Lyell's hypothesis

As the *Beagle* sailed out of Devonport in gloomy winter weather, Darwin lay in his hammock, 22 years old, miserably seasick and homesick, knowing that he would not see his family and friends for many years. To take his mind away from his troubles, Darwin read a new book, which Henslow had recommended: Sir Charles Lyell's *Principles of Geology.* "Read it by all means", Henslow had written, "for it is very interesting; but do not pay any attention to it except in regard to facts, for it is altogether wild as far as theory goes."

Reading Lyell's book with increasing excitement and absorption, Darwin could easily see what Henslow found objectionable: Lyell, a follower of the great Scottish geologist, James Hutton (1726-1797), introduced a revolutionary hypothesis into geology. According to Lyell, "No causes whatever have, from the earliest times to which we can look back, to the present, ever acted, but those now acting; and they have never acted with different degrees of energy from those which they now exert".

This idea seemed dangerous and heretical to deeply religious men like Henslow and Sedgwick. They believed that the earth's geology had been shaped by Noah's flood, and perhaps by other floods and catastrophes which had occurred before the time of Noah. The great geological features of the earth, its mountains, valleys and planes, they viewed as marks left behind by the various catastrophes through which the earth had passed.

All this was now denied by Lyell. He believed the earth to be enormously old - thousands of millions of years old. Over this vast period of time, Lyell believed, the long-continued action of slow forces had produced the geological features of the earth. Great valleys had been carved out by glaciers and by the slow action of rain and frost; and gradual changes in the level of the land, continued over enormous periods of time, had built up towering mountain ranges.

Lyell's belief in the immense age of the earth, based on geological evidence, made the evolutionary theories of Darwin's grandfather suddenly seem more plausible. Given such vast quantities of time, the long-continued action of small forces might produce great changes in biology as well as in geology!

By the time the *Beagle* had reached San Thiago in the Cape Verde Islands, Darwin had thoroughly digested Lyell's book, with its dizzying prospects. Looking at the geology of San Thiago, he realized "the wonderful superiority of Lyell's manner of treating geology". Features of the island which would have been incomprehensible on the basis of the usual Catastrophist theories were clearly understandable on the basis of Lyell's hypothesis.

As the *Beagle* slowly made its way southward along the South American coast, Darwin went on several expeditions to explore the interior. On one of these trips, he discovered some fossil bones in the red mud of a river bed. He carefully excavated the area around them, and found the remains of nine huge extinct quadrupeds. Some of them were as large as elephants, and yet in structure they seemed closely related to living South American species. For example, one of the extinct animals which Darwin discovered resembled an armadillo except for its gigantic size.

The *Beagle* rounded Cape Horn, lashed by freezing waves so huge that it almost floundered. After the storm, when the brig was anchored safely in the channel of Tierra del Fuego, Darwin noticed how a Fuegan woman stood for hours and watched the ship, while sleet fell and melted on her naked breast, and on the new-born baby she was nursing. He was struck by the remarkable degree to which the Fuegans had adapted to their frigid environment, so that they were able to survive with almost no shelter, and with no clothes except a few stiff animal skins, which hardly covered them, in weather which would have killed ordinary people.

Fig. 9.2 *Plate showing Fuegans from the voyage of the Beagle. Wellcome Images, Wikimedia Commons*

In 1835, as the *Beagle* made its way slowly northward, Darwin had many chances to explore the Chilean coast - a spectacularly beautiful country, shadowed by towering ranges of the Andes. One day, near Concepcion Bay, he experienced the shocks of a severe earthquake.

"It came on suddenly, and lasted two minutes", Darwin wrote, "The town of Concepcion is now nothing more than piles and lines of bricks, tiles and timbers."

Measurements which Darwin made showed him that the shoreline near Concepcion had risen at least three feet during the quake; and thirty miles away, Fitzroy, the captain of the *Beagle*, discovered banks of mussels ten feet above the new high-water mark. This was dramatic confirmation of Lyell's theories! After having seen how much the level of the land was changed by a single earthquake, it was easy for Darwin to imagine that similar events, in the course of many millions of years, could have raised the huge wall of the Andes mountains.

In September, 1835, the *Beagle* sailed westward to the Galapagos Islands, a group of small rocky volcanic islands off the coast of Peru. On these islands, Darwin found new species of plants and animals which did not exist anywhere else in the world. In fact, he discovered that each of the islands had its own species, similar to the species found on the other islands, but different enough to be classified separately.

The Galapagos Islands contained thirteen species of finches, found nowhere else in the world, all basically alike in appearance, but differing in certain features especially related to their habits and diet. As he turned these facts over in his mind, it seemed to Darwin that the only explanation was that the thirteen species of Galapagos finches were descended from a single species, a few members of which had been carried to the islands by strong winds blowing from the South American mainland.

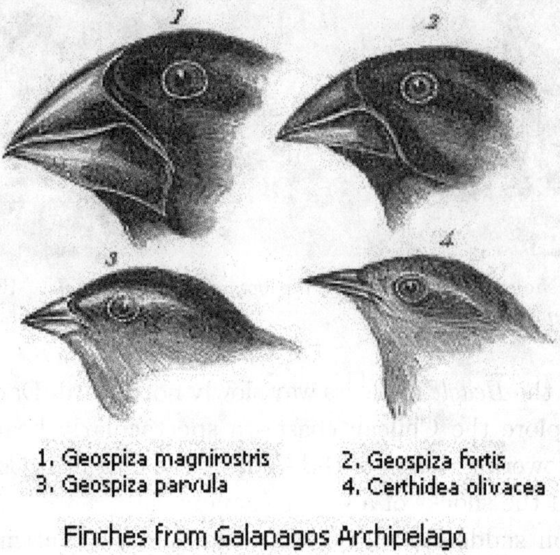

Fig. 9.3 *Darwin's finches. Public domain, Wikimedia Commons*

"Seeing this gradation and diversity of structure in one small, intimately related group of birds", Darwin wrote, "one might really fancy that from an original paucity of birds in this archipelago, one species had been taken and modified for different ends... Facts such as these might well undermine the stability of species."

As Darwin closely examined the plants and animals of the Galapagos Islands, he could see that although they were not quite the same as the corresponding South American species, they were so strongly similar that it seemed most likely that all the Galapagos plants and animals had reached the islands from the South American mainland, and had since been modified to their present form.

The idea of the gradual modification of species could also explain the fact, observed by Darwin, that the fossil animals of South America were more closely related to African and Eurasian animals than were the living South American species. In other words, the fossil animals of South America formed a link between the living South American species and the corresponding animals of Europe, Asia and Africa. The most likely explanation for this was that the animals had crossed to America on a land bridge which had since been lost, and that they had afterwards been modified.

The *Beagle* continued its voyage westward, and Darwin had a chance to study the plants and animals of the Pacific Islands. He noticed that there were no mammals on these islands, except bats and a few mammals brought by sailors. It seemed likely to Darwin that all the species of the Pacific Islands had reached them by crossing large stretches of water after the volcanic islands had risen from the ocean floor; and this accounted for the fact that so many classes were missing. The fact that each group of islands had its own particular species, found nowhere else in the world, seemed to Darwin to be strong evidence that the species had been modified after their arrival. The strange marsupials of the isolated Australian continent also made a deep impression on Darwin.

The Origin of Species

Darwin had left England on the *Beagle* in 1831, an immature young man of 22, with no real idea of what he wanted to do with his life. He returned from the five-year voyage in 1836, a mature man, confirmed in his dedication to science, and with formidable powers of observation, deduction and generalization. Writing of the voyage, Darwin says:

"I have always felt that I owe to the voyage the first real education of my mind... Everything about which I thought or read was made to bear directly on what I had seen, or was likely to see, and this habit was continued during the five years of the voyage. I feel sure that it was this training which has enabled me to do whatever I have done in science."

Darwin returned to England convinced by what he had seen on the voyage that plant and animal species had not been independently and miraculously created, but that they had been gradually modified to their present form over millions of years of geological time.

Darwin was delighted to be home and to see his family and friends once again. To his uncle, Josiah Wedgwood, he wrote:

"My head is quite confused from so much delight, but I cannot allow my sister to tell you first how happy I am to see all my dear friends again... I am most anxious once again to see Maer and all its inhabitants."

In a letter to Henslow, he said:

"My dear Henslow, I do long to see you. You have been the kindest friend to me that ever man possessed. I can write no more, for I am giddy with joy and confusion."

In 1837, Darwin took lodgings at Great Marlborough Street in London, where he could work on his geological and fossil collections. He was helped in his work by Sir Charles Lyell, who became Darwin's close friend. In 1837 Darwin also began a notebook on *Transmutation of Species*. His *Journal of researches into the geology and natural history of the various countries visited by the H.M.S. Beagle* was published in 1839, and it quickly became a best-seller. It is one of the most interesting travel books ever written, and since its publication it has been reissued more than a hundred times.

These were very productive years for Darwin, but he was homesick, both for his father's home at the Mount and for his uncle's nearby estate at Maer, with its galaxy of attractive daughters. Remembering his many happy visits to Maer, he wrote:

"In the summer, the whole family used often to sit on the steps of the old portico, with the flower-garden in front, and with the steep, wooded bank opposite the house reflected in the lake, with here and there a fish rising, or a water-bird paddling about. Nothing has left a more vivid picture in my mind than these evenings at Maer."

In the summer of 1838, tired of his bachelor life in London, Darwin wrote in his diary:

"My God, it is intolerable to think of spending one's whole life like a neuter bee, working, working, and nothing after all! Imagine living all one's days in smoky, dirty London! Only picture to yourself a nice soft wife on a sofa with a good fire, and books and music perhaps.. Marry! Marry! Marry! Q.E.D."

Having made this decision, Darwin went straight to Maer and proposed to his pretty cousin, Emma Wedgwood, who accepted him at once, to the joy of both families. Charles and Emma Darwin bought a large and pleasant country house at Down, fifteen miles south of London; and there, in December, 1839, the first of their ten children was born.

Darwin chose this somewhat isolated place for his home because he was beginning to show signs of a chronic illness, from which he suffered for the rest of his life. His strength was very limited, and he saved it for his work by avoiding social obligations. His illness was never accurately diagnosed during his own lifetime, but the best guess of modern doctors is that he had Chagas' disease, a trypanasome infection transmitted by the bite of a South American blood-sucking bug.

Darwin was already convinced that species had changed over long periods of time, but what were the forces which caused this change? In 1838 he found the answer:

"I happened to read for amusement Malthus on *Population*", he wrote, "and being well prepared to appreciate the struggle for existence which everywhere goes on from long-continued observation of the habits of animals and plants, it at once struck me that under these circumstances favorable variations would tend to be preserved, and unfavorable ones destroyed. The result would be the formation of new species"

"Here, then, I had at last got a theory by which to work; but I was so anxious to avoid prejudice that I determined not for some time to write down even the briefest sketch of it. In June, 1842, I first allowed myself the satisfaction of writing a very brief abstract of my theory in pencil in 33 pages; and this was enlarged during the summer of 1844 into one of 230 pages".

All of Darwin's revolutionary ideas were contained in the 1844 abstract, but he did not publish it! Instead, in an incredible Copernicus-like procrastination, he began a massive treatise on barnacles, which took him eight years to finish! Probably Darwin had a premonition of the furious storm of hatred and bigotry which would be caused by the publication of his heretical ideas.

Finally, in 1854, he wrote to his friend, Sir Joseph Hooker (the director of Kew Botanical Gardens), to say that he was at last resuming his work on the origin of species. Both Hooker and Lyell knew of Darwin's work on evolution, and for many years they had been urging him to publish it. By 1835, he had written eleven chapters of a book on the origin of species through natural selection; but he had begun writing on such a vast scale that the book might have run to four or five heavy volumes, which could have taken Darwin the rest of his life to complete.

Fortunately, this was prevented by the arrival at Down House of a bombshell in the form of a letter from a young naturalist named Alfred Russell Wallace. Like Darwin, Wallace had read Malthus' book *On Population*, and in a flash of insight during a period of fever in Malaya, he had arrived at a theory of evolution through natural selection which was precisely the same as the theory on which Darwin had been working for twenty years!

Wallace enclosed with his letter a short paper entitled *On the Tendency of Varieties to Depart Indefinitely From the Original Type.* It was a perfect summary of Darwin's theory of evolution!

"I never saw a more striking coincidence", the stunned Darwin wrote to Lyell, "If Wallace had my MS. sketch, written in 1842, he could not have made a better short abstract! Even his terms now stand as heads of my chapters... I should be extremely glad now to publish a sketch of my general views in about a dozen pages or so; but I cannot persuade myself that I can do so honourably... I would far rather burn my whole book than that he or any other man should think that I have behaved in a paltry spirit."

Both Lyell and Hooker acted quickly and firmly to prevent Darwin from suppressing his own work, as he was inclined to do. In the end, they found a happy solution: Wallace's paper was read to the Linnean Society together with a short abstract of Darwin's work, and the two papers were published together in the proceedings of the society. The members of the Society listened in stunned silence. As Hooker wrote to Darwin the next day, the subject was "too novel and too ominous for the old school to enter the lists before armouring."

Lyell and Hooker then persuaded Darwin to write a book of moderate size on evolution through natural selection. As a result, in 1859, he published *The Origin of Species*, which ranks, together with Newton's *Principia* as one of the two greatest scientific books of all time. What Newton did for physics, Darwin did for biology: He discovered the basic theoretical principle which brings together all the experimentally-observed facts and makes them comprehensible; and he showed in detail how this basic principle can account for the facts in a very large number of applications.

Darwin's *Origin of Species* can still be read with enjoyment and fascination by a modern reader. His style is vivid and easy to read, and almost all of his conclusions are still believed to be true. He begins by discussing the variation of plants and animals under domestication, and he points out that the key to the changes produced by breeders is selection: If we want to breed fast horses, we select the fastest in each generation, and use them as parents for the next generation.

Darwin then points out that a closely similar process occurs in nature: Every plant or animal species produces so many offspring that if all of them survived and reproduced, the population would soon reach astronomical numbers. This cannot happen, since the space and food supply are limited; and therefore, in nature there is always a struggle for survival. Accidental variations which increase an organism's chance of survival are more likely to be propagated to subsequent generations than are harmful variations. By this mechanism, which Darwin called "natural selection", changes in plants and animals occur in nature just as they do under domestication.

If we imagine a volcanic island, pushed up from the ocean floor and completely uninhabited, we can ask what will happen as plants and animals begin to arrive. Suppose, for example, that a single species of bird arrives on the island. The population will first increase until the environment cannot support larger numbers, and it will then remain constant at this level. Over a long period of time, however, variations may accidentally occur in the bird population which allow the variant individuals to make use of new types of food; and thus, through variation, the population may be further increased. In this way, a single species "radiates" into a number of sub-species which fill every available ecological niche. The new species produced in this way will be similar to the original ancestor species, although they may be greatly modified in features which are related to their new diet and habits. Thus, for example, whales, otters and seals retain the general structure of land-going mammals, although they are greatly modified in features which are related to their aquatic way of life. This is the reason, according to Darwin, why vestigial organs are so useful in the classification of plant and animal species.

The classification of species is seen by Darwin as a geneological classification. All living organisms are seen, in his theory, as branches of a single family tree! This is a truly remarkable assertion, since the common ancestors of all living things must have been extremely simple and primitive; and it follows that the marvellous structures of the higher animals and plants, whose complexity and elegance utterly surpasses the products of human intelligence, were all produced, over thousands of millions of years, by random variation and natural selection!

Each structure and attribute of a living creature can therefore be seen as having a long history; and a knowledge of the evolutionary history of the organs and attributes of living creatures can contribute much to our understanding of them. For instance, studies of the evolutionary history of the brain and of instincts can contribute greatly to our understanding of psychology, as Darwin pointed out.

Among the many striking observations presented by Darwin to support his theory, are facts related to morphology and embryology. For example, Darwin includes the following quotation from the naturalist, von Baer:

"In my possession are two little embryos in spirit, whose names I have omitted to attach, and at present I am quite unable to say to what class they belong. They may be lizards or small birds, or very young mammalia, so complete is the similarity in the mode of formation of the head and trunk in these animals. The extremities, however, are still absent in these embryos. But even if they had existed in the earliest stage of their development, we should learn nothing, for the feet of lizards and mammals, the wings and feet of birds, no less than the hands and feet of man, all arise from the same fundamental form."

Darwin also quotes the following passage from G.H. Lewis: "The tadpole of the common Salamander has gills, and passes its existence in the water; but the *Salamandra atra*, which lives high up in the mountains, brings forth its young full-formed. This animal never lives in the water. Yet if we open a gravid female, we find tadpoles inside her with exquisitely feathered gills; and when placed in water, they swim about like the tadpoles of the common Salamander or water-newt. Obviously this aquatic organization has no reference to the future life of the animal, nor has it any adaption to its embryonic condition; it has solely reference to ancestral adaptations; it repeats a phase in the development of its progenitors."

Darwin points out that, "...As the embryo often shows us more or less plainly the structure of the less modified and ancient progenitor of the group, we can see why ancient and extinct forms so often resemble in their adult state the embryos of existing species."

No abstract of Darwin's book can do justice to it. One must read it in the original. He brings forward an overwhelming body of evidence to support his theory of evolution through natural selection; and he closes with the following words:

"It is interesting to contemplate a tangled bank, clothed with many plants of many different kinds, with birds singing on the bushes, with various insects flitting about, and with worms crawling through the damp earth, and to reflect that these elaborately constructed forms, so different from each other, and dependant upon each other in so complex a manner, have all been produced by laws acting around us... There is grandeur in this view of life, with its several powers, having been originally breathed by the Creator into a few forms or into one; and that whilst this planet has gone cycling on according to the fixed law of gravity, from so simple a beginning, endless forms most beautiful and wonderful have been and are being evolved."

Fig. 9.4 *Charles Darwin in 1880. The photograph is by Elliott and Fry. Public domain, Wikimedia Commons*

Fig. 9.5 *"Man is But a Worm", a cartoon, published in Punch in 1882. Public domain, Wikimedia Commons*

Suggestions for further reading

(1) Sir Julian Huxley and H.B.D. Kettlewell, *Charles Darwin and his World*, Thames and Hudson, London (1965).
(2) Allan Moorehead, *Darwin and the Beagle*, Penguin Books Ltd. (1971).
(3) Francis Darwin (editor), *The Autobiography of Charles Darwin and Selected Letters*, Dover, New York (1958).
(4) Charles Darwin, *The Voyage of the Beagle*, J.M. Dent and Sons Ltd., London (1975).
(5) Charles Darwin, *The Origin of Species*, Collier MacMillan, London (1974).
(6) Charles Darwin, *The Expression of Emotions in Man and Animals*, The University of Chicago Press (1965).
(7) D.W. Forest, *Francis Galton, The Life and Work of a Victorian Genius*, Paul Elek, London (1974).
(8) Ruth Moore, *Evolution*, Time-Life Books (1962).

Chapter 10

VICTORY OVER DISEASE

Jenner

If the Europeans and Americans of the 19th century felt that their scientific civilization had something to offer to humanity as a whole, they may have had in mind not only factories, steamships, railways and telegraphs, but also great victories won against disease. The first of these victories was won against smallpox, a disease which at one time was so common that almost everyone was sure of getting it. In the more severe epidemics, one person out of three who contracted smallpox died of the disease. Those who recovered were often so severely disfigured that their faces were hardly human.

Since smallpox was so common that people scarcely hoped to avoid it entirely, they hoped instead to have a mild case. It had been noticed that anyone who survived an attack of smallpox could never be attacked again. In Turkey and China, people sometimes inoculated themselves with pus taken from the blisters of patients sick with smallpox in a mild form. The Turkish and Chinese custom of inoculation was introduced into Europe in the 18th century, and Diderot, the editor of the Encyclopedia, did much to make this practice popular. However, this type of inoculation was dangerous: It gave protection against future attacks, but often the inoculated person became severly ill or died. It was like "Russian roulette".

The story of safe immunization against smallpox began when an English physician named Edward Jenner (1749-1823) treated a dairymaid. He suspected that she might have smallpox; but when he told her this, she replied: "I cannot take the smallpox sir, because I have had the cowpox". She told him that it was common knowledge among the people of her district that anyone who had been ill with cowpox (a mild disease of cattle

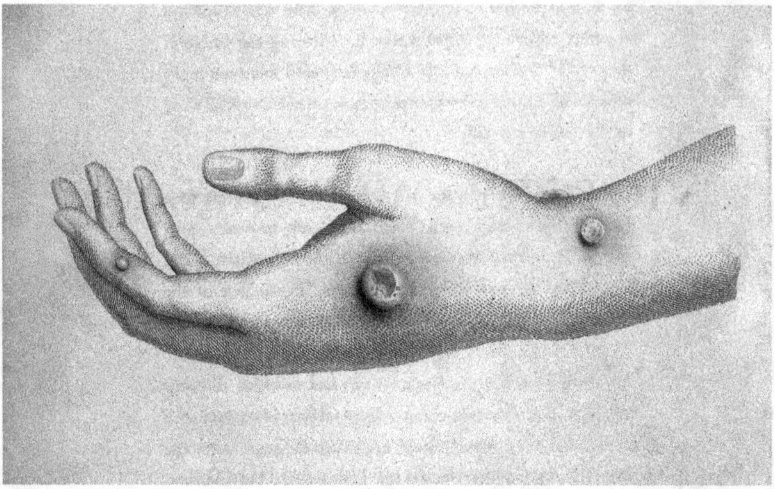

Fig. 10.1 *The hand of Sarah Nelmes infected with the cowpox. Wellcome Images, [CC BY 4.0], Wikimedia Commons*

which sometimes affected farmers and dairymaids), would never be attacked by smallpox.

Jenner realized that if her story were true, it might offer humanity a safe method of immunization against one of its most feared diseases. On May 14, 1796, he found a dairymaid with active cowpox, and taking a little fluid from a blister on her hand, he injected it into a boy. The boy became ill with cowpox, but he recovered quickly, because the disease is always mild.

Jenner then took the dangerous step of inoculating the boy with smallpox. If the boy had died, Jenner would have been a criminal - but he was immune! It took Jenner two years to find the courage and the opportunity to try the experiment again; but when he repeated it in 1798 with the same result, he decided to publish his findings.

So great was the terror of smallpox, that Jenner was immediately besieged with requests for immunization by inoculation with cowpox (which he called "vaccination" after *vacca*, the Latin word for "cow"). The practice quickly became accepted: The English Royal Family was vaccinated, and Parliament voted Jenner rewards totalling thirty thousand pounds - in those days an enormous sum.

In 1807, Bavaria made vaccination compulsory, and celebrated Jenner's birthday as a holiday. Russia also enthusiastically adopted vaccination. The first child in Russia to be vaccinated was given the name "Vaccinov",

Fig. 10.2 *Edward Jenner. Pastel by John Raphael Smith. Edward Jenner pioneered vaccination against smallpox using the cowpox vaccine. Wellcome Images, [CC BY 4.0], Wikimedia Commons*

and was educated at the expense of the state. Thanks to Jenner and the dairymaid, smallpox began to disappear from the earth.

Pasteur

In 1800, when vaccination began to be used against smallpox, no one understood why it worked. No one, in fact, understood what caused infectious diseases. It had been more than a century since Anton van Leeuwenhoek had studied bacteria with his home-made microscopes and described them in long letters to the Royal Society. However, the great Swedish naturalist, Carolus Linnaeus, left microscopic organisms out of his classification of all living things on the grounds that they were too insignificant and chaotic to be mentioned.

This was the situation when Louis Pasteur was born in 1822, in the Jura region of France, near the Swiss border. His father was a tanner in the small town of Arbois. Pasteur's parents were not at all rich, but they were very sincere and idealistic, and they hoped that their son would one day become a teacher.

As a boy, Louis Pasteur was considered to be a rather slow student, but he was artistically gifted. Between the ages of 13 and 19, he made many realistic and forceful portraits of the people of his town. His ambition was to become a professor of the fine arts; and with this idea he studied to qualify for the entrance examination of the famous École Normale of Paris, supporting himself with a part-time teaching job, and sometimes enduring semi-starvation when the money sent by his father ran out.

The earnest, industrious and artistically gifted boy would certainly have succeeded in becoming an excellent professor of the fine arts if he had not suddenly changed his mind and started on another path. This new path was destined to win Louis Pasteur a place among the greatest benefactors of humanity.

The change came when Pasteur attended some lectures by the famous chemist Jean Baptiste Dumas. Professor Dumas was not only a distinguished researcher; he was also a spellbinding speaker, whose lectures were always attended by six or seven hundred excited students. "I have to go early to get a place", Pasteur wrote to his parents, "just as in the theatre". Inspired by these lectures, Pasteur decided to become a chemist. He put away his brushes, and never painted again.

While he was still a student, Pasteur attracted the attention of Antoine Jerome Balard, the discoverer of the element bromine. Instead of being sent to teach at a high-school in the provinces after his graduation, Pasteur became an assistant in the laboratory of Balard, where he had a chance to work on a doctor's degree, and where he could talk with the best chemists in Paris. Almost every Thursday, he was invited to the home of Professor Dumas, where the conversation was always about science.

Pasteur's first important discovery came when he was 25. He had been studying the tartarates - a group of salts derived from tartaric acid. There was a mystery connected with these salts because, when polarized light was passed through them, they rotated the direction of polarization. On the other hand, paratartaric acid (now called racemic acid), did not exhibit this effect at all, nor did its salts. This was a mystery, because there seemed to be no chemical difference between tartaric acid and racemic acid.

Studying tiny crystals of paratartaric acid under his microscope, Pasteur noticed that there were two kinds, which seemed to be mirror images of one another. His vivid imagination leaped to the conclusion that the two types of crystals were composed of different forms of tartaric acid, the molecules of one form being mirror images of the other. Therefore the crystals too were mirror images, since, as Pasteur guessed, the shapes of the crystals resulted from the shapes of the molecules.

By painstakingly separating the tiny right-handed crystals from the left-handed ones, Pasteur obtained a pure solution of right-handed molecules, and this solution rotated polarized light. The left-handed crystals, when dissolved, produced the opposite rotation! Pasteur ran from the laboratory, embraced the first person that he met in the hall, and exclaimed: "I have just made a great discovery! I am so happy that I am shaking all over, and I am unable to set my eyes again to the polarimeter."

Jean Baptiste Biot, the founder of the field of polarimetry, was sceptical when he heard of Pasteur's results, and he asked the young man to repeat the experiments so that he could see the results with his own eyes. Under Biot's careful supervision, Pasteur separated the two types of crystals of racemic acid, and put a solution of the left-handed crystals into the polarimeter.

"At the first sight of the color tints presented by the two halves of the field", Pasteur wrote, "and without having to make a reading, Biot recognized that there was a strong rotation to the left. Then the illustrious old man, who was visibly moved, seized me by the hand and said: 'My dear son, all my life I have loved science so deeply that this stirs my heart!'"

As he continued his work with right- and left-handed molecules, Pasteur felt that he was coming close to an understanding of the mysteries of life itself, since, as Biot had shown, the molecules which rotate polarized light are almost exclusively molecules produced by living organisms. He soon discovered that he could make an optically active solution of tartaric acid in another way: When he let the mould *penicillium glaucum* grow in a solution of racemic acid, the left-handed form disappeared, and only the right-handed form remained. In this way, Pasteur became interested in the metabolism of microscopic organisms.

Pasteur's work on crystallography and optical activity had made him famous among chemists, and he was appointed Professor of Chemistry at the University of Strasbourg. He soon fell in love with and married the daughter of the Rector of the university, Marie Laurent. This marriage was very fortunate for Pasteur. In the words of Pasteur's assistant, Emil Roux, "Madame Pasteur loved her husband to the extent of understanding his studies... She was more than an incomparable companion for her husband: She was his best collaborator". She helped him in every way that she could - protecting him from everyday worries, taking dictation, copying his scientific papers in her beautiful handwriting, discussing his experiments and asking intelligent questions which helped him to clarify his thoughts.

After a few years at Strasbourg, Pasteur was appointed Dean of the Faculty of Sciences at the University of Lille. In appointing him, the French government explained to Pasteur that they expected him to place the Faculty of Sciences of the university at the service of the industry and agriculture of the district.

Pasteur took this commission seriously, and he soon put his studies of microorganisms to good use in the service of a local industry which produced alcohol from beet juice. He was able to show that whenever the vats of juice contained bacteria, they spoiled; and he showed the local manufacturers how eliminate harmful bacteria from their vats. As a result of this work, the industry was saved.

His work on fermentation put Pasteur into conflict with the opinions of the most famous chemists of his time. He believed that it was the action of the living yeast cells which turned sugar into alcohol, since he had observed that the yeasts were alive and that the amount of alcohol produced was directly proportional to the number of yeasts present. On the other hand, the Swedish chemist, Jöns Jakob Berzelius (1779-1848), had considered fermentation to be an example of catalysis, while Justus von Liebig (1805-1875) thought that the yeasts were decaying during fermentation, and that the breakdown of the yeast cells somehow assisted the conversion of sugar to alcohol. (Both Pasteur and Berzelius were right! Although the fermentation observed by Pasteur was an example of the action of living yeasts, it is possible to extract an enzyme from the yeasts which can convert sugar to alcohol without the presence of living cells.)

Pasteur studied other fermentation processes, such as the conversion of sugar into lactic acid by the bacilli which are found in sour milk, and the fermentation which produces butyric acid in rancid butter. He discovered that each species of microorganism produces its own specific type of fermentation; and he learned to grow pure cultures of each species.

At the suggestion of Napoleon III, Pasteur turned his attention to the French wine industry, which was in serious difficulties. He began to look for ways to get rid of the harmful bacteria which were causing spoilage of the wine. After trying antiseptics, and finding them unsatisfactory, Pasteur finally found a method for killing the bacteria, without affecting the taste of the wine, by heating it for several minutes to a temperature between 50 and 60 degrees centigrade. This process ("Pasteurization") came to be applied, not only to wine, but also to milk, cheese, butter, beer and many other kinds of food.

Pasteur developed special machines for heat-treating large volumes of liquids. He patented these, to keep anyone else from patenting them, but he made all his patents available to the general public, and refused to make any money from his invention of the Pasteurization process. He followed the same procedure in patenting an improved process for making vinegar, but refusing to accept money for it.

Pasteur was now famous, not only in the world of chemists and biologists, but also in the larger world. He was elected to membership by the French Academy of Sciences, and he was awarded a prize by the Academy for his research refuting the doctrine of spontaneous generation.

The germ theory of disease

In 1873, Louis Pasteur was elected to membership by the French Academy of Medicine. Many conservative physicians felt that he had no right to be there, since he was really a chemist, and had no medical "union card". However, some of the younger doctors recognized Pasteur as the leader of the most important revolution in medical history; and a young physician, Emil Roux, became one of Pasteur's devoted assistants.

When he entered the Academy of Medicine, Pasteur found himself in the middle of a heated debate over the germ theory of disease. According to Pasteur, every contagious disease is caused by a specific type of microorganism. To each specific disease there corresponds a specific germ.

Pasteur was not alone in advocating the germ theory, nor was he the first person to propose it. For example, Varro (117 B.C. - 26 B.C.), believed that diseases are caused by tiny animals, too small to be seen, which are carried by the air, and which enter the body through the mouth and nose.

In 1840, Jacob Henle, a distinguished Bavarian anatomist, had pointed out in an especially clear way what one has to do in order to prove that a particular kind of germ causes a particular disease: The microorganism must be found consistently in the diseased tissue; it must be isolated from the tissue and cultured; and it must then be able to induce the disease consistently. Finally, the newly-diseased animal or human must yield microorganisms of the same type as those found originally.

Henle's student, Robert Koch (1843-1910), brilliantly carried out his teacher's suggestion. In 1872, Koch used Henle's method to prove that anthrax is due to rodlike bacilli in the blood of the infected animal. Koch's pioneering contributions to microbiology and medicine were almost as great as those of Pasteur. Besides being the first person to prove beyond doubt that a specific disease was caused by a specific microorganism, Koch introduced a number of brilliant technical improvements which paved the way for rapid progress in bacteriology and medicine.

Instead of using liquids as culture media,Koch and his assistant, Petri, pioneered the use of solid media. Koch developed a type of gel made from agar-agar (a substance derived from seaweed). On the surface of this gel, bacteria grew in tiny spots. Since the bacteria could not move about on the solid surface, each spot represented a pure colony of a single species, derived from a single parent. Koch also pioneered techniques for staining bacteria, and he introduced the use of photography in bacteriology. He was later to isolate the bacillus which causes tuberculosis, and also the germ which causes cholera.

When Koch's work was attacked in the French Academy of Medicine, Pasteur rushed to his defense. In order to demonstrate that it was living bacilli in the blood of a sheep with anthrax which transmitted the disease, and not something else in the blood, Pasteur took a drop of infected blood and added it to a large flask full of culture medium. He let this stand until the bacteria had multiplied; and then he took a tiny drop from the flask and transferred it to a second flask of nutrient broth. He did this a hundred times, so that there was no possibility that anything whatever remained from the original drop of sheep's blood. Nevertheless, a tiny amount of liquid from the hundredth flask was just as lethal as fresh blood drawn from a sheep with anthrax.

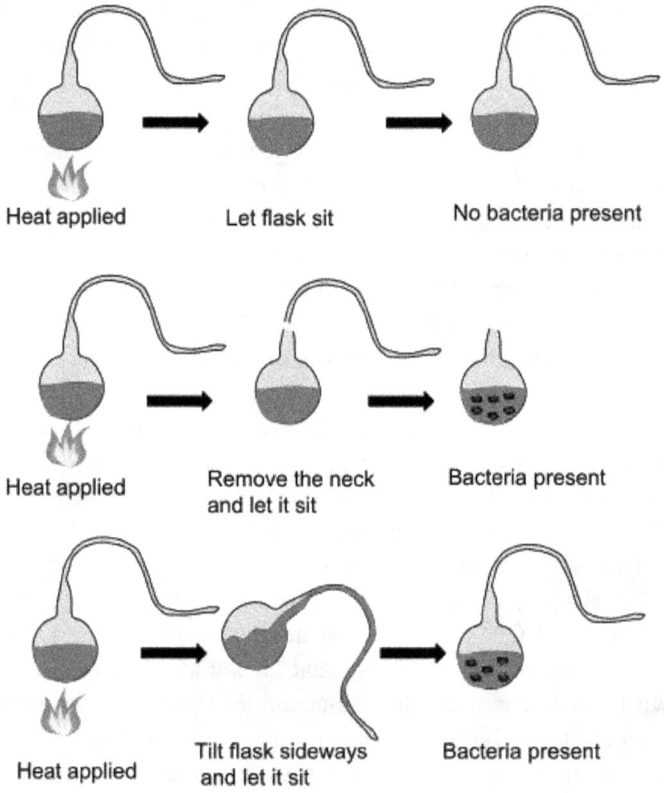

Fig. 10.3 *Louis Pasteur's pasteurization experiment illustrates the fact that the spoilage of the liquid was caused by particles in the air rather than than the air itself. These experiments were important pieces of evidence supporting the idea of Germ Theory of Disease. By Kgerow16, [CC BY-SA 4.0], Wikimedia Commons*

Vaccines

Pasteur read and reread the papers of Jenner on immunization against smallpox. He searched continually for something analogous to smallpox vaccination which could be applied to other diseases. Finally, the answer came by chance.

Pasteur and his assistants had been studying chicken cholera, an invariably fatal disease of chickens. Roux and Chamberland were carrying out a series of experiments where they made a fresh culture of chicken cholera bacteria every day. When they injected a bit of liquid from any of these cultures into a chicken, the chicken always died.

It was summer, and the young men went off for two weeks of vacation. When they came back, they took their two-week-old culture of chicken cholera out of the cupboard and injected it into a hen; but the hen didn't die. They decided that while they had been on vacation, the culture must have lost its strength; and after some effort, they obtained a new specimen of active chicken cholera bacteria, which they injected into their hens. All the hens died except one. The hen which had previously been inoculated with two-week-old culture didn't even get sick!

When Pasteur returned to his laboratory, the two young men hesitated to tell him about this strange result because they were afraid that he might be angry with them for going off on a holiday and breaking off the series of experiments. However, they finally confessed what had happened, and added the strange detail about the chicken which had not died. In the middle of their apologies, Pasteur raised his hand. "Please be quiet for a moment", he said, "I want to think". After a few moments of silence, Pasteur looked at Roux and Chamberland and said, "That's it! The hen that didn't die was *vaccinated* by the old culture!"

This was the big breakthrough - a turning point in medical history. Pasteur, Roux and Chamberland had discovered by chance a method of weakening a culture of bacteria so that it would not produce the fatal disease with which it was usually associated; but on the other hand, it was still able to alert the body's defense mechanisms, so that the inoculated animal became immune. This great discovery was made by chance, but, as Pasteur was fond of saying, "In research, chance favors the prepared mind".

Pasteur, Roux and Chamberland dropped everything else and began a series of experiments to find the best way of weakening their cultures of chicken cholera. They found that the critical factor was the proper amount of exposure to air. (Probably the culture contained a few mutant bacteria, able to grow well in air, but not able to produce chicken cholera; and during the exposure of a culture, these mutants multiplied rapidly, until the entire population was composed of mutants.)

Pasteur now began research on a vaccine against anthrax - a disease which was causing serious economic loss to farmers, and which could affect humans as well as animals. With anthrax, the problem was to keep the bacilli from forming spores. After much experimentation, the group found that if they held their anthrax cultures at a temperature between 42 °C and 43 °C, the bacilli would still grow, but they did not form spores.

Pasteur and his coworkers allowed their cultures to grow at 42 °C in shallow dishes, where there was good contact with the air. They found that

Fig. 10.4 *Photograph of Louis Pasteur. Public domain, Wikimedia Commons*

after two weeks, the cultures were weakened to the point where they would make a sheep sick, but not kill it. They developed a method for inoculating animals in two stages - first with a very much weakened culture, and later with a stronger one. After the second inoculation, the animals could stand an injection of even the most virulent anthrax bacilli without becoming ill.

When Pasteur published these results, there was much sarcasm among veterinarians. The editor of the *Veterinary Press*, a surgeon named Rossignol, wrote: "Monsieur Pasteur's discovery, *if it were genuine*, should not be kept in the laboratory". Rossignol proposed a public trial of the anthrax vaccine, and he started a campaign to collect money for the purchase of experimental animals.

Pasteur's friends warned him against accepting the risk of a public trial at such an early stage. He had not tested his vaccine sufficiently, and a failure would make him the laughing stock of Europe. However, Pasteur saw the trial as a chance to focus public attention on microorganisms and vaccines. Like Galileo, Pasteur had a flair for dramatic gestures and public debate; and the impact of his career was greatly enhanced by his ability to attract widespread attention.

A farm near Melun called Pouilly le Fort was chosen as the site for the experiment; and sixty sheep, together with several cows, were put at Pasteur's disposal. Thousands of people made the journey from Paris to Melun to watch the first injections, which were made on May 5, 1881. Twelve days later, the same sheep were inoculated with a stronger vaccine. Then, on May 31, the big test was made - both the vaccinated and unvaccinated animals were inoculated with a highly lethal culture of anthrax. Pasteur went back to Paris. There was nothing to do but wait.

The next afternoon, a telegram from Rossignol shattered Pasteur's confidence: It said that one of the vaccinated sheep was dying. Pasteur spent a sleepless night. The following morning, however, at nine o'clock, another telegram arrived from Rossignol: All the vaccinated sheep were well, even the one which had seemed to be dying; and all the unvaccinated sheep were either dying or already dead! Rossignol, who had been Pasteur's enemy, was completely converted; and his telegram ended with the words, "Stunning success!" When the aging Pasteur limped onto the field at Pouilly le Fort that afternoon, a great cheer went up from the thousands of people present.

Rabies

The next disease which Pasteur attempted to conquer was rabies, the terrifying and invariably fatal disease which often follows the bite of a mad dog. The rabies virus travels slowly through the body from the wounds to the spinal cord, where, after one or two months, it attacks the nervous system. If a victim is offered water and attempts to swallow, his head jerks back in terrible spasms, which make rabies extremely frightening, both for the victim and for the onlooker. For this reason, the disease is sometimes called hydrophobia - fear of water.

Pasteur and his coworkers soon discovered that even with their best microscopes, they were unable to see the organism which causes rabies. In fact, the disease is caused by a virus, much too small to be seen with an optical microscope. Thus the aging Pasteur was confronted with an entirely new technical problem, never before encountered in microbiology.

He soon found that it was impossible to culture the rabies virus in a flask or dish, as he was in the habit of doing with bacteria. Absorbed in his research, he forgot his wedding anniversary. Marie Pasteur, however, remembered; and she wrote in a letter to her daughter:

"Your father is absorbed in his thoughts. He talks little, sleeps little, rises at dawn, and in a word, continues the life which which I began with him this day thirty-five years ago."

Besides being technically difficult, the work on rabies was also dangerous. When Pasteur, Roux and Chamberland took samples of saliva from the foaming jaws of mad dogs, they risked being bitten by accident and condemned to an agonizing death from the convulsions of rabies. Since they could not culture the rabies virus in a dish or a flask of nutrient fluid, they were forced to grow it inside the nervous systems of experimental animals. After four years of difficult and hazardous work, they finally succeeded in developing a vaccine against rabies.

In the method which finally proved successful, they took a section of spinal cord from a rabbit with rabies and exposed it to air inside a germproof bottle. If the section of spinal cord remained in the bottle for a long time, the culture was very much weakened or "attenuated", while when it was exposed to air for a shorter time, it was less attenuated. As in the case of anthrax, Pasteur built up immunity by a series of injections, beginning with a very much attenuated culture, and progressing to more and more virulent cultures.

At last, Pasteur had a method which he believed could be used to save the lives of the victims of mad dogs and wolves; and he found himself faced with a moral dilemma: Everyone who developed rabies died of it; but not everyone who was bitten by a mad dog developed rabies. Therefore if Pasteur gave his vaccine to a human victim of a mad dog, he might harm someone who would have recovered without treatment.

He had published the results of his research, and he was inundated with requests for treatment, but still he hesitated. If he treated someone, and the person afterward died, he might be accused of murder; and all the work which he had done to build up public support for the new movement in medicine might be ruined.

Finally, on July 6, 1885, Pasteur's indecision was ended by the sight of a man and woman who had come to him with their frightened nine-year-old son. The boy, whose name was Joseph Meitner, had been severely bitten by a mad dog. It was one thing to write letters refusing requests for treatment, and another thing to look at a doomed and frightened child and turn him away.

Pasteur felt that he had to help the boy. He consulted Alfred Vulpian, a specialist in rabies, and Vulpian assured him that Joseph Meitner had been bitten so severly that without treatment, he would certainly develop rabies and die. Pasteur also consulted Dr. Granchier, a young physician who had joined his staff, and together the three men agreed that there was no time to lose - they would have to begin inoculations immediately if they were to save the boy's life. They decided to go ahead. To Pasteur's great joy, Joseph Meitner remained completely well.

The second rabies victim to be treated by Pasteur was a fourteen-year-old shepard named Jupille. He had seen a mad dog about to attack a group of small children, and he had bravely fought with the maddened animal so that the children could escape. Finally he had managed to tie its jaws together, but his hands were so badly bitten that without treatment, he was certain to die. Like Joseph Meitner, Jupille was saved by the Pasteur treatment. A statue of Jupille in front of the Pasteur Institute commemorates his bravery.

Pasteur had now grown so old, and was so worn out by his labors that he could do no more. The task of winning a final victory over infectious diseases was not finished - it was barely begun; but at least the feet of researchers had been placed on the right road; and there were younger men and women enthusiastically taking up the task which Pasteur laid down.

On December 27, 1892, physicians and scientists from many countries assembled in Paris to celebrate Pasteur's seventieth birthday. The old man was so weak that he was unable to reply in his own words to the address of Sir Joseph Lister and to the cheers of the crowd; but his words were read by his son. Pasteur spoke to the young men and women who would take his place in the fight against disease:

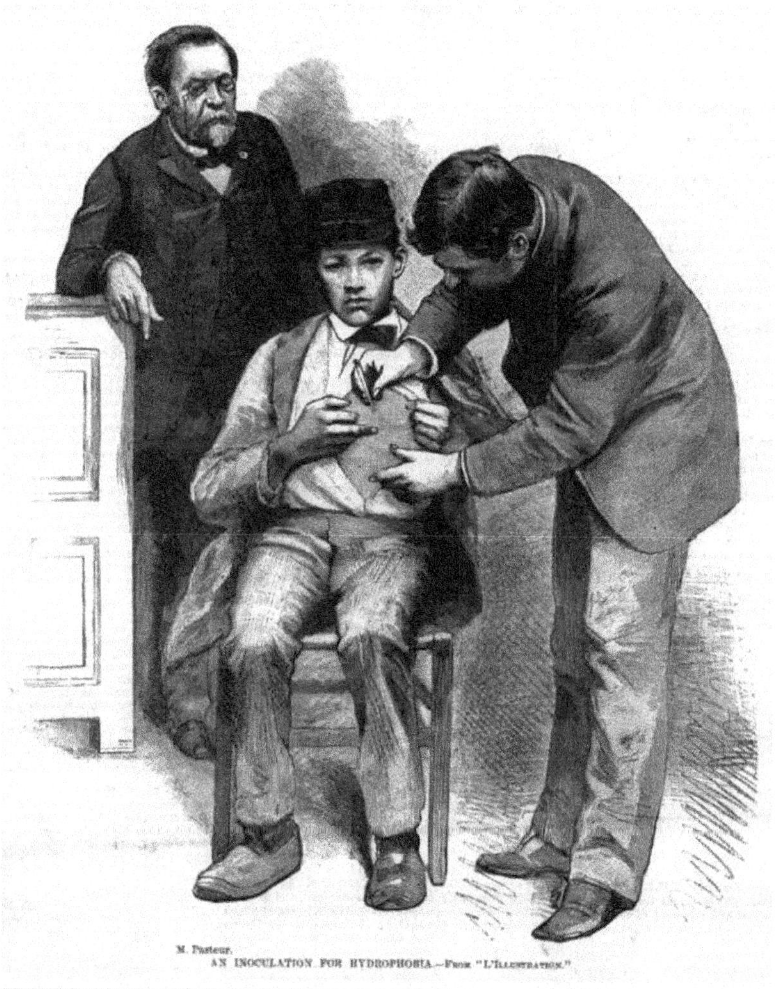
M. Pasteur.
AN INOCULATION FOR HYDROPHOBIA.

Fig. 10.5 *The innoculation of Jean-Baptiste Jupille. Public domain, Wikimedia Commons*

"Do not let yourselves be discouraged by the sadness of certain hours which pass over nations. Live in the serene peace of your laboratories and libraries. Say to yourselves first, 'What have I done for my instruction?', and as you gradually advance, 'What have I done for my country?', until the time comes when you may have the intense happiness of thinking that you have contributed in some way to the progress and good of humanity."

Suggestions for further reading

(1) Clifford Dobell (editor), *Antony van Leeuwenhoek and his Little Animals*, Dover, New York (1960).
(2) Paul de Kruif, *Microbe Hunters*, Pocket Books Inc., New York (1959).
(3) René Dubos, *Pasteur and Modern Science*, Heinemann, London (1960).
(4) A.P. Waterson and Lise Wilkinson, *An Introduction to the History of Virology*, Cambridge University Press (1978).
(5) P.E. Baldry, *The Battle Against Bacteria*, Cambridge University Press (1965).
(6) L. Wilkinson, *Animals and Disease; An Introduction to the History of Comparative Medicine*, Cambridge University Press, (1992).
(7) Arthur Rook (editor), *The Origins and Growth of Biology*, Penguin Books Ltd. (1964).

Chapter 11

ATOMS IN CHEMISTRY

Dalton

As we saw in an earlier chapter, atomism was originated by the Greek philosopher, Leucippus, in the 5th century B.C., and it was developed by his student Democritus. The atomists believed that all matter is composed of extremely small, indivisible particles (atoms).They believed that all the changes which we observe in matter are changes in the groupings of atoms, the atoms themselves being eternal.

The rational philosophy of Democritus was not very popular in his own time, but it was saved from being lost entirely by the Athenian philosopher Epicurus. Later, the Roman poet, Lucretius, published a long, philosophical poem, *De Natura Rerum*, in which he maintained that all things (even the gods!) are composed of atoms. In 1417, a single surviving manuscript copy of *De Natura Rerum* was discovered and printed.

The poem became very popular, and in this way, the ideas of Democritus were transmitted to the experimental scientists of the 17th century, almost all of whom were believers in the atomic theory of matter. Christian Huygens, for example, believed that light radiating from a flame is a wave-like disturbance produced by the violent motion of atoms in the flame. Sir Isaac Newton was also a believer in the atomic theory of matter. He believed (correctly) that chemical compounds are composed of atoms bonded together by forces which are fundamentally electrical in nature. The universally talented Robert Hooke came near to developing a kinetic theory of gases based on atomic ideas; but he lacked the mathematical power needed for such a theory.

At the beginning of the 19th century, an honest, ingenious, color-blind, devout, unmarried English provincial schoolteacher named John Dalton

(1766-1814) gave the atomic theory of matter new force by relating it to the observed facts of chemistry. Dalton was born in Cumberland, the son of a Quaker weaver, and he remained in the North of England all his life. At the early age of 12, he became a teacher; and he remained a teacher in various Quaker schools until 1800, when he became the Secretary of the Manchester Literary and Philosophical Society.

One of Dalton's early scientific interests was in meteorology, and he recorded the capricious weather of the Lake District in a diary which ultimately contained more than 200,000 entries. In speculating about water vapor in the atmosphere, John Dalton began to wonder why the various gases in the atmosphere did not separate into layers, since some of the gases in the mixture were less dense than others.

The only way that Dalton could explain the failure of the atmosphere to stratify was to imagine it as composed mainly of empty space through which atoms of the various gases moved almost independently, seldom striking one another. In this picture, he imagined each of the gases in the atmosphere as filling the whole available volume, almost as though the other gases in the mixture were not there.

Dalton believed the pressure on the walls of a vessel containing a mixture of gases to be due to the force of the atoms striking the walls; and he believed that each of the gases behaved as though the other gases were not there. Therefore he concluded that the total pressure must be the sum of the partial pressures, i.e. the sum of the pressures which would be exerted by each of the gases in the mixture if it occupied the whole volume by itself. This law, which he confirmed by experiment, is known as "Dalton's Law of Partial Pressures".

Convinced of the atomic picture by his studies of gases, John Dalton began to think about chemical reactions in terms of atoms. Here he made a bold guess - that all the atoms of a given element are of the same weight. He soon found that this hypothesis would explain one of the most important fact in chemistry, the fixed ratio of weights in which chemical elements combine to form compounds. (The law of definite proportions by weight in chemical reactions is known as "Proust's Law", after the French chemist, Joseph Louis Proust (1754-1826), who first proposed and defended it.)

In Dalton's view, molecules of the simplest compound formed from two elements ought to consist of one atom of the first element, united with one atom of the second element. For example, the simplest compound of carbon and oxygen should consist of one atom of carbon, bonded to one atom of oxygen (carbon monoxide). Dalton believed that besides such

Fig. 11.1 *Engraving of a painting of John Dalton. Public domain, Wikimedia Commons*

simple compounds, others with more complicated structure could also exist, (e.g. carbon dioxide).

By studying the weights of the elements which combined to form what he believed to be the simplest chemical compounds, Dalton was able to construct a table of the relative atomic weights of the elements. For example,

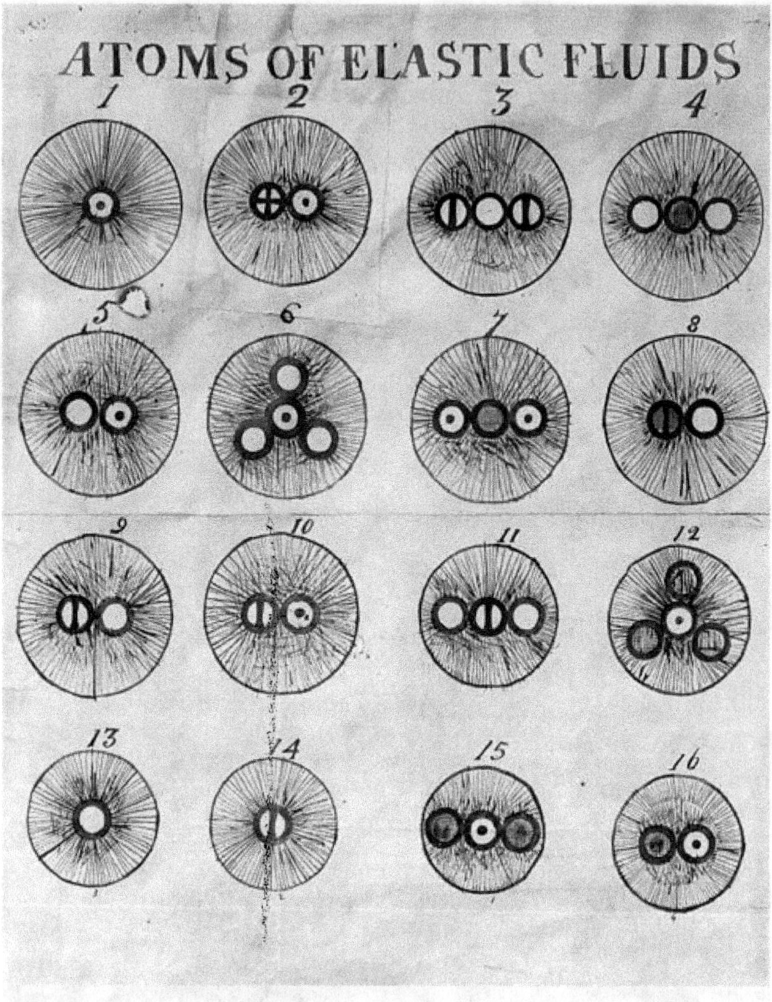

Fig. 11.2 *John Dalton: Composition and size of atoms. 1806-1809. Wellcome Images, Wikimedia Commons*

knowing that 12 ounces of carbon combine with 16 ounces of oxygen to form carbon monoxide, Dalton could deduce that the ratio of the weight of a carbon atom to the weight of an oxygen atom must be 12/16. His table of relative atomic weights contained some errors, but the principle which he used in constructing it was not only correct, but also very important.

Gay-Lussac and Avogadro

In 1808, John Dalton published his table of atomic weights in a book entitled *A New System of Chemical Philosophy.* A year later, in 1809, the celebrated French chemist and balloonist, Joseph Louis Gay-Lussac (1778-1850), made public an important law concerning the chemical reactions of gases: Gay-Lussac's experiments showed that the volumes of the reactants and the volumes of the products were related to each other by the ratios of simple whole numbers.

This law was strikingly similar to Proust's law of definite proportions by weight, on which Dalton had based his table of relative atomic weights. Gay-Lussac stated that his results were "very favorable to Dalton's ingenious ideas"; but there were problems in linking Dalton's ideas with Gay-Lussac's experiments.

Observation showed, for example, that one volume of hydrogen gas would unite with exactly the same volume of chlorine gas to form the gas of hydrochloric acid. The problem was that, if the temperature and pressure were kept constant, the resulting total volume of gas was the same after the reaction as before, although according to Dalton's ideas the number of particles should be cut in half!

This was a mystery which Dalton and Gay-Lussac failed to solve; but it was completely cleared up a year later, in 1810, by Amadeo Avogadro (1776-1856), Count of Quaregna and Professor of Philosophy at the University of Vercelli in Italy. Avogadro introduced a bold hypothesis - that a standard volume of any gas whatever, at room temperature and atmospheric pressure, contains a number of particles which is the same for every gas.

(Avogadro himself did not have any idea how many gas particles there are in a litre of gas; but we now know that at room temperature and atmospheric pressure, 22.4 litres contain 602,600,000,000,000,000,000,000 particles. This is the same as the number of atoms in a gram of hydrogen. To get some imaginative idea of the size of "Avogardo's number", we can think of the fact that the number of atoms in a drop of water is roughly the same as the number of drops of water in all the oceans in the world!)

Avogadro believed that the particles of a gas need not be single atoms, even if the gas contains only a single element. In this way, he could explain the mysterious proportions of volume observed by Gay-lussac. for example, in the reaction where hydrogen and chlorine combine to form hydrochloric acid, Avogadro assumed that every molecule of hydrogen gas consist of two atoms joined together, and similarly, every molecule of chlorine gas consist of two atoms. Then, in the reaction in which hydrochloric acid is formed, the total number of molecules is not changed by the reaction, which fits with Gay-lussac's observation that the volume occupied by the gasses is unchanged.

Although Avogadro completely solved the problem of reconciling Dalton's atomic ideas with Gay-lussac's volume ratios, there was a period of 50 years during which most chemists ignored the atomic theories of Dalton and Avogadro. However, it hardly mattered that the majority of chemists where unconvinced, since the greatest chemist of the period, Jöns Jakob Berzelius (1779-1849), was an ardent disciple of Dalton's atomism. His belief more than made up for the other chemists' disbelief!

After studying medicine at the University of Uppsala in Sweden, Berzelius became a chemist; and over a period of ten years, between 1807 and 1817, he analysed more than two thousand different chemical reactions. He showed that all these reactions follow Proust's law of definite proportions by weight. He also continued Dalton's work on relative atomic weights; and in 1828 he published the first reasonably accurate table of these weights.

Unfortunately, although Berzelius was a follower of Dalton, he did not appreciate the value of Avogadro's ideas; and therefore confusion about the distinction between atoms and molecules remained to plague chemistry until 1860. In that year, the first international scientific congress in history was held at Karlsruhe, Baden, to try to clear up the confusion about atomic weights. By that time, Dalton's atomic theory was widely accepted, but without Avogadro's clarifying ideas, it led to much confusion. In fact, the chemists of the period were almost at one another's throats, arguing about the correct chemical formulas for various compounds.

Among the delegates at the Karlsruhe Congress was the fiery Italian chemist, Stanislao Cannizzaro (1826-1910). He had been a revolutionist in 1848, and he was later to fight in the army of Garibaldi for the unification of Italy. Cannizzaro had read Avogadro's almost-forgotten papers; and he realized that Avogadro's hypothesis, together with Gay-Lussac's volume ratios, could be used to determine atomic weights unambiguously. He went to the congress filled with missionary zeal; and as a result of his efforts, most of the other delegates saw the light. One of the delegates, Lothar Meyer, said later: "The scales suddenly fell from my eyes, and they were replaced by a feeling of peaceful certainty."

Neither John Dalton nor Amadeo Avogadro lived to see the triumph of their theories at Karlsruhe, but towards the end of his life, John Dalton was much honored. He was given an honorary degree by Oxford University, invited to soirées by the Duke of Sussex, and presented to King William IV of England.

The presentation to the king involved some difficulty, since Dalton was forbidden by his Quaker religion to wear the sword required for court dress. Therefore it was arranged that he should be presented to the king wearing crimson academic robes from Oxford; but here again there was a difficulty: Bright colors were inconsistent with the simple clothes required by the Quakers. Dalton solved this problem by wearing the crimson robes anyway, and saying that he was colorblind, which was perfectly true!

Mendeléev

Among the distinguished delegates listening to Cannizzaro at the Karlsruhe Congress in 1860, was the brilliant young Russian chemist, Dmitri Ivanovich Mendeléev (1834-1907). He had been born in Tobolsk, Siberia, the youngest child in a family of 14 (some accounts even say 17!). His grandfather had brought the first printing press to Siberia, and had published Siberia's first newspaper. His father had been the principal of the high-school in Tobolsk, before blindness forced his retirement. Mendeléev's mother, a part-Mongol woman of incredible energy, then set up a glass factory to support her large family.

When Mendeléev was in his teens, two disasters struck the family: His father died and the glass factory burned down. His mother then gathered her last remaining strength, and traveled to St. Petersburg, where a friend of her dead husband obtained a university place for her favorite son, Dmitri. Soon afterward, she died.

After graduating from the university at the top of his class, Dmitri Mendeléev went to Germany to do postgraduate work under Bunsen, (the inventor of the spectroscope and the "Bunsen burner"). In 1860, he attended the First International Congress of Chemistry at Karlsruhe; and like Lothar Meyer, he was profoundly impressed by Cannizzaro's views on atomic weights.

Returning to St. Petersburg, (where he became a professor of chemistry in 1866), Mendeléev began to arrange the elements in order of their atomic weights. He soon noticed that when the elements were arranged in this way, their chemical properties showed a periodic variation. Arranged in order of their atomic weights, the first few elements were hydrogen, (helium was then unknown), lithium, beryllium, boron, carbon, nitrogen, oxygen and fluorine. Mendeléev noticed that lithium was a very active metal, with a valence (combining power) of 1; beryllium was a metal, with valence 2; boron had valence 3; and carbon had valence 4. Next came the non-metals: nitrogen with valence 3; oxygen with valence 2; and finally came fluorine, a very active non-metal with valence 1.

Continuing along the list of elements, arranged in order of their atomic weights, Mendeléev came next to sodium, a very active metal with valence 1; magnesium, a metal with valence 2; aluminium, with valence 3; silicon, with valence 4; phosphorus, a non-metal, with valence 3; sulphur, a non-metal with valence 2; and finally chlorine, a very active non-metal with valence 1. Mendeléev realized that there is a periodicity in the chemical properties of the elements: The elements of the first period, arranged in order of increasing atomic weight, had the valences 1,2,3,4,3,2,1. The second period exhibited the same pattern: 1,2,3,4,3,2,1.

Fig. 11.3 *Dmetri Mendeleev, (1834-1907). Public domain, Wikimedia Commons*

When he arranged all of the known elements in a table which exhibited the periodicity of their chemical properties, Mendeléev could see that there were some gaps. These gaps, he reasoned, must correspond to undiscovered elements! By studying the rows and columns of his periodic table, he calculated the chemical properties and the approximate atomic weights which these yet-unknown elements ought to have.

Mendeléev's predictions, made in 1869, were dramatically confirmed a decade later, when three of the elements whose discovery he had prophesied were actually found, and when their atomic weights and chemical properties turned out to be exactly as he had predicted! The discovery of these elements made Mendeléev world-famous, and it was clear that his periodic table contained some deep truth. However, the underlying meaning of the periodic table was not really understood; and it remained a mystery until it was explained by quantum theory in 1926.

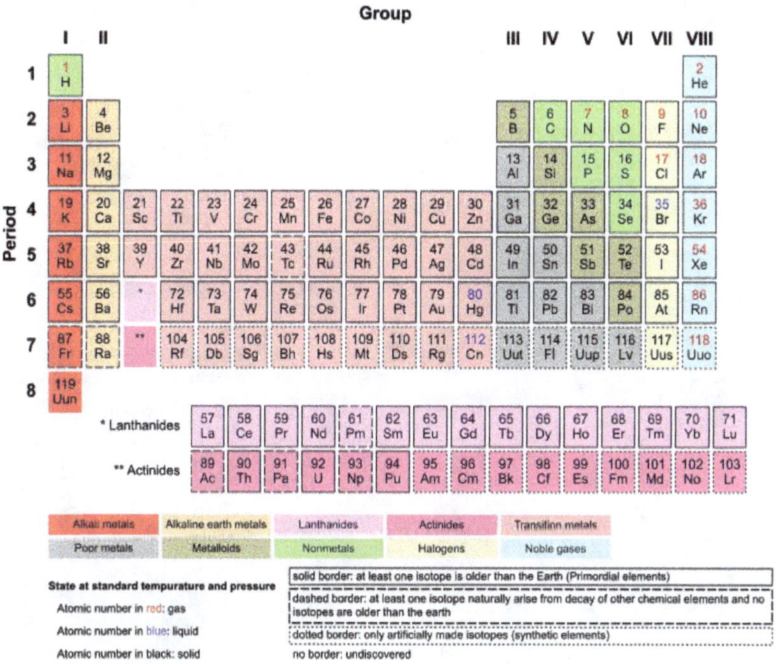

Fig. 11.4 *Periodic Table based on HTML Table found in Wikipedia. Armtuk, [CC BY-SA 4.0], Wikimedia Commons*

Suggestions for further reading

(1) Frank Greenaway, *John Dalton and the Atom*, Heinemann, London (1966).

(2) Sir Basil Schonland, *The Atomists, 1805-1933*, Clarendon Press, Oxford (1968).

(3) F.J. Moore, *A History of Chemistry*, McGraw-Hill (1939).
(4) W.G. Palmer, *A History of the Concept of Valency to 1930*, Cambridge University Press (1965).

Chapter 12

ELECTRICITY AND MAGNETISM

Galvani and Volta

While Dalton's atomic theory was slowly gaining ground in chemistry, the world of science was electrified (in more ways than one) by the discoveries of Franklin, Galvani, Volta, Ørsted, Ampère, Coulomb and Faraday.

A vogue for electrical experiments had been created by the dramatic experiments of Benjamin Franklin (1706-1790), who drew electricity from a thundercloud, and thus showed that lightning is electrical in nature. Towards the end of the 18th century, almost every scientific laboratory in Europe contained some sort of machine for generating static electricity. Usually these static electricity generators consisted of a sphere of insulating material which could be turned with a crank and rubbed, and a device for drawing off the accumulated static charge. Even the laboratory of the Italian anatomist, Luigi Galvani (1737-1798), contained such a machine; and this was lucky, since it led indirectly to the invention of the electric battery.

In 1771, Galvani noticed that some dissected frog's legs on his work table twitched violently whenever they were touched with a metal scalpel while his electrostatic machine was running. Since Franklin had shown lightning to be electrical, it occurred to Galvani to hang the frog's legs outside his window during a thunderstorm. As he expected, the frog's legs twitched violently during the thunderstorm, but to Galvani's surprise, they continued to move even after the storm was over. By further experimentation, he found that what made the frog's legs twitch was a closed electrical circuit, involving the brass hook from which they were hanging, and the iron lattice of the window.

Fig. 12.1 *Luigi Galvani (1737-1798). Public domain, Wikimedia Commons*

Galvani mentioned these experiments to his friend, the physicist Alessandro Volta (1745-1827). Volta was very much interested, but he could not agree with Galvani about the source of the electrical current which was making the frog's legs move. Galvani thought that the current was "animal electricity", coming from the frog's legs themselves, while Volta thought that it was the two different metals in the circuit which produced the current.

The argument over this question became bitter, and finally destroyed the friendship between the two men. Meanwhile, to prove his point, Volta constructed the first electrical battery. This consisted of a series of dishes containing salt solution, connected with each other by bridges of metal. One end of each bridge was made of copper, while the other end was made of zinc. Thus, as one followed the circuit, the sequence was: copper, zinc,

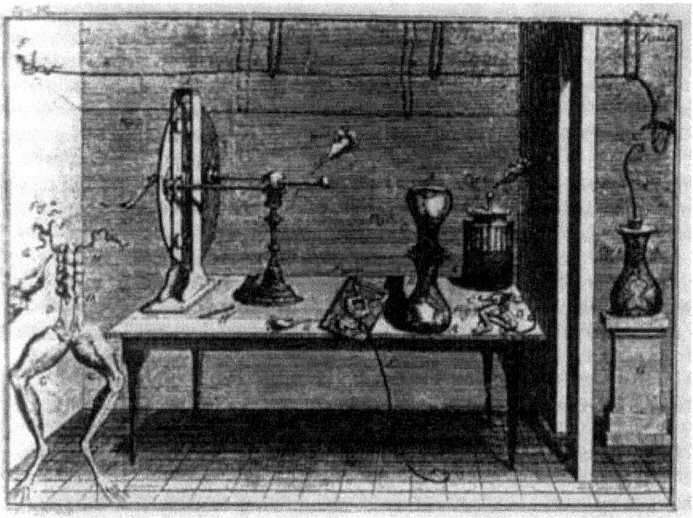

Fig. 12.2 *Apparatus used by Galvani. Public domain, Wikimedia Commons*

salt solution, copper, zinc, salt solution, and so on.

Volta found that when a closed circuit was formed by such an arrangement, a steady electrical current flowed through it. The more units connected in series in the battery, the stronger was the current. He next constructed a more compact arrangement, which came to be known as the "Voltaic pile". Volta's pile consisted of a disc of copper, a disc of zinc, a disc of cardboard soaked in salt solution, another disc of copper, another disc of zinc, another disc of cardboard soaked in salt solution, and so on. The more elements there were in the pile, the greater was the electrical potential and current which it produced.

The invention of the electric battery lifted Volta to a peak of fame where he remained for the rest of his life. He was showered with honors and decorations, and invited to demonstrate his experiments to Napoleon, who made him a count and a senator of the Kingdom of Lombardy. When Napoleon fell from power, Volta adroitly shifted sides, and he continued to receive honors as long as he lived.

News of the Voltaic pile spread like wildfire throughout Europe and started a series of revolutionary experiments both in physics and in chemistry. On March 20, 1800, Sir Joseph Banks, the President of the Royal Society, received a letter from Volta explaining the method of constructing batteries. On May 2 of the same year, the English chemist, William Nichol-

Fig. 12.3 *Alessandro Volta (1745-1827). Public domain, Wikimedia Commons*

son (1755-1815), (to whom Banks had shown the letter), used a Voltaic pile to separate water into hydrogen and oxygen.

Shortly afterwards, the brilliant young English chemist, Sir Humphrey Davy (1778-1829), constructed a Voltaic pile with more than two hundred and fifty metal plates. On October 6, 1807, he used this pile to pass a current through molten potash, liberating a previously unknown metal, which he called potassium. During the year 1808, he isolated barium, strontium, calcium, magnesium and boron, all by means of Voltaic currents.

Ørsted, Ampère and Faraday

In 1819, the Danish physicist, Hans Christian Ørsted (1777-1851), was demonstrating to his students the electrical current produced by a Voltaic pile. Suspecting some connection between electricity and magnetism, he brought a compass needle near to the wire carrying the current. To his astonishment, the needle turned from north, and pointed in a direction perpendicular to the wire. When he reversed the direction of the current, the needle pointed in the opposite direction.

Ørsted's revolutionary discovery of a connection between electricity and magnetism was extended in France by André Marie Ampère (1775-1836). Ampère showed that two parallel wires, both carrying current, repel each other if the currents are in the same direction, but they attract each other if the currents are opposite. He also showed that a helical coil of wire carrying a current produces a large magnetic field inside the coil; and the more turns in the coil, the larger the field.

The electrochemical experiments of Davy, and the electromagnetic discoveries of Ørsted and Ampère, were further developed by the great experimental physicist and chemist, Michael Faraday (1791-1867). He was one of ten children of a blacksmith, and as a boy, he had little education. At the age of 14, he was sent out to work, apprenticed to a London bookbinder. Luckily, the bookbinder sympathized with his apprentice's desire for an education, and encouraged him to read the books in the shop (outside of working hours). Faraday's favorites were Lavoisier's textbook on chemistry, and the electrical articles in the Encyclopedia Britannica.

In 1812, when Michael Faraday was 21 years old, a customer in the bookshop gave him tickets to attend a series of lectures at the Royal Institution, which were to be given by the famous chemist Humphry Davy. At that time, fashionable London socialites (particularly ladies) were flocking to the Royal Institution to hear Davy. Besides being brilliant, he was also extremely handsome, and his lectures, with their dramatic chemical demonstrations, were polished to the last syllable.

Michael Faraday was, of course, thrilled to be present in the glittering audience, and he took careful notes during the series of lectures. These notes, to which he added beautiful colored diagrams, came to 386 pages. He bound the notes in leather and sent them to Sir Joseph Banks, the President of the Royal Society, hoping to get a job related to science. He received no reply from Banks, but, not discouraged, he produced another version of his notes, which he sent to Humphry Davy.

Faraday accompanied his notes with a letter saying that he wished to work in science because of "the detachment from petty motives and the unselfishness of natural philosophers". Davy told him to reserve judgement on that point until he had met a few natural philosophers, but he gave Faraday a job as an assistant at the Royal Institution.

In 1818, Humphry Davy was knighted because of his invention of the miner's safety lamp. He married a wealthy and fashionable young widow, resigned from his post as Director of the Royal Institution, and set off on a two-year excursion of Europe, taking Michael Faraday with him. Lady Davy regarded Faraday as a servant; but in spite of the humiliations which she heaped on him, he enjoyed the tour of Europe and learned much from it. He met, and talked with, Europe's most famous scientists; and in a sense, Europe was his university.

Returning to England, the modest and devoted Faraday finally rose to outshine Sir Humphry Davy, and he became Davy's successor as Director of the Royal Institution. Faraday showed enormous skill, intuition and persistence in continuing the electrical and chemical experiments begun by Davy.

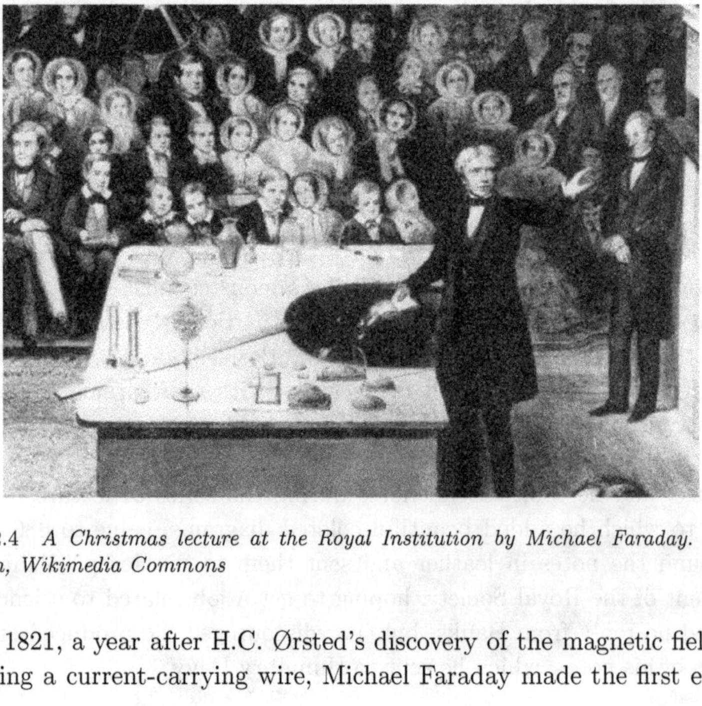

Fig. 12.4 *A Christmas lecture at the Royal Institution by Michael Faraday. Public domain, Wikimedia Commons*

In 1821, a year after H.C. Ørsted's discovery of the magnetic field surrounding a current-carrying wire, Michael Faraday made the first electric motor. His motor was simply a current-carrying wire, arranged so that it could rotate around the pole of a magnet; but out of this simple device, all modern electrical motors have developed. When asked what use his motor was, Faraday replied: "What use is a baby?"

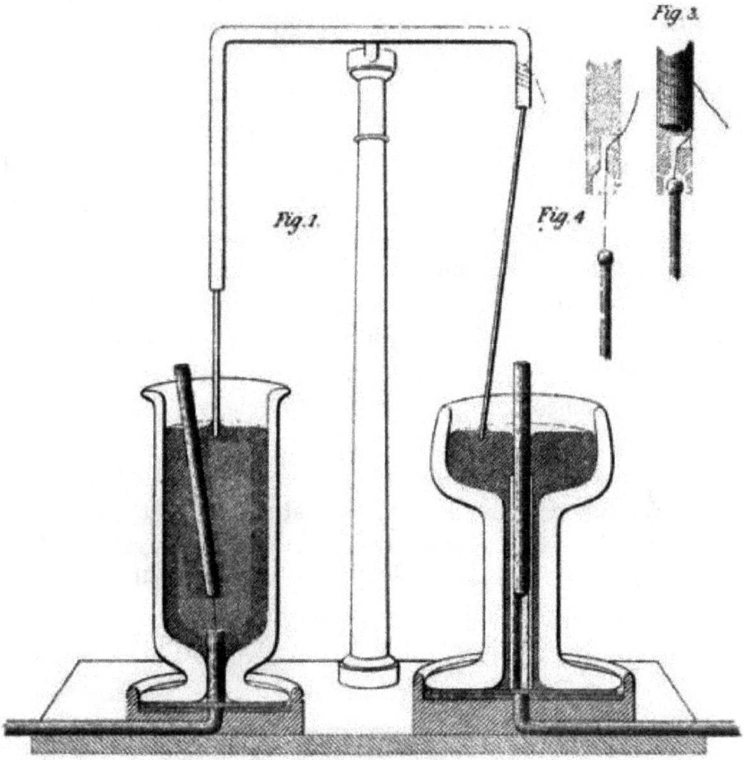

Fig. 12.5 *Faraday's experiment showing that an electric current could produce mechanical rotation in a magnetic field. This was the first electric motor! On the right side of the figure, a current-carrying rod rotates about a fixed magnet in a pool of mercury. On the left, the rod is fixed and the magnet rotates. Public domain, Wikimedia Commons*

Ørsted had shown that electricity could produce magnetism; and Faraday, with his strong intuitive grasp of the symmetry of natural laws, believed that the relationship could be reversed. He believed that magnetism could be made to produce electricity. In 1822, he wrote in his notebook: "Convert magnetism to electricity". For almost ten years, he tried intermittently to produce electrical currents with strong magnetic fields, but without success. Finally, in 1831, he discovered that a *changing* magnetic field would produce a current.

Faraday had wrapped two coils of wire around a soft iron ring; and he discovered that at precisely the instant when he started a current flowing in one of the coils, a momentary current was induced in the other coil. When he stopped the current in the first coil, so that the magnetic field collapsed, a momentary current in the opposite direction was induced in the second coil.

Next, Faraday tried pushing a permanent magnet in and out of a coil of wire; and he found that during the time when the magnet was in motion, so that the magnetic field in the coil was changing, a current was induced in the coil. Finally, Faraday made the first dynamo in history by placing a rotating copper disc between the poles of a magnet. He demonstrated that when the disc rotated, an electrical current flowed through a circuit connecting the center with the edge. He also experimented with static electricity, and showed that insulating materials become polarized when they are placed in an electric field.

Faraday continued the experiments on electrolysis begun by Sir Humphry Davy. He showed that when an electrical current is passed through a solution, the quantities of the chemical elements liberated at the anode and cathode are directly proportional to the total electrical charge passed through the cell, and inversely proportional to the valence of the elements. He realized that these laws of electrolysis supported Dalton's atomic hypothesis, and that they also pointed to the existence of an indivisible unit of electrical charge.

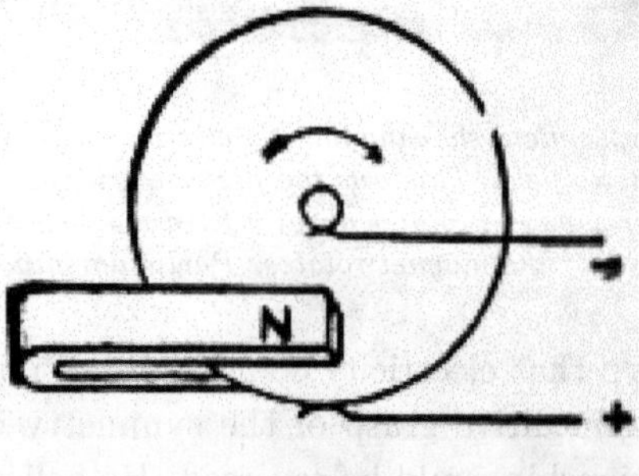

Fig. 12.6 *Faraday also showed that a copper disc, rotating between the poles of a magnet could produce an electric current. Public domain, Wikimedia Commons*

Faraday believed (correctly) that light is an electromagnetic wave; and to prove the connection of light with the phenomena of electricity and magnetism, he tried for many years to change light by means of electric and magnetic fields. Finally, towards the end of his career, he succeeded in rotating the plane of polarization of a beam of light passing through a piece of heavy glass by placing the glass in a strong magnetic field. This phenomenon is now known as the "Faraday effect".

Because of his many contributions both to physics and to chemistry (including the discovery of benzene and the first liquefaction of gases), and especially because of his contributions to electromagnetism and electrochemistry, Faraday is considered to be one of the greatest masters of the experimental method in the history of science. He was also a splendid lecturer. Fashionable Londoners flocked to hear his discourses at the Royal Institution, just as they had flocked to hear Sir Humphry Davy. Prince Albert, Queen Victoria's husband, was in the habit of attending Faraday's lectures, bringing with him Crown Prince Edward (later Edward VII).

As Faraday grew older, his memory began to fail, probably because of mercury poisoning. Finally, his unreliable memory forced him to retire from scientific work. He refused both an offer of knighthood and the Presidency of the Royal Society, remaining to the last the simple, modest and devoted worker who had first gone to assist Davy at the Royal Institution.

Maxwell and Hertz

Michael Faraday had no mathematical training, but he made up for this lack with his powerful physical intuition. He visualized electric and magnetic fields as "lines of force" in the space around the wires, magnets and electrical condensers with which he worked. In the case of magnetic fields, he could even make the lines of force visible by covering a piece of cardboard with iron filings, holding it near a magnet, and tapping the cardboard until the iron filings formed themselves into lines along the magnetic lines of force.

In this way, Faraday could actually see the magnetic field running from the north pole of a magnet, out into the surrounding space, and back into the south pole. He could also see the lines of the magnetic field forming circles around a straight current-carrying wire. Similarly, Faraday visualized the lines of force of the electric field as beginning at the positive charges of the system, running through the intervening space, and ending at the negative charges.

Meanwhile, the German physicists (especially the great mathematician and physicist, Johann Karl Friedrich Gauss (1777-1855)), had utilized the similarity between Coulomb's law of electrostatic force and Newton's law of gravitation. Coulomb's law states that the force between two point charges varies as the inverse square of the distance between them - in other words, it depends on distance in exactly the same way as the gravitational force. This allowed Gauss and the other German mathematicians to take over the whole "action at a distance" formalism of theoretical astronomy, and to apply it to electrostatics.

Faraday was unhappy with the idea of action at a distance, and he expressed his feelings to James Clerk Maxwell (1831-1879), a brilliant young mathematician from Edinburgh who had come to visit him. The young Scottish mathematical genius was able to show Faraday that his idea of lines of force did not in any way contradict the German conception of action at a distance. In fact, when put into mathematical form, Faraday's picture of lines of force fit beautifully with the ideas of Gauss.

During the nine years from 1864 to 1873, Maxwell worked on the problem of putting Faraday's laws of electricity and magnetism into mathematical form. In 1873, he published *A Treatise on Electricity and Magnetism*, one of the truly great scientific classics. Maxwell achieved a magnificent synthesis by expressing in a few simple equations the laws governing electricity and magnetism in all its forms. His electromagnetic equations have withstood the test of time; and now, a century later, they are considered to be among the most fundamental laws of physics.

Maxwell's equations not only showed that visible light is indeed and electromagnetic wave, as Faraday had suspected, but they also predicted the existence of many kinds of invisible electromagnetic waves, both higher and lower in frequency than visible light. We now know that the spectrum of electromagnetic radiation includes (starting at the low-frequency end) radio waves, microwaves, infra-red radiation, visible light, ultraviolet rays, X-rays and gamma rays. All these types of radiation are fundamentally the same, except that their frequencies and wave lengths cover a vast range. They all are oscillations of the electromagnetic field; they all travel with the speed of light; and they all are described by Maxwell's equations.

Maxwell's book opened the way for a whole new category of inventions, which have had a tremendous impact on society. However, when the *Treatise on Electricity and Magnetism* was published, very few scientists could understand it. Part of the problem was that the scientists of the 19th century would have liked a mechanical explanation of electromagnetism.

Fig. 12.7 *James Clerk Maxwell (1831-1879). When he put Faraday's experimental observations into mathematical form, Maxwell's equations predicted the existence of radio waves, and these were soon discovered by Heinrich Hertz. Public domain, Wikimedia Commons*

Even Maxwell himself, in building up his ideas, made use of mechanical models, "..replete with ropes passing over pulleys, rolled over drums, pulling weights, or at times comprising tubes pumping water into other elastic tubes which expanded and contracted, the whole mass of machinery noisy with the grinding of interlocked gear wheels". In the end, however, Maxwell abandoned as unsatisfactory the whole clumsy mechanical scaffolding which he had used to help his intuition; and there is no trace of mechanical ideas in his final equations. As Synge has expressed it, "The robust body of the Cheshire cat was gone, leaving in its place only a sort of mathematical grin".

Lord Kelvin (1824-1907), a prominent English physicist of the time, was greatly disappointed because Maxwell's theory could offer no mechanical explanation for electromagnetism; and he called the theory "a failure - the hiding of ignorance under the cover of a formula". In Germany, the eminent physicist, Hermann von Helmholtz (1821-1894), tried hard to understand Maxwell's theory in mechanical terms, and ended by accepting Maxwell's equations without ever feeling that he really understood them.

In 1883, the struggles of von Helmholtz to understand Maxwell's theory produced a dramatic proof of its correctness: Helmholtz had a brilliant student named Heinrich Hertz (1857-1894), whom he regarded almost as a son. In 1883, the Berlin Academy of Science offered a prize for work in the field of electromagnetism; and von Helmholtz suggested to Hertz that he should try to win the prize by testing some of the predictions of Maxwell's theory.

Hertz set up a circuit in which a very rapidly oscillating electrical current passed across a spark gap. He discovered that electromagnetic waves were indeed produced by this rapidly-oscillating current, as predicted by Maxwell! The waves could be detected with a small ring of wire in which there was a gap. As Hertz moved about the darkened room with his detector ring, he could see a spark flashing across the gap, showing the presence of electromagnetic waves, and showing them to behave exactly as predicted by Maxwell.

The waves detected by Hertz were, in fact, radio waves; and it was not long before the Italian engineer, Guglielmo Marconi (1874-1937), turned the discovery into a practical means of communication. In 1898, Marconi used radio signals to report the results of the boat races at the Kingston Regatta, and on December 12, 1901, using balloons to lift the antenae as high as possible, he sent a signal across the Atlantic Ocean from England to Newfoundland.

In 1904, a demonstration of a voice-carrying radio apparatus developed by Fessenden was the sensation of the St. Louis World's Fair; and in 1909, Marconi received the Nobel Prize in physics for his development of radio communications. In America, the inventive genius of Alexander Graham Bell (1847-1922) and Thomas Alva Edison (1847-1931) turned the discoveries of Faraday and Maxwell into the telephone, the electric light, the cinema and the phonograph.

Suggestions for further reading

(1) F.K. Richtmeyer and E.H. Kennard, *Introduction to Modern Physics*, McGraw-Hill (1947).
(2) E.T. Whittaker, *A History of the Aether and Electricity*, Cambridge University Press (1953).
(3) D.K.C. Macdonald, *Faraday, Maxwell and Kelvin*, Heinemann, London (1964).
(4) Otto Glasser, *Wilhelm Conrad Röntgen and the Early History of Röntgen Rays*, Charles C. Thomas, Springfield Illinois (1934).

Chapter 13

ATOMIC AND NUCLEAR PHYSICS

The discovery of electrons

In the late 1880's and early a 1890's, a feeling of satisfaction, perhaps even smugness, prevailed in the international community of physicists. It seemed to many that Maxwell's electromagnetic equations, together with Newton's equations of motion and gravitation, were the fundamental equations which could explain all the phenomena of nature. Nothing remained for physicists to do (it was thought) except to apply these equations to particular problems and to deduce the consequences. The inductive side of physics was thought to be complete.

However, in the late 1890's, a series of revolutionary discoveries shocked the physicists out of their feeling of complacency and showed them how little they really knew. The first of these shocks was the discovery of a subatomic particle, the electron. In Germany, Julius Plücker (1801-1868), and his friend, Heinrich Geisler (1814-1879), had discovered that an electric current could be passed through the gas remaining in an almost completely evacuated glass tube, if the pressure were low enough and the voltage high enough. When this happened, the gas glowed, and sometimes the glass sides of the tube near the cathode (the negative terminal) also glowed. Plücker found that the position of the glowing spots on the glass near the cathode could be changed by applying a magnetic field.

In England, Sir William Crookes (1832-1919) repeated and improved the experiments of Plücker and Geisler: He showed that the glow on the glass was produced by rays of some kind, streaming from the cathode; and he demonstrated that these "cathode rays" could cast shadows, that they could turn a small wheel placed in their path, and that they heated the glass where they struck it.

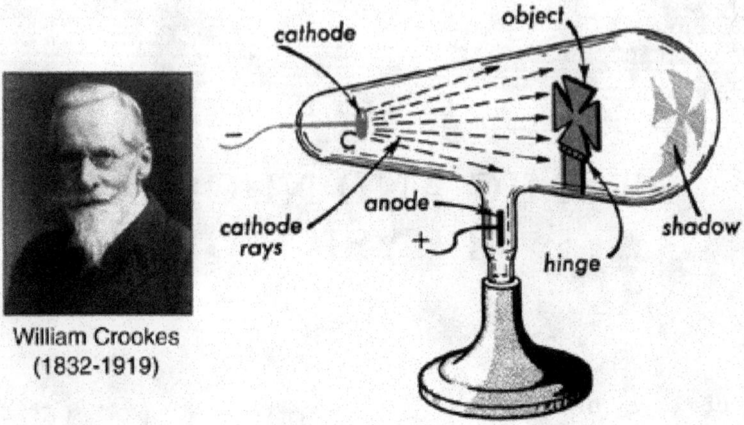

Fig. 13.1 *Sir William Crookes showed that cathode rays could cast shadows. Source: nau.edu*

Sir William Crookes believed that the cathode rays were electrically charged particles of a new kind - perhaps even a "fourth state of matter". His contemporaries laughed at these speculations; but a few years later a brilliant young physicist named J.J.Thomson (1856-1940), working at Cambridge University, entirely confirmed Crookes' belief that the cathode rays were charged particles of a new kind.

Thomson, an extraordinarily talented young scientist, had been appointed full professor and head of the Cavendish Laboratory at Cambridge at the age of 27. His predecessors in this position had been James Clerk Maxwell and the distinguished physicist, Lord Rayleigh, so the post was quite an honor for a man as young as Thomson. However, his brilliant performance fully justified the expectations of the committee which elected him. Under Thomson's direction, and later under the direction of his student, Ernest Rutherford, the Cavendish Laboratory became the world's greatest center for atomic and subatomic research; and it maintained this position during the first part of the twentieth century.

J.J. Thomson's first achievement was to demonstrate conclusively that the "cathode rays" observed by Plücker, Geisler and Crookes were negatively charged particles. He and his students also measured their ratio of charge to mass. If the charge was the same as that on an ordinary negative ionthen the mass of the particles was astonishingly small - almost two thousand times smaller than the mass of a hydrogen atom! Since the hydrogen atom is the lightest of all atoms, this indicated that the cathode rays were *subatomic* particles.

The charge which the cathode rays particles carried was recognized to be the fundamental unit of electrical charge, and they were given the name "electrons". All charges observed in nature were found to be integral multiples of the charge on an electron. The discovery of the electron was the first clue that the atom, thought for so long to be eternal and indivisible, could actually be torn to pieces.

X-rays

In 1895, while the work leading to the discovery of the electron was still going on, a second revolutionary discovery was made. In the autumn of that year, Wilhelm Konrad Roentgen (1845-1923), the head of the department of physics at the University of Würtzburg in Bavaria, was working with a discharge tube, repeating some of the experiments of Crookes.

Roentgen was especially interested in the luminescence of certain materials when they were struck by cathode rays. He darkened the room, and turned on the high voltage. As the current surged across the tube, a flash of light came from an entirely different part of the room! To Roentgen's astonishment, he found that a piece of paper which he had coated with barium platinocyanide was glowing brightly, even though it was so far away from the discharge tube that the cathode rays could not possibly reach it!

Roentgen turned off the tube, and the light from the coated paper disappeared. He turned on the tube again, and the bright glow on the screen reappeared. He carried the coated screen into the next room. Still it glowed! Again he turned off the tube, and again the screen stopped glowing. Roentgen realized that he had discovered something completely strange and new. Radiation of some kind was coming from his discharge tube, but the new kind of radiation could penetrate opaque matter!

Years later, when someone asked Roentgen what he thought when he discovered X-rays, he replied: "I didn't think. I experimented!" During the next seven weeks he experimented like a madman; and when he finally announced his discovery in December, 1895, he was able to report all of the most important properties of X-rays, including their ability to ionize gases and the fact that they cannot be deflected by electric or magnetic fields. Roentgen correctly believed X-rays to be electromagnetic waves, just like light waves, but with very much shorter wavelength.

Fig. 13.2 *Wilhelm Konrad Roentgen (1845-1923) Wellcome Images, [CC BY-SA 4.0], Wikimedia Commons*

It turned out that X-rays were produced by electrons from the cathode of the discharge tube. These electrons were accelerated by the strong electric field as they passed across the tube from the cathode (the negative terminal) to the anode (the positive terminal). They struck the platinum anode with very high velocity, knocking electrons out of the inner parts of the platinum atoms. As the outer electrons fell inward to replace these lost inner electrons, electromagnetic waves of very high frequency were emitted.

On January 23, 1896, Roentgen gave the first public lecture on X-rays; and in this lecture he demonstrated to his audience that X-ray photographs could be used for medical diagnosis. When Roentgen called for a volunteer from the audience, the 79 year old physiologist, Rudolf von Kölliker stepped up to the platform, and an X-ray photograph was taken of the old man's hand. The photograph, still in existence, shows the bones beautifully.

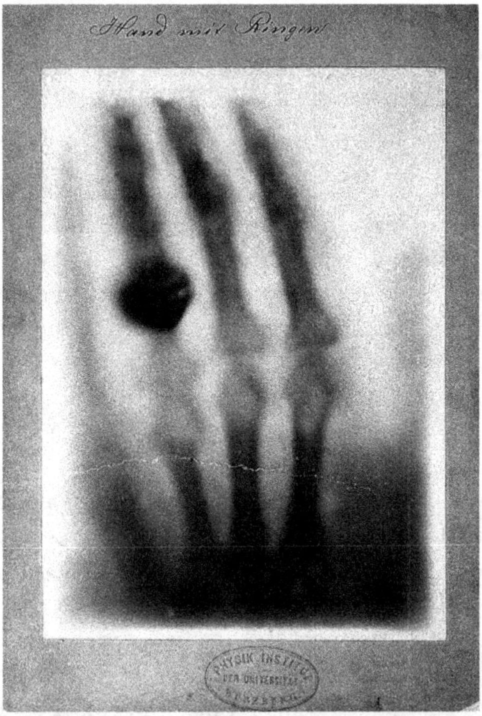

Fig. 13.3 *X-ray photograph by W.K. Roentgen. Wellcome Images, [CC BY-SA 4.0], Wikimedia Commons*

Wild enthusiasm for Roentgen's discovery swept across Europe and America, and soon many laboratories were experimenting with X-rays. The excitement about X-rays led indirectly to a third revolutionary discovery - radioactivity.

Radioactivity

On the 20th of January, 1896, only a month after Roentgen announced his discovery, an excited crowd of scientists gathered in Paris to hear the mathematical physicist Henri Poincaré lecture on Roentgen's X-rays. Among them was Henri Becquerel (1852-1908), a professor of physics working at the Paris Museum of Natural History and the École Polytechnique. Becquerel, with his neatly clipped beard, looked the very picture of a 19th century French professor; and indeed, he came from a family of scientists.

His grandfather had been a pioneer of electrochemistry, and his father had done research on fluorescence and phosphorescence.

Like his father, Henri Becquerel was studying fluorescence and phosphorescence; and for this reason he was especially excited by the news of Roentgen's discovery. He wondered whether there might be X-rays among the rays emitted by fluorescent substances. Hurrying to his laboratory, Becquerel prepared an experiment to answer this question.

He wrapped a large number of photographic plates in black paper, so that ordinary light could not reach them. Then he carried the plates outdoors into the sunlight, and on each plate he placed a sample of a fluorescent compound from his collection. After several hours of exposure, he developed the plates. If X-rays were present in the fluorescent radiation, then the photographic plates should be darkened, even though they were wrapped in black paper.

When he developed the plates, he found, to his excitement, that although most of them were unaffected, one of the plates was darkened! This was the plate on which he had placed the compound, potassium uranium sulfate. Experimenting further, Becquerel found other compounds which would darken the photographic plates - sodium uranium sulfate, ammonium uranium sulfate and uranium nitrate. All were compounds of uranium!

At the end of February, Becquerel made his first report to the French Academy of Sciences; and until the end of March, he brought a new report every week, describing new properties of the remarkable radiation from uranium compounds. Then the weather turned against him, and for many weeks, Paris was covered with thick clouds. Too impatient to wait for sunshine, Becquerel continued his experiments in cloudy weather, hoping that even without direct sunlight there would be some slight effect.

To his astonishment, the plates were blackened as much as before, although without direct sunlight the fluorescence of the uranium compounds was much diminished! Could it be that the mysterious penetrating radiation from the uranium compounds was independent of fluorescence? To answer this question, Becquerel next tried placing the uranium-containing compounds on photographic plates in a completely darkened room. Still the plates were blackened! The effect was completely independent of exposure to sunlight!

This was indeed something completely new and strange: The radiation seemed to come from the uranium atoms themselves, rather than from chemical changes in the compounds to which the atoms belonged. If the energy of Becquerel's rays did not come from sunlight, what was its source?

Two of the most basic assumptions of classical science seemed to be challenged - the indivisibility of the atom and the conservation of energy.

Marie and Pierre Curie

Among Henri Becquerel's colleagues in Paris were two dedicated and talented scientists, Marie and Pierre Curie. As a boy, Pierre Curie (1859-1906), the son of an intellectual Parisian doctor, had never been to school. His father had educated him privately, recognizing that his son's original and unworldly mind was unsuited for an ordinary education.

At the age of 16, Pierre Curie had become a Bachelor of Science, and at 18, he had a Master's degree in physics. Together with his brother, Jacques, Pierre Curie had discovered the phenomenon of piezoelectricity - the electrical potential produced when certain crystals, such as quartz, are compressed. He had also discovered a law governing the temperature-dependence of magnetism, "Curie's Law".

Although Pierre Curie had an international reputation as a physicist, his position as chief of the laboratory at the School of Physics and Chemistry of the City of Paris was miserably paid; and his modest, unworldly character prevented him from seeking a better position. He only wanted to be allowed to continue his research.

In 1896, when Becquerel announced his revolutionary discovery of radioactivity, Pierre Curie was newly married to a Polish girl, much younger than himself, but equally exceptional in character and ability. Marie Sklodowska Curie (1867-1934) had been born in Warsaw, in a Poland which did not officially exist, since it had been partitioned between Germany, Austria and Russia. Her father was a teacher of mathematics and physics and her mother was the principal of a girl's school.

Marie Sklodowska's family was a gifted one, with strong intellectual traditions; but it was difficult for her to obtain a higher education in Poland. Her mother died, and her father's job was withdrawn by the government. Marie Sklodowska was forced to work in a humiliating position as a governess in an uncultured family, meanwhile struggling to educate herself by reading books of physics and mathematics. She had a romance with the son of a Polish landowning family; but in the end, he rejected her because of her inferior social position.

Marie Sklodowska transmuted her unhappiness and humiliation into a fanatical devotion to science. She once wrote to her brother: "You must believe yourself to be born with a gift for some particular thing; and you must achieve that thing, no matter what the cost." Although she could not know it at the time, she was destined to become the greatest woman scientist in history.

Marie Sklodowska's chance for a higher education came at last when her married sister, who was studying medicine in Paris, invited Marie to live with her there and to enroll in the Sorbonne. After living in Paris with her sister for a year while studying physics, Marie found her sister's household too distracting for total concentration. She moved to a tiny, comfortless garret room, where she could be alone with her work.

Rejecting all social life, enduring freezing temperatures in winter, and sometimes fainting from hunger because she was too poor to afford proper food, Marie Sklodowska was nevertheless completely happy because at last she had the chance to study and to develop her potentialities. She graduated from the Sorbonne at the top of her class.

Pierre Curie had decided never to marry. He intended to devote himself totally to science; but when he met Marie, he recognized in her a person with whom he could share his ideals and his devotion to his work. After some hesitation by Marie, to whom the idea of leaving Poland forever seemed like treason, they were married. They spent a happy honeymoon touring the countryside of France on a pair of bicycles.

The next step for the young Polish student, who had now become Madame Curie, was to begin research for a doctor's degree; and she had to decide on a topic of research. The year was 1896, and news of Becquerel's remarkable discovery had just burst upon the scientific world. Marie Curie decided to make Becquerel's rays the topic of her thesis.

Using a sensitive electrometer invented by Pierre and Jacques Curie, she systematically examined all the elements to see whether any others besides uranium produced the strange penetrating rays. Almost at once, she made an important discovery: Thorium was also radioactive; but besides uranium and thorium, none of the other elements made the air of her ionization chamber' conduct electricity, discharging the electrometer. Among the known elements, only uranium and thorium were radioactive.

Next, Marie Curie tested all the compounds and minerals in the collection at the School of Physics. One of the minerals in the collection was pitchblend, an ore from which uranium can be extracted. She of course expected this uranium-containing ore to be radioactive; but to her astonishment, her measurements showed that the pitchblende was much *more* radioactive than could be accounted for by its content of uranium and thorium!

Since both Marie Curie's own work, and that of Becquerel, had shown radioactivity to be an atomic property, and since, among the known elements, the only two radioactive ones were uranium and thorium, she and her husband were forced to the inescapable conclusion that the pitchblende must contain small traces of a new, undiscovered, highly radioactive element, which had escaped notice in the chemical analysis of the ore.

At this point, Pierre Curie abandoned his own research and joined Marie in an attempt to find the unknown element which they believed must exist in pitchblende. By July, 1898, they had isolated a tiny amount of a new element, a hundred times more radioactive than uranium. They named it "polonium" after Marie's native country.

By this time, however, they had discovered that the extra radioactivity of pitchblende came from not one, but at least two new elements. The second undiscovered element, however, was enormously radioactive, and present only in infinitesimal concentrations. They realized that, in order to isolate a weighable amount of it, they would have to begin with huge amounts of raw pitchblende ore.

The Curies wrote to the directors of the mines at St. Joachimsthal in Bohemia, where silver was extracted from pitchblende, and begged for a few tons of the residue left after the extraction process. When they received a positive reply, they spent their small savings to pay the transportation costs.

The only place the Curies could find to work with the pitchblende ore was an old shed with a leaky roof - a chillingly cold place in the winter. Remembering the four years which she and her husband spent in this shed, Marie Curie wrote:

"This period was, for my husband and myself, the heroic period of our common existence... It was in this miserable old shed that the best and happiest years of our lives were spent, entirely consecrated to work. I sometimes passed the whole day stirring a boiling mass of material with an iron rod nearly as big as myself. In the evening, I was broken with fatigue... I came to treat as many as twenty kilograms of matter at a time, which had the effect of filling the shed with great jars full of precipitates and liquids. It was killing work to carry the receivers, to pour off the liquids and to stir for hours at a stretch the boiling matter in a smelting basin."

Marie and Pierre Curie began by separating the ore into fractions by various chemical treatments. After each treatment, they tested the fractions by measuring their radioactivity. They could easily see which fraction contained the highly radioactive unknown element. The new element, which

Fig. 13.4 *Pierre Curie, (1859-1906). He shared the 1903 Nobel Prize in Physics with his wife Marie. Public domain, Wikimedia Commons*

they named "radium", had chemical properties almost identical to those of barium; and the Curies found that it was almost impossible to separate radium from barium by ordinary chemical means.

In the end, they resorted to fractional crystalization, repeated several thousand times. At each step, the radium concentration of the active fraction was slightly enriched, and the radioactivity became progressively stronger. Finally it was two million times as great as the radioactivity of uranium. One evening, when Marie and Pierre Curie entered their laboratory without lighting the lamps, they saw that all their concentrated samples were glowing in the dark.

After four years of backbreaking labor, the Curies isolated a small amount of pure radium and measured its atomic weight. This achievement, together with their other work on radioactivity, brought them the 1903 Nobel Prize in Physics (shared with Becquerel), as well as worldwide fame. Madame Curie, the first great woman scientist in history, became a symbol of what women could do. The surge of public enthusiasm, which had started with Roentgen's discovery of X-rays, reached a climax with Madame Curie's isolation of radium.

It had been discovered that radium was helpful in treating cancer; and Madame Curie was portrayed by newspapers of the period as a great humanitarian. Indeed, the motives which inspired Marie and Pierre Curie to their heroic labors were both humanitarian and idealistic. They believed that only good could come from any increase in human knowledge. They did not know that radium is also a dangerous element, capable of causing cancer as well as curing it; and they could not forsee that research on radioactivity would eventually lead to nuclear weapons.

Fig. 13.5 *Marie Curie. Nobel Prize in Physics photo (1903). Later, she also won the Nobel Prize in Chemistry. Public domain, Wikimedia Commons*

Rutherford's model of the atom

In 1895, the year during which Roentgen made his revolutionary discovery of X-rays, a young New Zealander named Ernest Rutherford was digging potatoes on his father's farm, when news reached him that he had won a scholarship for advanced study in England. Throwing down his spade, Rutherford said, "That's the last potato I'll dig!" He postponed his marriage plans and sailed for England, where he enrolled as a research student at Cambridge University. He began work at the Cavendish Laboratory, under the leadership of J.J. Thomson, the discoverer of the electron.

In New Zealand, Rutherford had done pioneering work on the detection of radio waves, and he probably would have continued this work at Cambridge, if it had not been for the excitement caused by the discoveries of Roentgen and Becquerel. Remembering this period of his life, Rutherford wrote:

"Few of you can realize the enormous sensation caused by the discovery of X-rays by Roentgen in 1895. It interested not only the scientific man, but also the man in the street, who was excited by the idea of seeing his own insides and his bones. Every laboratory in the world took out its old Crookes' tubes to produce X-rays, and the Cavendish was no exception."

J.J. Thomson, who was interested in studying ions (charged atoms or molecules) in gases, soon found that gaseous ions could be produced very conveniently by means of X-rays. Rutherford abandoned his research on radio waves, and joined Thomson in this work.

"When I entered the Cavendish Laboratory", Rutherford remembered later, "I began to work on the ionization of gases by means of X-rays. After reading the paper of Becquerel, I was curious to know whether the ions produced by the radiation from uranium were of the same nature as those produced by X-rays; and in particular, I was interested because Becquerel thought that his radiation was somehow intermediate between light and X-rays."

"I therefore proceeded to make a systematic examination of the radiation, and I found that it was of two types - one which produced intense ionization, and which was absorbed by a few centimeters of air, and the other, which produced less intense ionization, but was more penetrating. I called these alpha rays and beta rays respectively; and when, in 1898, Villard discovered a still more penetrating type of radiation, he called it gamma-radiation."

Rutherford later showed that the alpha-rays were actually ionized helium atoms thrown out at enormous velocities by the decaying uranium, and that beta-rays were high-speed electrons. The gamma-rays turned out to be electromagnetic waves, just like light waves, but of extremely short wavelength.

Fig. 13.6 *Rutherford receivng the 1908 Nobel Prize in Chemistry. Public domain, Wikimedia Commons*

Rutherford returned briefly to New Zealand to marry his sweetheart, Mary Newton; and then he went to Canada, where he had been offered a post as Professor of Physics at McGill University. In Canada, with the collaboration of the chemist, Frederick Soddy (1877-1956), Rutherford continued his experiments on radioactivity, and worked out a revolutionary theory of transmutation of the elements through radioactive decay.

During the middle ages, alchemists had tried to change lead and mercury into gold. Later, chemists had convinced themselves that it was impossible to change one element into another. Rutherford and Soddy now claimed that radioactive decay involves a whole series of transmutations, in which one element changes into another!

Returning to England as head of the physics department at Manchester University, Rutherford continued to experiment with alpha-particles. He was especially interested in the way they were deflected by thin metal foils. Rutherford and his assistant, Hans Geiger (1886-1945), found that most of the alpha-particles passed through a metal foil with only a very slight deflection, of the order of one degree.

In 1911, a young research student named Ernest Marsden joined the group, and Rutherford had to find a project for him. What happened next, in Rutherford's own words, was as follows:

"One day, Geiger came to me and said, 'Don't you think that young Marsden, whom I'm training in radioactive methods, ought to begin a small research?' Now I had thought that too, so I said, 'Why not let him see if any alpha-particles can be scattered through a large angle?' I may tell you in confidence that I did not believe that they would be, since we knew that the alpha-particle was a very fast, massive particle, with a great deal of energy; and you could show that if the scattering was due to the accumulated effect of a number of small scatterings, the chance of an alpha-particle's being scattered backward was very small."

"Then I remember two or three days later, Geiger coming to me in great excitement and saying, 'We have been able to get some of the alpha-particles coming backwards'. It was quite the most incredible event that has ever happened to me in my life. It was almost as incredible as if you fired a 15-inch shell at a piece of tissue paper and it came back and hit you."

"On consideration, I realized that this scattering backwards must be the result of a single collision, and when I made calculations, I found that it was impossible to get anything of that order of magnitude unless you took a system in which the greater part of the mass of the atom was concentrated in a minute nucleus."

"It was then that I had the idea of an atom with a minute massive center carrying a charge. I worked out mathematically what laws the scattering should obey, and found that the number of particles scattered through a given angle should be proportional to the thickness of the scattering foil, the square of the nuclear charge, and inversely proportional to the fourth power of the velocity. These deductions were later verified by Geiger and Marsden in a series of beautiful experiments."

Planck, Einstein and Bohr

According to the model proposed by Rutherford in 1911, every atom has an extremely tiny nucleus, which contains almost all of the mass of the atom. Around this tiny but massive nucleus, Rutherford visualized light, negatively-charged electrons circulating in orbits, like planets moving around the sun. Rutherford calculated that the diameter of the whole atom had to be several thousand times as large as the diameter of the nucleus.

Rutherford's model of the atom explained beautifully the scattering experiments of Geiger and Marsden, but at the same time it presented a serious difficulty: According to Maxwell's equations, the electrons circulating in their orbits around the nucleus ought to produce electromagnetic waves. It could easily be calculated that the electrons in Rutherford's atom ought to lose all their energy of motion to this radiation, and spiral in towards the nucleus. Thus, according to classical physics, Rutherford's atom could not be stable. It had to collapse.

The paradox was solved by Niels Bohr (1885-1962), a gifted young theoretical physicist from Copenhagen who had come to Manchester to work with Rutherford. Bohr was not at all surprised by the failure of classical concepts when applied to Rutherford's nuclear atom. Since he had been educated in Denmark, he was more familiar with the work of German physicists than were his English colleagues at Manchester. In particular, Bohr had studied the work of Max Planck (1858-1947) and Albert Einstein (1879-1955).

Just before the turn of the century, the German physicist, Max Planck, had been studying theoretically the electromagnetic radiation coming from a small hole in an oven. The hole radiated as though it were an ideally black body. This "black body radiation" was very puzzling to the physicists of the time, since classical physics failed to explain the frequency distribution of the radiation and its dependence on the temperature of the oven.

In 1901, Max Planck had discovered a formula which fitted beautifully with the experimental measurements of the frequency distribution of black body radiation; but in order to derive his formula, he had been forced to make a radical assumption which broke away completely from the concepts of classical physics.

Planck had been forced to assume that light (or, more generally, electromagnetic radiation of any kind) can only be emitted or absorbed in amounts of energy which Planck called "quanta". The amount of energy in each of these "quanta" was equal to the frequency of the light multiplied by a constant, h, which came to be known as "Planck's constant".

This was indeed a strange assumption! It seemed to have been pulled out of thin air; and it had no relation whatever to anything that had been discovered previously in physics. The only possible justification for Planck's quantum hypothesis was the brilliant success of his formula in explaining the puzzling frequency distribution of the black body radiation. Planck himself was greatly worried by his own radical break with classical concepts, and he spent many years trying unsuccessfully to relate his quantum hypothesis to classical physics.

In 1905, Albert Einstein published a paper in the *Annalen der Physik* in which he applied Planck's quantum hypothesis to the photoelectric effect. (At that time, Einstein was 25 years old, completely unknown, and working as a clerk at the Swiss Patent Office.) The photoelectric effect was another puzzling phenomenon which could not in any way be explained by classical physics. The German physicist Lenard had discovered in 1903 that light with a frequency above a certain threshold could knock electrons out of the surface of a metal; but below the threshold frequency, nothing at all happened, no matter how long the light was allowed to shine.

Using Planck's quantum hypothesis, Einstein offered the following explanation for the photoelectric effect: A certain minimum energy was needed to overcome the attractive forces which bound the electron to the metal surface. This energy was equal to the threshold frequency multiplied by Planck's constant. Light with a frequency equal to or higher than the threshold frequency could tear an electron out of the metal; but the quantum of energy supplied by light of a lower frequency was insufficient to overcome the attractive forces.

Einstein later used Planck's quantum formula to explain the low-temperature behavior of the specific heats of crystals, another puzzling phenomenon which defied explanation by classical physics. These contributions by Einstein were important, since without this supporting evidence it could be maintained that Planck's quantum hypothesis was an *ad hoc* assumption, introduced for the sole purpose of explaining black body radiation.

As a student, Niels Bohr had been profoundly impressed by the radical ideas of Planck and Einstein. In 1912, as he worked with Rutherford at Manchester, Bohr became convinced that the problem of saving Rutherford's atom from collapse could only be solved by means of Planck's quantum hypothesis.

Fig. 13.7 *Niels Bohr and Albert Einstein in a photo by Paul Ehrenfest. Public domain, Wikimedia Commons*

Returning to Copenhagen, Bohr continued to struggle with the problem. In 1913, he found the solution: The electrons orbiting around the nucleus of an atom had "angular momentum". Assuming circular orbits, the angular momentum was given by the product of the mass and velocity of the electron, multiplied by the radius of the orbit. Bohr introduced a quantum hypothesis similar to that of Planck: He assumed that the angular momentum of an electron in an allowed orbit, (multiplied by 2 pi), had to be equal to an integral multiple of Planck's constant. The lowest value of the integer, n=1, corresponded to the lowest allowed orbit. Thus, in Bohr's model, the collapse of Rutherford's atom was avoided.

Bohr calculated that the binding energies of the various allowed electron orbits in a hydrogen atom should be a constant divided by the square of the integer n; and he calculated the value of the constant to be 13.5 electron-Volts. This value fit exactly the observed ionization energy of hydrogen. After talking with the Danish spectroscopist, H.M. Hansen, Bohr realized with joy that by combining his formula for the allowed orbital energies with the Planck-Einstein formula relating energy to frequency, he could explain the mysterious line spectrum of hydrogen.

Fig. 13.8 *Another photo of Bohr and Einstein by Ehrenfest. Public domain, Wikimedia Commons*

When Niels Bohr published all this in 1913, his paper produced agonized cries of "foul!" from the older generation of physicists. When Lord Rayleigh's son asked him if he had seen Bohr's paper, Rayleigh replied: "Yes, I have looked at it; but I saw that it was of no use to me. I do not say that discoveries may not be made in that sort of way. I think very likely they may be. But it does not suit me." However, as more and more atomic spectra and properties were explained by extensions of Niels Bohr's theories, it became clear that Planck, Einstein and Bohr had uncovered a whole new stratum of phenomena, previously unsuspected, but of deep and fundamental importance.

Atomic numbers

Bohr's atomic theory soon received strong support from the experiments of one of the brightest of Rutherford's bright young men - Henry Moseley (1887-1915). Moseley came from a distinguished scientific family. Not only his father, but also both his grandfathers, had been elected to the Royal Society. After studying at Oxford, where his father had once been a professor, Moseley found it difficult to decide where to do his postgraduate work. Two laboratories attracted him: the great J.J. Thomson's Cavendish Laboratory at Cambridge, and Rutherford's laboratory at Manchester. Finally, he decided on Manchester, because of the revolutionary discoveries of Rutherford, who two years earlier had won the 1908 Nobel Prize for Chemistry.

Rutherford's laboratory was like no other in the world, except J.J. Thomson's. In fact, Rutherford had learned much about how to run a laboratory from his old teacher, Thomson. Rutherford continued Thomson's tradition of democratic informality and cheerfulness. Like Thomson, he had a gift for infecting his students with his own powerful scientific curiosity, and his enthusiastic enjoyment of research.

Thomson had also initiated a tradition for speed and ingenuity in the improvisation of experimental apparatus - the so-called "sealing-wax and string" tradition - and Rutherford continued it. Niels Bohr, after working with Rutherford, was later to continue the tradition of informality and enthusiasm at the Institute for Theoretical Physics which Bohr founded in Copenhagen in 1920.

Most scientific laboratories of the time offered a great contrast to the informality, enthusiasm, teamwork and speed of the Thomson-Rutherford-

Bohr tradition. E.E. da C. Andrade, who first worked in Lenard's laboratory at Heidelberg, and later with Rutherford at Manchester, has given the following description of the contrast between the two groups:

"At the Heidelberg colloquium, Lenard took the chair, very much like a master with his class. He had the habit, if any aspect of his work was being treated by the speaker, of interrupting with, 'And who did that first?' The speaker would reply with a slight bow, 'Herr Geheimrat, you did that first', to which Lenard answered, 'Yes, I did that first'."

"At the Manchester colloquium, which met on Friday afternoons, Rutherford was, as in all his relations with the research workers, the boisterous, enthusiastic, inspiring friend, undoubtedly the leader but in close community with the led, stimulating rather than commanding, 'gingering up', to use a favourite expression of his, his team."

Although Rutherford occasionally swore at his "lads", his affection for them was very real. He had no son of his own, and he became a sort of father to the brilliant young men in his laboratory. Their nickname for him was "Papa". Such was the laboratory which Harry Moseley joined in 1910. At almost the same time, Moseley's childhood friend, Charles Darwin (the grandson of the "right" Charles Darwin), also joined Rutherford's team.

After working on a variety of problems in radioactivity which were given to him by Rutherford, Moseley asked whether he and Charles Darwin might be allowed to study the spectra of X-rays. At first, Rutherford said no, since no one at Manchester had any experience with X-rays; "and besides", Rutherford added with a certain amount of bias, "all science is either radioactivity or else stamp-collecting".

However, after looking more carefully at what was being discovered about X-rays, Rutherford gave his consent. In 1912, a revolutionary discovery had been made by the Munich physicist, Max von Laue (1879-1960): It had long been known that because of its wavelike nature, white light can be broken up into the colors of the spectrum by means of a "diffraction grating" - a series of parallel lines engraved very closely together on a glass plate.

For each wavelength of light, there are certain angles at which the new wavelets produced by the lines of the diffraction grating reinforce each other instead of cancelling. The angles of reinforcement are different for each wavelength, and thus the different colors are separated by the grating.

Max von Laue's great idea was to do the same thing with X-rays, using a crystal as a diffraction grating. The regular lines of atoms in the crystal, von Laue reasoned, would act be fine enough to fit the tiny wavelength of the X-rays, believed to be less than one ten-millionth of a centimeter.

Von Laue's experiment, performed in 1912, had succeeded beautifully, and his new technique had been taken up in England by a father and son team, William Henry Bragg (1862-1942) and William Lawrence Bragg (1890-1971). The Braggs had used X-ray diffraction not only to study the spectra of X-rays, but also to study the structure of crystals. Their techniques were later to become one of the most valuable research tools available for studying molecular structure.

Having finally obtained Rutherford's permission, Moseley and Darwin threw themselves into this exciting field of study. Remembering his work with Harry Moseley, Charles Darwin later wrote:

"Working with Moseley was one of the most strenuous exercises I have ever undertaken. He was, without exception, the hardest worker I have ever known... There were two rules for his work: First, when you started to set up the apparatus for an experiment, you must not stop until it was set up. Second, when the apparatus was set up, you must not stop work until the experiment was done. Obeying these rules implied a most irregular life, sometimes with all-night sessions; and indeed, one of Moseley's experteses was the knowledge of where in Manchester one could get a meal at three in the morning."

After about a year, Charles Darwin left the experiments to work on the theoretical aspects of X-ray diffraction. (He was later knighted for his distinguished contributions to theoretical physics.) Moseley continued the experiments alone, systematically studying the X-ray spectra of all the elements in the periodic system.

Niels Bohr had shown that the binding energies of the allowed orbits in a hydrogen atom are equal to Rydberg's constant , R (named after the distinguished Swedish spectroscopist, Johannes Robert Rydberg), divided by the square of an integral "quantum number", n. He had also shown that for heavier elements, the constant, R, is equal to the square of the nuclear charge, Z, multiplied by a factor which is the same for all elements. The constant, R, could be observed in Moseley's studies of X-ray spectra: Since X-rays are produced when electrons are knocked out of inner orbits and outer electrons fall in to replace them, Moseley could use the Planck-Einstein relationship between frequency and energy to find the energy difference between the orbits, and Bohr's theory to relate this to R.

Moseley found complete agreement with Bohr's theory. He also found that the nuclear charge, Z, increased regularly in integral steps as he went along the rows of the periodic table: Hydrogen had Z=1, helium Z=2, lithium Z=3, and so on up to uranium with Z=92. The 92 electrons of a uranium atom made it electrically neutral, exactly balancing the charge of the nucleus. The number of electrons of an element, and hence its chemical properties, Moseley found, were determined uniquely by its nuclear charge, which Moseley called the "atomic number".

Moseley's studies of the nuclear charges of the elements revealed that a few elements were missing. In 1922, Niels Bohr received the Nobel Prize for his quantum theory of the atom; and he was able to announce at the presentation ceremony that one of Moseley's missing elements had been found at his institute. Moseley, however, was dead. He was one of the ten million young men whose lives were needlessly thrown away in Europe's most tragic blunder - the First World War.

A wave equation for matter

In 1926, the difficulties surrounding the "old quantum theory" of Max Planck, Albert Einstein and Niels Bohr were suddenly solved, and its true meaning was understood. Two years earlier, a French aristocrat, Prince Louis de Broglie, writing his doctoral dissertation at the Sorbonne in Paris, had proposed that very small particles, such as electrons, might exhibit wavelike properties. The ground state and higher excited states of the electron in Bohr's model of the hydrogen atom would then be closely analogous to the fundamental tone and higher overtones of a violin string.

Almost the only person to take de Broglie's proposal seriously was Albert Einstein, who mentioned it in one of his papers. Because of Einstein's interest, de Broglie's matter-waves came to the attention of other physicists. The Austrian theoretician, Erwin Schrödinger, working at Zürich, searched for the underlying wave equation which de Broglie's matter-waves obeyed.

Schrödinger's gifts as a mathematician were so great that it did not take him long to solve the problem. The Schrödinger wave equation for matter is now considered to be more basic than Newton's equations of motion. The wavelike properties of matter are not apparent to us in our daily lives because the wave-lengths are extremely small in comparison with the sizes of objects which we can perceive. However, for very small and light particles, such as electrons moving in their orbits around the nucleus of an atom, the wavelike behavior becomes important.

Schrödinger was able to show that Niels Bohr's atomic theory, including Bohr's seemingly arbitrary quantization of angular momentum, can be derived by solving the wave equation for the electrons moving in the attractive field of the nucleus. The allowed orbits of Bohr's theory correspond in Schrödinger's theory to harmonics, similar to the fundamental harmonic and higher overtones of an organ pipe or a violin string. (If Pythagoras had been living in 1926, he would have rejoiced to see the deepest mysteries of matter explained in terms of harmonics!)

Bohr himself believed that a complete atomic theory ought to be able to explain the chemical properties of the elements in Mendeléev's periodic system. Bohr's 1913 theory failed to pass this test, but the new de Broglie-Schrödinger theory succeeded! Through the work of Pauli, Heitler, London, Slater, Pauling, Hund, Mulliken, Hückel and others, who applied Schrödinger's wave equation to the solution of chemical problems, it became apparent that the wave equation could indeed (in principle) explain all the chemical properties of matter.

Strangely, the problem of developing the fundamental quantum theory of matter was solved not once, but three times in 1926! At the University of Göttingen in Germany, Max Born (1882-1970) and his brilliant young students Werner Heisenberg and Pascal Jordan solved the problem in a completely different way, using matrix methods. At the same time, a theory similar to the "matrix mechanics" of Heisenberg, Born and Jordan was developed independently at Cambridge University by a 24 year old mathematical genius named Paul Adrien Maurice Dirac. At first, the Heisenberg-Born-Jordan-Dirac quantum theory seemed to be completely different from the Schrödinger theory; but soon the Göttingen mathematician David Hilbert (1862-1943) was able to show that the theories were really identical, although very differently expressed.

Fig. 13.9 *Bust of Erwin Schrödinger in the courtyard arcade of the main building, University of Vienna. Daderot at the English Language Wikipedia, [CC BY-SA 3.0], Wikimedia Commons*

Suggestions for further reading

(1) Alfred Romer (editor), *The Discovery of Radioactivity and Transmutation*, Dover, New York (1964).

(2) Marie Curie, *Pierre Curie*, Dover, New York (1963).
(3) Eve Curie, *Madame Curie*, Pocket Books Inc., New York (1958).
(4) Joseph Needham and Walter Pagel (editors), *Background to Modern Science*, Cambridge University Press (1938).
(5) R. Harre (editor), *Scientific Thought 1900-1960, A Selective Survey*, Clarendon Press, Oxford (1969).
(6) J.B. Birks (editor), *Rutherford at Manchester*, Heywood and Company Ltd., London (1962).
(7) Niels Bohr, *On the Constitution of Atoms and Molecules*, W.A. Benjamin Inc., New York (1963).
(8) S. Rosenthal (editor), *Niels Bohr, His Life and Work*, North Holland (1967).
(9) Ruth Moore, *Niels Bohr, the Man and the Scientist*, Hodder and Stoughten, London (1967).
(10) Bernard Jaffe, *Moseley and the Numbering of the Elements*, Heinemann, London (1971).
(11) E.A. Hylleraas, *Remaniscences from Early Quantum Mechanics of Two-Electron Atoms*, Reviews of Modern Physics, *35*, 421 (1963).

Chapter 14

RELATIVITY

Einstein

Albert Einstein was born in Ulm, Germany, in 1879. He was the son of middle-class, irreligious Jewish parents, who sent him to a Catholic school. Einstein was slow in learning to speak, and at first his parents feared that he might be retarded; but by the time he was eight, his grandfather could say in a letter:

"Dear Albert has been back in school for a week. I just love that boy, because you cannot imagine how good and intelligent he has become."

Remembering his boyhood, Einstein himself later wrote:

"When I was 12, a little book dealing with Euclidian plane geometry came into my hands at the beginning of the school year. Here were assertions, as for example the intersection of the altitudes of a triangle in one point, which - though by no means self-evident - could nevertheless be proved with such certainty that any doubt appeared to be out of the question. The lucidity and certainty made an indescribable impression on me."

When Albert Einstein was in his teens, the factory owned by his father and uncle began to encounter hard times. The two Einstein families moved to Italy, leaving Albert alone and miserable in Munich, where he was supposed to finish his course at the gymnasium. Einstein's classmates had given him the nickname "Beidermeier", which means something like "Honest John"; and his tactlessness in criticizing authority soon got him into trouble. In Einstein's words, what happened next was the following:

"When I was in the seventh grade at the Lutpold Gymnasium, I was summoned by my home-room teacher, who expressed the wish that I leave the school. To my remark that I had done nothing wrong, he replied only,

Fig. 14.1 *Albert Einstein at the age of three years. This is believed to be the oldest known photograph of Einstein. Public domain, Wikimedia Commons*

'Your mere presence spoils the respect of the class for me'."

Einstein left gymnasium without graduating, and followed his parents to Italy, where he spent a joyous and carefree year. He also decided to change his citizenship. "The over-emphasized military mentality of the German State was alien to me, even as a boy", Einstein wrote later. "When my father moved to Italy, he took steps, at my request, to have me released from German citizenship, because I wanted to be a Swiss citizen."

The financial circumstances of the Einstein family were now precarious, and it was clear that Albert would have to think seriously about a practical career. In 1896, he entered the famous Zürich Polytechnic Institute with the intention of becoming a teacher of mathematics and physics. However, his undisciplined and nonconformist attitudes again got him into trouble. His mathematics professor, Hermann Minkowski (1864-1909), considered Einstein to be a "lazy dog"; and his physics professor, Heinrich Weber, who originally had gone out of his way to help Einstein, said to him in anger and exasperation: "You're a clever fellow, but you have one fault: You won't let anyone tell you a thing! You won't let anyone tell you a thing!"

Einstein missed most of his classes, and read only the subjects which interested him. He was interested most of all in Maxwell's theory of electromagnetism, a subject which was too "modern" for Weber. There were two major examinations at the Zürich Polytechnic Institute, and Einstein would certainly have failed them had it not been for the help of his loyal friend, the mathematician Marcel Grossman.

Grossman was an excellent and conscientious student, who attended every class and took meticulous notes. With the help of these notes, Einstein managed to pass his examinations; but because he had alienated Weber and the other professors who could have helped him, he found himself completely unable to get a job. In a letter to Professor F. Ostwald on behalf of his son, Einstein's father wrote: "My son is profoundly unhappy because of his present joblessness; and every day the idea becomes more firmly implanted in his mind that he is a failure, and will not be able to find the way back again."

From this painful situation, Einstein was rescued (again!) by his friend Marcel Grossman, whose influential father obtained for Einstein a position at the Swiss Patent Office - Technical Expert (Third Class). Anchored at last in a safe, though humble, position, Einstein married one of his classmates, a Serbian girl named Mileva Maric. He learned to do his work at the Patent Office very efficiently; and he used the remainder of his time on his own calculations, hiding them guiltily in a drawer when footsteps approached.

Fig. 14.2 *Albert Einstein with Zürich friends Habicht and Solovine, ca. 1903. They met informally to discuss physics, calling themselves the "Olympia Academy". Public domain, Wikimedia Commons*

Special relativity

In 1905, this Technical Expert (Third Class) astonished the world of science with five papers, written within a few weeks of each other, and published in the *Annalen der Physik.* Of these five papers, three were classics: One of these was the paper in which Einstein applied Planck's quantum hypothesis to the photoelectric effect. The second paper discussed "Brownian motion", the zig-zag motion of small particles suspended in a liquid and hit randomly by the molecules of the liquid. This paper supplied a direct proof of the validity of atomic ideas and of Boltzmann's kinetic theory.

The third paper was destined to establish Einstein's reputation as one of the greatest physicists of all time. It was entitled *On the Electrodynamics of Moving Bodies*, and in this paper, Albert Einstein formulated his special theory of relativity.

The theory of relativity grew out of problems connected with Maxwell's electromagnetic theory of light. Ever since the wavelike nature of light had first been demonstrated, it had been supposed that there must be some medium to carry the light waves, just as there must be some medium (for example air) to carry sound waves. A word was even invented for the medium which was supposed to carry electromagnetic waves: It was called the "ether".

By analogy with sound, it was believed that the velocity of light would depend on the velocity of the observer relative to the "ether". However, all attempts to measure differences in the velocity of light in different directions had failed, including an especially sensitive experiment which was performed in America in 1887 by A.A. Michelson and E.W. Morley.

Even if the earth had, by a coincidence, been stationary with respect to the "ether" when Michelson and Morley first performed their experiment, they should have found an "ether wind" when they repeated their experiment half a year later, with the earth at the other side of its orbit. Strangely, the observed velocity of light seemed to be completely independent of the motion of the observer!

In his famous 1905 paper on relativity, Einstein made the negative result of the Michelson-Morley experiment the basis of a far-reaching principle: He asserted that no experiment whatever can tell us whether we are at rest or whether we are in a state of uniform motion. With this assumption, the Michelson-Morley experiment of course had to fail, and the measured velocity of light had to be independent of the motion of the observer.

Einstein's Principle of Special Relativity had other extremely important consequences: He soon saw that if his principle were to hold, then Newtonian mechanics would have to be modified. In fact, Einstein's Principle of Special Relativity required that *all* fundamental physical laws exhibit a symmetry between space and time. The three space dimensions, and a fourth dimension, ict, had to enter every fundamental physical law in a symmetrical way. (Here i is the square root of -1, c is the velocity of light, and t is time.)

When this symmetry requirement is fulfilled, a physical law is said to be "Lorentz-invariant" (in honor of the Dutch physicist H.A. Lorentz, who anticipated some of Einstein's ideas). Today, we would express Einstein's principle by saying that every fundamental physical law must be Lorentz-invariant (i.e. symmetrical in the space and time coordinates). The law will then be independent of the motion of the observer, provided that the observer is moving uniformly.

Einstein was able to show that, when properly expressed, Maxwell's equations are already Lorentz-invariant; but Newton's equations of motion have to be modified. When the needed modifications are made, Einstein found, then the mass of a moving particle appears to increase as it is accelerated. A particle can never be accelerated to a velocity greater than the velocity of light; it merely becomes heavier and heavier, the added energy being converted into mass.

From his 1905 theory, Einstein deduced his famous formula equating the energy of a system to its mass multiplied by the square of the velocity of light. As we shall see, his formula was soon used to explain the source of the energy produced by decaying uranium and radium; and eventually it led to the construction of the atomic bomb. Thus Einstein, a lifelong pacifist, who renounced his German citizenship as a protest against militarism, became instrumental in the construction of the most destructive weapon ever invented - a weapon which casts an ominous shadow over the future of humankind.

Just as Einstein was one of the first to take Planck's quantum hypothesis seriously, so Planck was one of the first physicists to take Einstein's relativity seriously. Another early enthusiast for relativity was Hermann Minkowski, Einstein's former professor of mathematics. Although he once had characterized Einstein as a "lazy dog", Minkowski now contributed importantly to the mathematical formalism of Einstein's theory; and in 1907, he published the first book on relativity. In honor of Minkowski's contributions to relativity, the 4-dimensional space-time continuum in which we live is sometimes called "Minkowski space".

In 1908, Minkowski began a lecture to the Eightieth Congress of German Scientists and Physicians with the following words:

"From now on, space by itself, and time by itself, are destined to sink completely into the shadows; and only a kind of union of both will retain an independent existence."

General relativity

Gradually, the importance of Einstein's work began to be realized, and he was much sought after. He was first made Assistant Professor at the University of Zürich, then full Professor in Prague, then Professor at the Zürich Polytechnic Institute; and finally, in 1913, Planck and Nernst persuaded Einstein to become Director of Scientific Research at the Kaiser Wilhelm Institute in Berlin. He was at this post when the First World War broke out.

While many other German intellectuals produced manifestos justifying Germany's invasion of Belgium, Einstein dared to write and sign an anti-war manifesto. Einstein's manifesto appealed for cooperation and understanding among the scholars of Europe for the sake of the future; and it proposed the eventual establishment of a League of Europeans. During the war, Einstein remained in Berlin, doing whatever he could for the cause of peace, burying himself unhappily in his work, and trying to forget the agony of Europe, whose civilization was dying in a rain of shells, machine-gun bullets, and poison gas.

The work into which Einstein threw himself during this period was an extension of his theory of relativity. He already had modified Newton's equations of motion so that they exhibited the space-time symmetry required by his Principle of Special Relativity. However, Newton's law of gravitation remained a problem. Obviously it had to be modified, since it was not Lorentz-invariant; but how should it be changed?

What principles could Einstein use in his search for a more correct law of gravitation? Certainly whatever new law he found would have to give results very close to Newton's law, since Newton's theory could predict the motions of the planets with almost perfect accuracy. This was the deep problem with which he struggled.

In 1907, Einstein had found one of the principles which was to guide him - the Principle of Equivalence of inertial and gravitational mass. After turning Newton's theory over and over in his mind, Einstein realized that Newton had used mass in two distinct ways: His laws of motion stated that the force acting on a body is equal to the mass of the body multiplied by its acceleration; but according to Newton, the gravitational force on a body is also proportional to its mass.

In Newton's theory, gravitational mass, by a coincidence, is equal to inertial mass; and this holds for all bodies. Einstein wondered - can the equality between the two kinds of mass be a coincidence? Why not make a theory in which they necessarily have to be the same?

He then imagined an experimenter inside a box, unable to see anything outside it. If the box is on the surface of the earth, the person inside it will feel the pull of the earth's gravitational field. If the experimenter drops an object, it will fall to the floor with an acceleration of 32 feet per second per second. Now suppose that the box is taken out into empty space, far away from strong gravitational fields, and accelerated by exactly 32 feet per second per second. Will the enclosed experimenter be able to tell the difference between these two situations? Certainly no difference can be detected by dropping an object, since in the accelerated box, the object will fall to the floor in exactly the same way as before.

With this "thought experiment" in mind, Einstein formulated a general Principle of Equivalence: He asserted that no experiment whatever can tell an observer enclosed in a small box whether the box is being accelerated, or whether it is in a gravitational field. According to this principle, gravitation and acceleration are locally equivalent, or, to say the same thing in different words, gravitational mass and inertial mass are equivalent.

Einstein soon realized that his Principle of Equivalence implied that a ray of light must be bent by a gravitational field. This conclusion followed because, to an observer in an accelerated frame, a light beam which would appear straight to a stationary observer, must necessarily appear very slightly curved. If the Principle of Equivalence held, then the same slight bending of the light ray would be observed by an experimenter in a stationary frame in a gravitational field.

Another consequence of the Principle of Equivalence was that a light wave propagating upwards in a gravitational field should be very slightly shifted to the red. This followed because in an accelerated frame, the wave crests would be slightly farther apart than they normally would be, and the same must then be true for a stationary frame in a gravitational field. It seemed to Einstein that it ought to be possible to test experimentally both the gravitational bending of a light ray and the gravitational red shift.

This seemed promising; but how was Einstein to proceed from the Principle of Equivalence to a Lorentz-invariant formulation of the law of gravitation? Perhaps the theory ought to be modeled after Maxwell's electromagnetic theory, which was a field theory, rather than an "action at a distance" theory. Part of the trouble with Newton's law of gravitation was that it allowed a signal to be propagated instantaneously, contrary to the Principle of Special Relativity. A field theory of gravitation might cure this defect, but how was Einstein to find such a theory? There seemed to be no way.

From these troubles Albert Einstein was rescued (a third time!) by his staunch friend Marcel Grossman. By this time, Grossman had become a professor of mathematics in Zürich, after having written a doctoral dissertation on tensor analysis and non-Euclidian geometry - the very things that Einstein needed. The year was 1912, and Einstein had just returned to Zürich as Professor of Physics at the Polytechnic Institute. For two years, Einstein and Grossman worked together; and by the time Einstein left for Berlin in 1914, the way was clear.

With Grossman's help, Einstein saw that the gravitational field could be expressed as a curvature of the 4-dimensional space-time continuum. The mathematical methods appropriate for describing the curvature of a many-

Fig. 14.3 *Albert Einstein and his first wife, his classmate Mileva Maric. Public domain, Wikimedia Commons*

dimensional space had already been developed in the early 19th century by Nickolai Ivanovich Lobachevski (1793-1856), Karl Friedrich Gauss (1777-1855) and Bernard Riemann (1826-1866).

As an example of a curved space, we might think of the 2-dimensional space formed by the surface of a sphere. The geometry of figures drawn on a sphere is non-Euclidian: Parallel lines meet, and the angles of a triangle add up to more than 180 degrees. Non-Euclidian spaces of higher dimension are hard to visualize, but they can be treated mathematically.

Gauss and Riemann had introduced a "metric tensor" which contained all the necessary information about the curvature of a non-Euclidian space; and Einstein saw that this metric tensor could be used to express the gravitational field. The orbits of the planets became "geodesics" in curved space. A geodesic is the shortest distance between two points, but in the curved space-time continuum of Einstein's theory, the geodesics were not straight lines.

By 1915, working by himself in Berlin, Einstein was able to show that the simplest theory of this form yielded Newton's law of gravitation as a first approximation, and in a higher approximation, it gave the correct movement of the perihelion of the orbit of Mercury. It had long been known that Mercury's point of closest approach to the sun (its perihelion) drifted slowly forward at the rate of between 40 and 50 seconds of arc per century. Einstein calculated that the change of Mercury's perihelion each century should be 43 seconds of arc. In January, 1916, he wrote to his friend Paul Ehrenfest:

"Imagine my joy at the feasibility of the general covariance, and at the result that the equations yield the correct perihelion of mercury. I was beside myself with ecstasy for days."

In 1919, a British expedition, headed by Sir Arthur Eddington, sailed to a small island off the coast of West Africa. Their purpose was to test Einstein's prediction of the bending of light in a gravitational field by observing stars close to the sun during a total eclipse. The observed bending agreed exactly with Einstein's predictions; and as a result he became world-famous.

The general public was fascinated by relativity, in spite of the abstruseness of the theory (or perhaps because of it). Einstein, the absent-minded professor, with long, uncombed hair, became a symbol of science. The world was tired of war, and wanted something else to think about.

Einstein met President Harding, Winston Churchill and Charlie Chaplin; and he was invited to lunch by the Archbishop of Canterbury. Although adulated elsewhere, he was soon attacked in Germany. Many Germans, looking for an excuse for the defeat of their nation, blamed it on the pacifists and Jews; and Einstein was both these things.

Albert Einstein's famous relativistic formula, relating energy to mass, soon yielded an understanding of the enormous amounts of energy released in radioactive decay. Marie and Pierre Curie had noticed that radium maintains itself at a temperature higher than its surroundings. Their measurements and calculations showed that a gram of radium produces roughly 100 gram-calories of heat per hour.

This did not seem like much energy until Rutherford found that radium has a half-life of about 1,000 years. In other words, after a thousand years, a gram of radium will still be producing heat, its radioactivity only reduced to one-half its original value. During a thousand years, a gram of radium produces about a million kilocalories - an enormous amount of energy in relation to the tiny size of its source! Where did this huge amount of energy come from? Conservation of energy was one of the most basic principles of physics. Would it have to be abandoned?

The mass defect

The source of the almost-unbelievable amounts of energy released in radioactive decay could be understood through Einstein's formula equating the energy of a system to its mass multiplied by the square of the velocity of light, and through accurate measurements of atomic weights. Einstein's formula asserted that mass and energy are equivalent. It was realized that in radioactive decay, neither mass nor energy is conserved, but only a quantity more general than both, of which mass and energy are particular forms.

The quantitative verification of the equivalence of mass and energy depended on very accurate measurements of atomic weights. Until 1912, the atomic weights of the elements were a puzzle. For some elements, the weights were very nearly integral multiples of the atomic weight of hydrogen, in units of which carbon was found to have an atomic weight almost exactly equal to 12, while nitrogen, oxygen and sodium were respectively 14, 16 and 23. This almost exact numerical correspondence made the English chemist, William Prout (1785-1850), propose that hydrogen might be the fundamental building-block of nature, and that atoms of all elements might be built up out of hydrogen.

Prout's hypothesis was destined to be killed several times, and revived several times. It was soon discovered that many elements have atomic weights which are not even nearly integral multiples of the weight of hydrogen. This discovery killed Prout's hypothesis for the first time. However, through their studies of radioactive decay, Rutherford and Soddy discovered isotopes; and isotopes revived Prout's hypothesis.

Rutherford and Soddy demonstrated that in the decay of uranium to its final product, lead, a whole chain of intermediates is involved, all of them radioactive, and each one changing spontaneously to the next. But what elements could these intermediate links of the decay chain be? After all, among the known elements, only uranium, polonium, radium, actinium and thorium were radioactive - and one could show that these elements could not represent all the intermediates of the Rutherford-Soddy decay chain.

In 1912, in Rutherford's Manchester laboratory, a young chemist named Georg von Hevesy was trying to separate by chemical means two radioactive decay products known to be different from each other because their half-lives were different. But no matter what he tried, von Hevesy could not separate them. All chemical methods failed.

Hevesy discussed his troubles with Niels Bohr, who suggested that the two decay products might be atoms with the same nuclear charges, but

Fig. 14.4 *Together with Rutherford, Frederick Soddy received the Nobel Prize in Chemistry in 1921 for the discovery of isotopes. Public domain, Wikimedia Commons*

different atomic weights. Since the number of electrons was determined by the nuclear charge, and since the chemical properties were determined by the number of electrons, it would be impossible to separate the two decay products by chemical means. They were, in fact, different varieties of the same element.

The same idea occurred simultaneously and independently to Frederick Soddy. In the autumn of 1912, he published a detailed paper explaining the concept, and introducing the word "isotope". Each chemical element, Soddy explained, is a mixture of isotopes. For those elements whose atomic weight is nearly an integral multiple of the atomic weight of hydrogen, a single isotope dominates the mixture. All the isotopes of a given element have the same nuclear charge (atomic number) and the same number of electrons; but two different isotopes of the same element have different atomic weights and different nuclear properties, some isotopes being radioactive, while others are stable.

When a nucleus emits a beta-particle (a high-speed electron carrying one unit of negative charge, but very little mass), the weight of the nucleus is almost unchanged, but its charge increases by one unit. Therefore beta-decay produces a product which is one place higher in the periodic table than its parent.

In alpha-decay, on the other hand, a helium ion, with two units of positive charge, and four units of mass, is thrown out of the decaying nucleus. Therefore, in alpha-decay, the product is two places lower in the periodic table, and four atomic mass units lighter than the parent atom.

The concept of isotopes allowed Frederick Soddy to identify clearly all the intermediate links in the decay chains which he and Rutherford had studied; and he later received the Nobel Prize in Chemistry for his work. Georg von Hevesy became the first scientist to use radioactive isotopes as tracers in biochemistry; and he also received the Nobel Prize in Chemistry.

Meanwhile, at the Cavendish Laboratory in Cambridge, J.J. Thomson and his student, Francis Aston (1877-1945), developed a "mass-spectrograph" - an instrument which could separate isotopes from one another by accelerating them with both electric and magnetic fields. In Aston's hands, the mass spectrograph became a precision instrument. Using it, he could not only separate isotopes from one another - he could also measure their masses very accurately. He found these masses to be almost exactly integral multiples of the mass of a hydrogen atom, but not quite! There was always a little mass missing!

The explanation for the missing mass - the mass defect - was found through Prout's hypothesis (newly revived) and Einstein's formula relating mass to energy. The nucleus of an atom was visualized as being composed of hydrogen nuclei (protons) and electrons bound tightly together. The mass defect, through Einstein's formula, was equivalent to the energy which would be needed to separate these elementary particles.

By observing the mass defects of isotopes, one could calculate their binding energies; and from these, the vast amounts of energy available for release through nuclear transmutation could also be calculated. For the first time, humans realized the enormous power which was potentially available in the atomic nucleus.

Suggestions for further reading

(1) Paul Arthur Schlipp (editor), *Albert Einstein: Philosopher-Scientist*, Open Court Publishing Co., Lasalle Illinois (1970).

(2) Banesh Hoffmann, *Albert Einstein, Creator and Rebel*, The Viking Press, New York (1972).

(3) Albert Einstein and Leopold Infeld, *The Evolution of Physics*, Cambridge University Press (1971).

Chapter 15

NUCLEAR FISSION

Artificial transmutations

During the First World War, Rutherford's young men had joined the army, and he had been forced to spend most of his own time working on submarine detection. In spite of this, he had found some spare time for his scientific passion - bombarding matter with alpha particles. Helped by his laboratory steward, Kay, Rutherford had studied the effects produced when alpha particles from a radium source struck various elements. In a letter to Niels Bohr, dated December 9, 1917, Rutherford wrote:

"I have got, I think, results that will ultimately have great importance. I wish that you were here to talk matters over with me. I am detecting and counting the lighter atoms set in motion by alpha particles, and the results, I think, throw a good deal of light on the character and distribution of forces near the nucleus... I am trying to break up the atom by this method. In one case, the results look promising, but a great deal of work will be required to make sure. Kay helps me, and is now an expert counter. Best wishes for a happy Christmas."

In July, 1919, Bohr was at last able to visit Manchester, and he heard the news directly from his old teacher: Rutherford had indeed produced artificial nuclear transmutations! In one of his experiments, an alpha-particle (i.e. a helium nucleus with nuclear charge 2) was absorbed by a nitrogen nucleus. Later, the compound nucleus threw out a proton with charge 1; and thus the bombarded nucleus gained one unit of charge. It moved up one place in the periodic table and became an isotope of oxygen.

Bohr later wrote: "I learned in detail about his great new discovery of controlled, or so-called artificial, nuclear transmutations, by which he gave birth to what he liked to call 'modern Alchemy', and which in the course of time, was to give rise to such tremendous consequences as regards man's mastery of the forces of nature."

Other scientists rushed to repeat and extend Rutherford's experiments. Particle accelerators were built by E.O. Lawrence (1901-1958) in California, by J.H. van de Graff (1901-1967) at the Massachusetts Institute of Technology and by John Cockcroft (1897-1967), working with Rutherford at the Cavendish Laboratory. These accelerators could hurl protons at energies of a million electron-volts. Thus, protons became another type of projectile which could be used to produce nuclear transmutations.

Neutrons

During the 1920s, nuclear transmutations could be achieved only with light elements. The charges on the nuclei of heavy elements were so large that, with the energies available, alpha particles and protons could not react with them. The positively charged projectiles were kept at a distance by the electrostatic repulsion of the heavy nuclei: They could not come close enough for the powerful but short-range nuclear attractive forces to become effective. However, in 1932, a new projectile was discovered - a projectile which was destined to unlock, with grave consequences, the colossal energies of the heavy nuclei. This new projectile was the neutron.

Rutherford and Bohr had for some time suspected that an electrically neutral particle with roughly the same mass as a proton might exist. The evidence for such a particle was as follows: Each isotope was characterized by a nuclear charge and by a nuclear weight. The nuclear charge was an integral multiple of the proton charge, while the nuclear weight was approximately an integral multiple of the proton weight. For example, the isotope carbon-12 had charge 6 and weight 12. This might be explained by supposing the carbon-12 nucleus to be composed of twelve protons and six electrons. However, there were theoretical objections to a model in which many electrons were concentrated within the tiny volume of a nucleus. Therefore, in 1920, Rutherford postulated the existence of neutrons - elementary particles with almost the same mass as protons, but no electrical charge. Then (for example) the carbon-12 nucleus could be thought of as being composed of six protons and six neutrons.

In 1930, the German physicist, Walter Bothe (1891-1957), discovered a strange, penetrating type of radiation coming from beryllium which had been bombarded with alpha particles. In 1931 and 1932, Bothe's experiments were repeated in Paris by Irène Joliot-Curie (1897-1956) and her husband Frédéric (1900-1958). The Joliot-Curies noticed that the mysterious rays emanating from the bombarded beryllium could easily penetrate lead. They also noticed that when the rays hit a piece of paraffin, hydrogen nuclei were knocked out.

The strange rays were, in fact, neutrons, as the Joliot-Curies would have realized immediately if they had been familiar with Rutherford's prediction of the neutron's existence. The Joliot-Curies might have made the correct identification of the rays given time; but Rutherford's assistant, James Chadwick (1891-1974), was faster. On February 17, 1932, he published a paper in *Nature* reporting a series of experiments.

Chadwick had studied not only the velocities of the hydrogen nuclei knocked out of paraffin by Bothe's rays but also the velocities of nuclei knocked out of many other materials. In every case, he found that the velocities were consistent with the identification of the rays as neutrons. Chadwick completed his proof by showing that the rays moved with one-tenth the velocity of light, so that they had to be material particles rather than radiation; and he showed that the rays could not be deflected by a magnet. Therefore they carried no charge.

Fig. 15.1 *Irène Joliot-Curie as a child, with her parents, Pierre and Marie Curie. Although she and her husband Frèdèric narrowly missed discovering the neutron, they soon made another discovery of major importance - artificial radioactivity. Public domain, Wikimedia Commons*

Fermi

Although Irène and Frédéric Joliot-Curie narrowly missed discovering the neutron, they soon made another discovery of major importance - artificial radioactivity. The Joliot-Curies had been bombarding an aluminum target with alpha-particles and studying the resulting radiation. One day in 1934, they noticed to their astonishment that the aluminum target continued to radiate even after they had stopped the alpha-particle bombardment. They discovered that some of the aluminum atoms in the target had been converted to a radioactive isotope of phosphorus!

In 1934, news of the startling discoveries of Bothe, Chadwick and the Joliot-Curies reached a brilliant young professor of theoretical physics in Rome. Although he was only 33 years old, Enrico Fermi (1901-1954) already had a worldwide reputation for his work in quantum theory. He also had attracted a school of extremely talented young students, the first physicists in Italy to enter the new fields of quantum mechanics and relativity: Persico,

Amaldi, Rasetti, Segrè, Pontecorvo, Majorana, Racah and Wick. It was a happy, informal group of young men.

Because of his reputation for scientific infallibility, Enrico Fermi was nicknamed "the Pope", while Franco Rasetti was "the Cardinal" and Emilio Segrè was "the Basilisk". A medical colleague, Professor Trabacci, who generously supplied the group with equipment and chemicals, was known as "the Divine Providence".

In 1934, Fermi was feeling somewhat discouraged with theoretical work, and in the mood to try something new. His paper on the theory of beta-decay (later regarded as one of his major achievements) had just been rejected by *Nature*. At that moment, he heard of Chadwick's neutrons and the Joliot-Curie's artificial radioactivity. Putting the two things together, Fermi decided to try to produce artificial radioactivity by bombarding elements with neutrons.

There were good theoretical reasons why Fermi's plan should work, as well as practical reasons why it should fail. The argument in favor of neutrons was that they had no charge. Therefore they should be able to approach the nuclei of even heavy elements without being repelled by the electrostatic potential. The practical argument against neutrons was that it was difficult to produce them in worthwhile numbers. The yield of neutrons was only one for every hundred thousand alpha-particles.

Although he had no experience in working with radioactivity, Fermi managed to make his own Geiger counter. He also made a neutron source for himself by condensing radon gas (donated by "the Divine Providence") into a small glass tube of powdered beryllium held at liquid air temperature.

Being a methodical person, Fermi began at the bottom of the periodic table and worked systematically upwards. The first eight elements which he bombarded with neutrons showed no artificial radioactivity, and Fermi almost became discouraged. Finally, he came to fluorine, and to his delight, he succeeded in making it strongly radioactive by neutron bombardment. He succeeded also with several other elements beyond fluorine; and realizing that the line of research was going to be very fruitful, he enlisted help from Segrè, Amaldi, and the chemist, d'Agostino. Fermi also sent a cable to Franco Rasetti, who was on vacation in Marocco.

In order that the source should not disturb the measurements, the room where the elements were irradiated was far from the room where their radioactivity was measured - at the other end of a long corridor. The half-life of the induced radioactivity was very short in some elements, which meant that Fermi and Amaldi had to run full tilt with their samples, from one end of the hallway to the other.

One day a visitor arrived from Spain and asked to see *"Sua Eccellenza Fermi"*. (Fermi was a member of the Royal Academy of Italy, and therefore had the title "Excellency", which much embarrassed him). "The Pope is upstairs", said Segrè, and then, realizing that the visitor did not know this nickname, he added: "I mean Fermi, of course." The Spanish visitor arrived on the second floor of the institute just in time to see *"Sua Eccellenza Fermi"* dash wildly down the length of the corridor.

After this fashion, Fermi and his group finally reached the top of the periodic table. They carefully purified uranium from its disintegration products and bombarded it with neutrons. A new radioactivity was induced, quite different from the ordinary activity of uranium. The question was: to what element or elements had the uranium been converted?

With the help of the chemist, d'Agostino, they analysed the uranium target, and proved definitely that neutron bombardment had not converted uranium to any of the nearby heavy elements at the top of the periodic table. It seemed most likely that what they had produced by bombarding uranium was a new, unstable element, which had never before existed - element number 93! However, they lacked definite proof; and Fermi, always cautious, refused to jump to such a sensational conclusion.

By this time, the summer of 1934 had begun. The university year ended, as was traditional, with a meeting of the Accademia dei Lincei, attended by the King of Italy. In 1934, the speaker at this meeting was Senator Corbino, who had been a talented physicist before he became a politician. Corbino had been responsible for raising money to support Fermi's group of young physicists; and he was justly proud of what they had achieved. In his 1934 speech before the king, Senator Corbino glowingly described their production of neutron-induced radioactivity; and he ended the speech with the words:

"The case of uranium, atomic number 92, is particularly interesting. It seems that, having absorbed the neutron, it converts rapidly by emission of an electron, into the element one place higher in the periodic system, that is, into a new element having atomic number 93... However, the investigation is so delicate that it justifies Fermi's prudent reserve and a continuation of the experiments before an announcement of the discovery. For what my own opinion on this matter is worth, and I have followed the investigations daily, I believe that production of this new element is certain."

Corbino had not cleared this announcement with Fermi. It was immediately picked up by both the Italian and international press and given great publicity. A new element had been made by man! The official newspapers

Fig. 15.2 *Enrico Fermi at the blackboard. Public domain, Wikimedia Commons*

of fascist Italy, in particular, made much of this "great discovery" which, they claimed, showed that Italy was regaining the glorious position which it had held in the days of the Roman Empire.

Fermi was thrown into a mood of deep despair by this premature publicity. He could not sleep, and woke his wife in the middle of the night to tell her that his reputation as a scientist was in jeopardy. Next morning, Fermi and Corbino prepared a statement attempting to halt the publicity: "The public is giving an incorrect interpretation to Senator Corbino's speech... Numerous and delicate tests must still be performed before the production of element 93 is actually proved."

Before the question of element 93 could be cleared up, the attention of Fermi's group was distracted by an accidental discovery of extreme importance. They had been obtaining inconsistent and inexplicable results. The radioactivity induced in a sample depended in what seemed to be a completely illogical way on the conditions under which the experiment was performed. For example, if the target was bombarded with neutrons while standing on a wooden table, the induced activity was much stronger than when the target was on a marble table.

Fermi suspected that these strange results were due to scattering of neutrons by surrounding objects. He prepared a lead wedge to insert between neutron source and the counter to measure the scattering. However, he did not use the lead wedge which he had so carefully prepared.

"I was clearly dissatisfied with something", Fermi remembered later, "I tried every excuse to postpone putting the piece of lead in its place. I said to myself, 'No, I do not want this piece of lead here; what I want is a piece of paraffin.' It was just like that, with no advance warning, no prior reasoning. I immediately took some odd piece of paraffin and placed it where the piece of lead was to have been."

The effect of the paraffin was amazing. The radioactivity increased a hundredfold! Puzzled, the group adjourned for lunch and siesta. When they reassembled a few hours later, Fermi had developed a theory to explain what was happening: The neutrons had almost the same mass as the hydrogen atoms in the paraffin. When they collided with the hydrogen atoms, the neutrons lost almost all their energy of motion, just as a billiard ball loses almost all its speed when it collides with another ball of equal mass. What Fermi and the others had discovered by accident was that slow neutrons are much more effective than fast ones in producing nuclear reactions.

"What we need", said Fermi, "is a large amount of water." The group excitedly took the neutron source and targets to Senator Corbino's nearby garden, where there was a goldfish pond. The hydrogen-containing water of the pond produced the same result: It slowed the neutrons, and greatly enhanced their effect.

That evening, at Edouardo Amaldi's house, they prepared a paper reporting their discovery. Fermi dictated, while Segrè wrote. Meanwhile, Rasetti, Amaldi and Pontecorvo walked up and down, all offering suggestions simultaneously. They made so much noise that when they left, the maid asked Mrs. Amaldi whether her guests had been drunk.

The happy and carefree days of the little group of physicists in Rome were coming to an end. They had thought that they could isolate themselves from politics; but in 1935, it became clear that this was impossible.

One day, in 1935, Segrè said to Fermi: "You are the Pope, and full of wisdom. Can you tell me why we are now accomplishing less than a year ago?"

Fermi answered without hesitation: "Go to the physics library. Pull out the big atlas that is there. Open it. You shall find your explanation." When Segrè did this, the atlas opened automatically to a much-thumbed map of Ethiopia.

In 1935, Mussolini's government had attacked Ethiopia, and Italy had been condemned by the League of Nations. For thinking Italians, this shock revealed the true nature of Mussolini's government. They could no longer ignore politics. Within a few years, Enrico Fermi and most of his group had decided that they could no longer live under the fascist government of Italy. By 1939, most of them were refugees in the United States.

Hahn, Meitner and Frisch

Without knowing it, Enrico Fermi and his group had split the uranium atom; but four years were to pass before this became apparent. All the experts agreed that Fermi's group had undoubtedly produced transuranic elements. There was only one dissenting voice - that of the German chemist, Ida Noddack, who was an expert in the chemistry of rare elements. Knowing no nuclear physics, but a great deal of chemistry, Ida Noddack saw the problem from a different angle; and in 1934 she wrote:

"It would be possible to assume that when a nucleus is demolished in this novel way by neutrons, nuclear reactions occur which may differ considerably from those hitherto observed in the effects produced on atomic nuclei by protons and alpha rays. It would be conceivable that when heavy nuclei are bombarded with neutrons, the nuclei in question might break into a number of larger pieces, which would, no doubt, be isotopes of known elements, but not neighbors of the elements subjected to radiation."

No one took Ida Noddack's suggestion seriously. The energy required to smash a heavy nucleus into fragments was believed to be so enormous that it seemed ridiculous to suggest that this could be accomplished by a slow neutron.

Many other laboratories began to bombard uranium and thorium with slow neutrons to produce "transuranic elements". In Paris, Irène Joliot-Curie and Paul Savitch worked on this problem, while at the Kaiser Wilhelm Institute in Berlin, Otto Hahn (1879-1968), Lise Meitner (1878-1968) and Fritz Strassmann (1902-) did the same.

Meanwhile, night was falling on Europe. In 1929, an economic depression, caused in part by the shocks of the First World War, began in the United States; and it soon spread to Europe. Without the influx of American capital, the postwar reconstruction of the German economy collapsed. The German middle class, which had been dealt a severe blow by the great inflation of 1923, now received a second heavy blow. The desperation produced by economic chaos drove the German voters into the hands of political extremists.

On January 30, 1933, Adolf Hitler was appointed Chancellor and leader of a coalition cabinet by President Hindenberg. Although Hitler was appointed legally to this post, he quickly consolidated his power by unconstitutional means: On May 2, Hitler's police seized the headquarters of all trade unions, and arrested labor leaders. The Communist and Socialist parties were also banned, their assets seized and their leaders arrested. Other political parties were also smashed. Acts were passed eliminating Jews from public service; and innocent Jewish citizens were boycotted, beaten and arrested.

On March 11, 1938, Nazi troops entered Austria. Lise Meitner, who was working with Otto Hahn in Berlin, was a Jew, but until Hitler's invasion of Austria, she had been protected by her Austrian citizenship. Now, she was forced to escape from Germany. Saying goodbye only to Otto Hahn and to a few other close friends, she went to Holland for a vacation, from which she did not plan to return. From there, she went to Stockholm, where she had been offered a post by the Nobel Institute.

Meanwhile, Hahn and Strassmann continued to work on what they believed to be production of transuranic elements. They had been getting results which differed from those of the Paris group, but they believed that Irène Joliot-Curie must be mistaken. When Strassmann tried to show Hahn one of the new papers from Paris, he continued to puff calmly on his cigar and replied: "I am not interested in our lady-friend's latest writings". However, Strassmann would not be deterred, and he quickly summarized the most recent result from Paris.

"It struck Hahn like a thunderbolt", Strassmann said later, "He never finished that cigar. He laid it down, still glowing, on his desk, and ran downstairs with me to the laboratory."

Hahn and Strassmann quickly repeated the experiments which Irène Joliot-Curie had reported. They now suspected that one of the products which she had produced was actually an isotope of radium. Since radium has almost the same chemical properties as barium, they tried percipitating it together with a barium carrier. This procedure worked: The new substance came down with the barium.

Otto Hahn was the most experienced radiochemist in the world, and many years previously he had developed a method for separating radium from barium. He and Strassmann now tried to apply this method. It did not work. No matter how they tried, they could not separate the active substance from barium.

Could it be that an isotope of barium had been produced by bombarding uranium with neutrons? Impossible! It would mean that the uranium nucleus had split roughly in half, against all the well-established rules of nuclear physics. It could not happen - and yet their chemical tests told them again and again that the product really was barium. Finally, they sat down and wrote a paper:

"We come to this conclusion", Hahn and Strassmann wrote, "Our 'radium' isotopes have the properties of barium. As chemists, we are in fact bound to affirm that the new bodies are not radium but barium; for there is no question of elements other than radium and barium being present... As nuclear chemists, we cannot decide to take this step, in contradiction to all previous experience in nuclear physics."

On December 22, 1938, Otto Hahn mailed the this paper to the journal, *Naturwissenschaften*. "After the manuscript was mailed", he said later, "the whole thing seemed so improbable to me that I wished I could get the document back out of the mail box."

After making this strange discovery, Otto Hahn's first act had been to write to Lise Meitner, who had worked by his side for so many years. She received his letter just as she was starting for her Christmas vacation, which was to be spent at the small Swedish town of Kungälv, near Göteborg.

It was even more clear to Lise Meitner than it had been to Hahn that something of tremendous importance had unexpectedly come to light. As it happened, Lise Meitner's nephew, O.R. Frisch, had come to Kungälv to spend Christmas with his aunt, hoping to keep her from being lonely during her first Christmas as a refugee. Frisch was a physicist, working at Niels

Fig. 15.3 *Otto Hahn and Lise Meitner. Public domain, Wikimedia Commons*

Bohr's institute in Copenhagen. He was one of the many scientists whom Bohr saved from the terror and persecution of Hitler's Germany by offering them refuge in Copenhagen.

When Frisch arrived, Lise Meitner immediately showed him Otto Hahn's letter. "I wanted to discuss with her a new experiment I was planning", Frisch said later, "but she wouldn't listen. I had to read the letter. Its content was indeed so startling that I was at first inclined to be sceptical."

Frisch put on his skis, and went out to get some air; but his aunt followed him over the snow, insisting that he think about the problem of uranium and barium. Lise Meitner knew the precision and thoroughness of Otto Hahn's methods so well that she could not imagine him making a mistake of that kind. If Hahn said that bombarding uranium with neutrons produced barium, then it *did* produce barium. She insisted that her nephew should try to explain this impossible result, rather than shrugging it off as an error.

Finally, aunt and nephew sat down on a log in the middle of the snow-filled Swedish forest and tried to make some calculations on the back of an envelope. They continued their calculations back at their hotel, consulting some tables of isotopic masses which Frisch had brought with him. Gradually, they formed a picture of what had happened:

The uranium nucleus was like a liquid drop. Although the powerfully attractive short-range nuclear forces produced a surface tension which tended to keep the drop together, there were also powerful electrostatic repulsive forces which tended to make it divide. Under certain conditions, the nucleus could become non-spherical in shape, with a narrow waist. If this happened, the electrostatic repulsion would split the nucleus into two fragments, and would drive the fragments apart with tremendous energy of motion.

Frisch and Meitner calculated that for a single uranium nucleus, the energy of motion would be roughly two hundred million electron volts. What was the source of this gigantic energy? By consulting tables of isotopic masses, the two scientists were able to show that in the splitting of uranium, a large amount of the mass is converted to energy. If one of the fragments was an isotope of barium, the other had to be an isotope of krypton. Using Einstein's formula relating energy to mass, they found that the lost mass was exactly equivalent to two hundred million electron volts. Everything checked. This had to be the explanation.

Meitner and Frisch were struck by the colossal size of the energy released in the fission of uranium. Ordinary combustion releases one or two electron volts per atom. They realized with awe that in the fission of uranium, a hundred million times as much energy is released!

When O.R. Frisch returned to Copenhagen, Niels Bohr was preparing to leave for a lecture tour in America. Frisch had only a few minutes to tell him what had happened, but Bohr was quick to understand. "I had hardly begun to tell him", Frisch said later, "when he struck his forehead and exclaimed, 'Oh what idiots we all have been! But this is wonderful! This is just as it must be!'"

There was no time to talk, but as Niels Bohr entered the taxi which would take him to the liner, *Drottningholm*, he asked Frisch whether he had written a paper. Frisch handed some rough notes to Bohr, and said that he would write a paper immediately. Bohr promised that he would not talk about the new discovery until the paper was ready.

Bohr's assistant, Rosenfeld, had accompanied him on the trip, and the long sea voyage to New York gave the two physicists a good opportunity to think about the revolutionary new discovery of nuclear fission. A blackboard was installed in Bohr's stateroom on the *Drottningholm*. Bohr and Rosenfeld covered this blackboard with calculations, and by the end of the voyage, they were convinced that Otto Frisch and Lise Meitner had correctly analysed the problem of nuclear fission.

At the harbor in New York, they were met by Professor John Wheeler of Princeton, together with Enrico Fermi and his wife, Laura, who had become refugees in America. Laura Fermi remembered later the tense and worried expression with which Bohr described the rapidly-deteriorating political situation in Europe. With her imperfect knowledge of English and the noise of the pier, she could only make out a few of the words - "Europe - war - Hitler - danger".

Rosenfeld accompanied Wheeler to Princeton, while Bohr and his 19 year old son, Erik, remained a few days in New York. At Princeton, Rosenfeld was invited to address the "Journal Club", a small, informal group of physicists. Bohr had neglected to tell Rosenfeld that he had promised not to talk about nuclear fission until the Hahn-Strassmann and Meitner-Frisch papers were out; and Rosenfeld spoke about the revolutionary new discovery to the physicists at Princeton.

The news spread with explosive speed. Telephone calls and letters went out to other parts of America. The physicist, I.I. Rabi, who happened to be at Princeton, returned to Colombia University, where Fermi was working, and told him the news. Fermi acted with characteristic speed and decisiveness. He devised an experiment to detect the high-energy fragments produced by uranium fission; and he suggested to his co-worker, Dunning, that the experiment should be performed as fast as possible. Fermi himself had to leave for a theoretical physics meeting in Washington, where Bohr would be present.

When Bohr heard that Rosenfeld had talked about fission, he was very upset, because he had promised Frisch to remain silent until the papers were out. He sent a telegram to Copenhagen urging Frisch to hurry with his manuscript, and urging him to perform an experiment to detect the fission fragments.

In fact, Otto Frisch had already performed this experiment, using a radium-lined ionization chamber containing a radium-beryllium neutron source. An amplifier connected with the chamber had shown giant bursts of ionization, which could only be due to the immensely energetic fission fragments.

On January 16, 1939, the same day that Rosenfeld had revealed the news about fission to the physicists at Princeton, Otto Frisch had mailed two papers to *Nature*. The first of these papers presented the theory of nuclear fission which he and Lise Meitner had developed, while the second described his experimental detection of the high-energy fragments.

Fig. 15.4 *Left to right: William Penney, Otto Frisch, Rudolf Peierls and John Douglas Cockcroft. Public domain, Wikimedia Commons*

On January 26, Bohr and Fermi arrived at the American capital to attend the Fifth Washington Conference on Theoretical Physics. The same day, Erik Bohr received a letter from his brother, Hans. The letter contained the news that Frisch had completed his experiment and had sent the paper to London. Simultaneously, Bohr learned from a reporter who was covering the conference that the Hahn-Strassmann paper had just been published in

Naturwissenschaften. At last, Bohr felt free to speak. He asked the chairman whether he might make an announcement of the utmost importance; and he told the astonished physicists the whole story.

While Bohr was speaking, Dr. Tuve of the Carnegie Institution whispered to his colleague, Halfstead, that he should quickly put a new filament in the Carnegie accelerator. Several physicists rushed for the door to make long-distance telephone calls. Fermi decided to leave the conference immediately, and to return to New York. On the way out, Fermi met Robert B. Potter, a reporter from *Science Service*, who asked: "What does it all mean?" Fermi explained as well as he could, and Potter wrote the following story, which was released to newspapers and magazines:

"New hope for releasing the enormous energy within the atom has arisen from German experiments that are now creating a sensation among eminent physicists gathered here for the Conference on Theoretical Physics. It is calculated that only five million electron volts of energy can release two hundred million electron volts of energy, forty times the amount shot into it by a neutron (neutral atomic particle). World famous Niels Bohr of Copenhagen and Enrico Fermi of Rome, both Nobel Prize winners, are among those who acclaim this experiment as one of the most important in recent years. American scientists join them in this acclaim."

Chapter 16

HIROSHIMA AND NAGASAKI

Chain reactions

Within hours of Bohr's announcement, scientists in various parts of America had begun to set up experiments to look for high-energy fission fragments. On the evening of January 26, Bohr watched, while giant pulses of ionization produced by the fission fragments were recorded on an oscilloscope at the Carnegie Institution's accelerator in Washington. Similar experiments were simultaneously being performed in New York and California.

At Columbia University, following Fermi's suggestion, Dunning had performed the experiment a day earlier, on January 25. The news spread rapidly. On the 9th of February, the Austrian physicists, Jentschke and Prankl, reported to the Vienna Academy that they too had observed fission fragments. By March 8, which was Otto Hahn's 60th birthday, an avalanche of papers on uranium fission had developed in the international scientific literature.

In the spring of 1939, Bohr and Wheeler published an important theoretical paper in which they showed that in nuclei with an even atomic mass numbers, the ground state energy is especially low because of pairing of the nuclear particles. For this reason, Bohr and Wheeler believed that it is the rare isotope, urnaium-235, which undergoes fission. They reasoned that when a slow neutron is absorbed by uranium-235, it becomes a highly-excited state of uranium-236. The extra energy of this excited state can deform the nucleus into a non-spherical shape, and the powerful electrostatic repulsive forces between the protons can then cause the nucleus to split.

During the early spring of 1939, a number of scientists, including Fermi,

Szilard and the Joliot-Curies, were becoming acutely aware of another question: Are neutrons produced in uranium fission? This was a question of critical importance, because if more than one neutron was produced, a chain reaction might be possible.

At Columbia University, Enrico Fermi and Leo Szilard began experiments to determine whether neutrons are produced; and similar experiments were performed by the Joliot-Curies in Paris. Both groups found that roughly two neutrons are released. This meant that a nuclear chain reaction might indeed be possible: It might be possible to arrange the uranium in such a way that each neutron released by the fission of a nucleus would have a good chance of causing a new fission.

The possibility of nuclear power became clear to the physicists, as well as the possibility of a nuclear bomb many millions of times more powerful than any ordinary bomb. Leo Szilard (who had seen the atrocities of Hitler's Germany at close range) became intensely worried that the Nazis would develop nuclear weapons. Therefore he proposed that the international community of physicists should begin a self-imposed silence concerning uranium fission, and especially concerning the neutrons produced in fission.

In Fermi's words, Szilard "...proceeded to startle physicists by proposing to them that, given the circumstances of the period - you see it was early 1939, and war was very much in the air - given the circumstances of the period, given the danger that atomic energy, and possibly atomic weapons, could become the chief tool of the Nazis to enslave the world, it was the duty of the physicists to depart from what had been the tradition of publishing significant results as soon as the *Physical Review* or other scientific journals might turn them out, and that instead one had to go easy, keep back some of the results until it was clear whether these results were potentially dangerous..."

"He sent in this vein a number of cables to Joliot in France, but he did not get a favorable response from him; and Joliot published his results more or less like results in physics had been published until that day. So the fact that neutrons are emitted in fission in some abundance - the order of magnitude one or two or three - became a matter of general knowledge; and of course that made the possibility of a chain reaction appear to most physicists as a vastly more real possibility than it had until that time."

On March 16, 1939, exactly two months after Bohr had arrived in America, he and Wheeler mailed their paper on uranium fission to a journal. On the same day, Enrico Fermi went to Washington to inform the Office of

Naval Operations that it might be possible to construct an atomic bomb; and on the same day, German troops poured into Czechoslovakia.

A few days later, a meeting of six German atomic physicists was held in Berlin to discuss the applications of uranium fission. Otto Hahn, the discoverer of fission, was not present, since it was known that he was opposed to the Nazi regime. He was even said to have exclaimed: "I only hope that you physicists will never construct a uranium bomb! If Hitler ever gets a weapon like that, I'll commit suicide."

The meeting of German atomic physicists was supposed to be secret; but one of the participants reported what had been said to Dr. S. Flügge, who wrote an article about uranium fission and about the possibility of a chain reaction. Flügge's article appeared in the July issue of *Naturwissenschaften*, and a popular version of it was printed in the *Deutsche Allgemeine Zeitung*. These articles greatly increased the alarm of American atomic scientists, who reasoned that if the Nazis permitted so much to be printed, they must be far advanced on the road to building an atomic bomb.

Einstein writes to Roosevelt

In the summer of 1939, while Hitler was preparing to invade Poland, alarming news reached the physicists in the United States: A second meeting of German atomic scientists had been held in Berlin, this time under the auspices of the Research Division of the German Army Weapons Department. Furthermore, Germany had stopped the sale of uranium from mines in Czechoslovakia.

The world's most abundant supply of uranium, however, was not in Czechoslovakia, but in Belgian Congo. Leo Szilard was deeply worried that the Nazis were about to construct atomic bombs; and it occurred to him that uranium from Belgian Congo should not be allowed to fall into their hands.

Szilard knew that his former teacher, Albert Einstein, was a personal friend of Elizabeth, the Belgian Queen Mother. Einstein had met Queen Elizabeth and King Albert of Belgium at the Solvay Conferences, and mutual love of music had cemented a friendship between them. When Hitler came to power in 1933, Einstein had moved to the Institute of Advanced Studies at Princeton; and Szilard decided to visit him there. Szilard reasoned that because of Einstein's great prestige, and because of his longstanding friendship with the Belgian Royal Family, he would be the proper person to warn the Belgians not to let their uranium fall into the hands of the Nazis.

It turned out that Einstein was vacationing at Peconic, Long Island, where he had rented a small house from a friend named Dr. Moore. Leo Szilard set out for Peconic, accompanied by the theoretical physicist, Eugene Wigner, who, like Szilard, was a Hungarian and a refugee from Hitler's Europe.

For some time, the men drove around Peconic, unable to find Dr. Moore's house. Finally Szilard, with his gift for foreseeing the future, exclaimed: "Let's give it up and go home. Perhaps fate never intended it. We should probably be making a frightful mistake in applying to any public authorities in a matter like this. Once a government gets hold of something, it never lets go." However, Wigner insisted that it was their duty to contact Einstein and to warn the Belgians, since they might thus prevent a world catastrophe. Finally they found the house by asking a small boy in the street if he knew where Einstein lived.

Einstein agreed to write a letter to the Belgians warning them not to let uranium from the Congo fall into the hands of the Nazis. Wigner suggested that the American State Department ought to be notified that such a letter was being written.

On August 2, 1939, Szilard again visited Einstein, this time accompanied by Edward Teller, who (like Szilard and Wigner) was a refugee Hungarian physicist. By this time, Szilard's plans had grown more ambitious; and he carried with him the draft of a letter to the American President, Franklin D. Roosevelt. Einstein made a few corrections, and then signed the fateful letter, which reads (in part) as follows:

"Some recent work of E. Fermi and L. Szilard, which has been communicated to me in manuscript, leads me to expect that the element uranium may be turned into an important source of energy in the immediate future. Certain aspects of the situation seem to call for watchfulness and, if necessary, quick action on the part of the Administration. I believe, therefore, that it is my duty to bring to your attention the following.."

"It is conceivable that extremely powerful bombs of a new type may be constructed. A single bomb of this type, carried by boat and exploded in a port, might very well destroy the whole port, together with some of the surrounding territory.."

"I understand that Germany has actually stopped the sale of uranium from Czechoslovakian mines which she has taken over. That she should have taken such an early action might perhaps be understood on the ground that

Fig. 16.1 *Albert Einstein and Leo Szilard with the fateful letter to Roosevelt. Einstein later bitterly regretted signing the letter. Source: priceeconomics.com*

the son of the German Under-Secretary of State, von Weizäcker, is attached to the Kaiser Wilhelm Institute in Berlin, where some of the American work is being repeated."

On October 11, 1939, three weeks after the defeat of Poland, Roosevelt's economic advisor, Alexander Sachs, personally delivered the letter to the President. After discussing it with Sachs, the President commented,"This calls for action." Later, when atomic bombs were dropped on civilian populations in an already virtually-defeated Japan, Einstein bitterly regretted having signed the letter to Roosevelt.

The first nuclear reactor

As a result of Einstein's letter, President Roosevelt set up an Advisory Committee on Uranium. On December 6, 1941, the day before the Japanese attack on Pearl Harbor, the Committee decided to make an all-out effort to develop atomic energy and atomic bombs. This decision was based in part on intelligence reports indicating that the Germans had set aside a large section of the Kaiser Wilhelm Institute for Research on uranium; and it was based in part on promising results obtained by Enrico Fermi's group at Columbia University.

Enrico Fermi and his group at Columbia University had been exploring the possibility of building a chain-reacting pile using natural uranium, together with a moderator to slow the neutrons. Fermi's own description of the research is as follows:

"...We soon reached the conclusion that in order to have any chance of success with natural uranium, we had to use slow neutrons. So there had to be a moderator. And this moderator could be water, or other substances. Water was soon discarded. It is very effective in slowing down the neutrons, but it absorbs a little bit too many of them, and we couldn't afford that. Then it was thought that graphite might be a better bet..."

"This brings us to the fall of 1939, when Einstein wrote his now famous letter to Roosevelt, advising him of what was the situation in physics - what was brewing, and that he thought that the government had the duty to take an interest and to help along the development. And in fact, help came along to the tune of six thousand dollars a few months later; and the six thousand dollars were used to buy huge amounts - or what seemed at the time, when the eyes of physicists had not yet been distorted - what seemed at the time a huge amount of graphite."

"So the physicists on the seventh floor of Pupin Laboratories started looking like coal miners, and the wives to whom these physicists came home tired at night were wondering what was happening. We know that there is smoke in the air, but after all..."

"We started to construct this structure that at that time looked again an order of magnitude larger than anything we had seen before. Actually, if anybody would look at this structure now, he would probably extract his magnifying glass and go close to see it. But for the ideas of the time, it looked really big. It was a structure of graphite bricks, and spread through these graphite bricks in some sort of pattern, were big cans, cubic cans, containing uranium oxide."

Fermi's results indicated that it would be possible to make a chain-reacting pile using graphite as a moderator, provided that enough very pure graphite and very pure uranium oxide could be obtained. Leo Szilard undertook the task of procuring the many tons of these substances which would be required.

Work on the pile was moved to the University of Chicago, and the number of physicists employed on the project was greatly enlarged. Work preceeded with feverish speed, because it was feared that the Nazis would win the race. Leona Woods, one of the few women employed on the project, recalled later: "We were told, day and night, that it was our duty to catch up with the Germans."

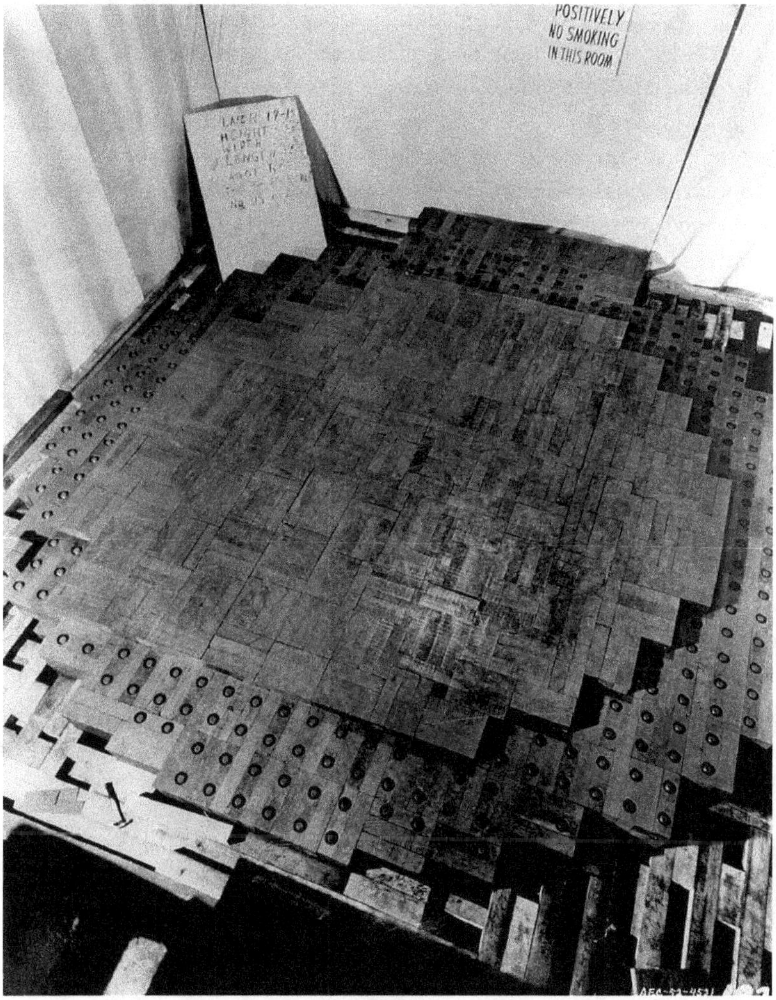

Fig. 16.2 *This is the only photograph made during the construction of the first nuclear reactor. Public domain, Wikimedia Commons*

During the summer of 1942, Fermi succeeded in constructing a uranium-graphite lattice with a neutron reproduction factor greater than unity. In other words, when he put a radium-beryllium neutron source into the lattice, more neutrons came out than were produced by the source. This meant that a chain-reacting pile could definitely be built. It was only a matter of obtaining sufficient amounts of very pure graphite and uranium.

Fermi calculated that a spherical pile, 26 feet in diameter, would be sufficiently large to produce a self-sustained chain reaction. At first, it was planned that the pile should be built at Argonne Laboratory, just outside Chicago. However, the buildings were not yet ready, and therefore Fermi suggested that the pile should instead be built in a squash court under the abandoned football stadium at the University of Chicago. (Football had been banned by the university's president, Robert Hutchens, who felt that it distracted students from their academic work.)

The squash court was not quite as high as Fermi would have liked it to be, and in case of a miscalculation of the critical size of the pile, it would be impossible to add extra layers. Therefore, Fermi and his young co-worker, Herbert Anderson, ordered an enormous cubical rubber balloon from the Goodyear Tyre Company, and the pile was built inside the balloon. The idea was that, if necessary, the air inside the pile could be pumped out to reduce the absorption of neutrons by nitrogen. This turned out not to be necessary; and the door of the balloon was never sealed.

The graphite-uranium lattice was spherical in shape, and it rested on blocks of wood. The physicists labored furiously, putting the tons of uranium and graphite into place, measuring and cutting the blocks of wood needed to support the pile, and swearing to ease the tension. Leona Woods, wearing goggles and overalls, was indistinguishable from the men as she worked on the pile. Everyone was covered from head to foot with black graphite dust, and graphite also made the floor treacherously slippery.

On December 1, 1942, Herbert Anderson stayed up all night putting the finishing touches on the pile. If he had pulled out the neutron absorbing cadmium control rods, Anderson would have been the first man in history to achieve a self-sustaining nuclear chain reaction. However, he had promised Fermi not to do so.

Enrico Fermi got a good night's sleep; and on the next morning, December 2, he was ready to conduct the historic experiment. About forty people were present. Most of them were scientists who had worked on the pile; but there were a few visitors, including a representative of the giant DuPont chemical company, which was undertaking a contract to build more chain-reacting piles.

Fermi, and all the spectators, stood on the balcony of the squash court. On the floor of the court stood a single physicist, George Weil, who was ready to pull out the final control rod. On the top of the pile, crouched in the cramped space under the top of the balloon, was a "suicide squad" - three young physicists who had volunteered to sit there during the experiment with containers of cadmium salt solution, which they would pour into the pile if anything went wrong.

Fermi was confident that nothing would go wrong. He had calculated that even if the last control rod were removed completely, the neutron flux within the pile would not jump rapidly to a high level. Instead, it would begin to increase slowly and steadily. The slow response of the pile was due to the fact that much time was required for the fast neutrons released by fission to be slowed by collisions with carbon atoms in the graphite moderator.

Although, according to theory, there was no danger, Fermi approached the chain reaction with great caution. He explained to the spectators that George Weil would pull out the final control rod by very slow stages; and at each stage, measurements would be made to make sure that the behavior of the pile checked with calculations. The neutron flux was measured by Geiger counters, and recorded by a pen on a roll of paper.

"Pull it out a foot, George", Fermi said; and he explained to the spectators: "Now the pen will move up to this point and then level off." The response was exactly as predicted.

Throughout the morning, this procedure was repeated. However, by lunchtime, much of the control rod still remained within the pile. Fermi was a man of fixed habits, and although no one else showed any signs of being hungry, he said: "Let's go to lunch."

Fig. 16.3 *Sketch of the world's first nuclear reactor, Chicago Pile 1 or CP-1, which was constructed under the football grandstands at the University of Chicago. Public domain, Wikimedia Commons*

After lunch, the experiment was continued; and by 2:30 in the afternoon, the critical point was reached. "Pull it out another foot, George", Fermi said, and then he added: "This will do it. Now the pile will chain-react." The Geiger counters began to click faster and faster, and the recording pen moved upward with no sign of leveling off. On top of the pile, the suicide squad waited tensely with their containers of cadmium solution.

Leona Woods whispered to Fermi: "When do we get scared?" However, the pile behaved exactly as predicted, and after 28 minutes, the control rod was reinserted. Eugene Wigner then produced a bottle of Chianti wine which he had kept concealed until that moment, and everyone drank a little, in silence, from paper cups.

The atomic bomb

The chain-reacting pile had a double significance: Its first meaning was a hopeful one - It represented a new source of energy for mankind. The second meaning was more sinister - It was a step on the road to the construction of atomic bombs.

According to the Bohr-Wheeler theory, it was predicted that plutonium-239 should be just as fissionable as uranium-235. Instead of trying to separate the rare isotope, uranium-235, from the common isotope, uranium-238, the physicists could just operate the pile until a sufficient amount of plutonium accumulated, and then separate it out by ordinary chemical means.

This was done on a very large scale by the Dupont chemical company. Four large chain-reacting piles were built beside the Colombia River at Hanford, Washington. Cold water from the river was allowed to flow through the piles to carry away the heat.

An alternative method for producing atomic bombs was to separate the rare fissionable isotope of uranium from the common isotope. Three different methods for isotope separation seemed possible: One could make a gaseous compound of uranium and allow it to diffuse through a porous barrier. (The lighter isotope would diffuse slightly faster.) Alternatively, one could use a high-speed gas centrifuge; or one could separate the isotopes in a mass spectrograph.

All three methods of isotope separation were tried, and all proved successful. Under Harold Urey's direction, a huge plant to carry out the gaseous separation methods was constructed at Oak Ridge Tennessee; and at the University of California in Berkeley, Ernest O. Lawrence and his group converted the new giant cyclotron into a mass spectrograph. Ultimately, 150,000 people were working at Hanford, Oak Ridge and Berkeley, producing material for atomic bombs. Of these, only a few knew the true purpose of the work in which they were engaged.

Calculations performed in England by Otto Frisch and Rudolf Peierels showed that the critical mass of fissionable material needed for a bomb was about two kilograms. If this mass of material were suddenly assembled, a chain-reaction would start spontaneously. An avalanche of neutrons would develop with almost-instantaneous speed, because no time would be needed for the neutrons to be slowed by a moderator. The lower efficiency of the fast neutrons would be offset by the high concentration of fissionable nuclei, and the result would be a nuclear explosion.

Following a joint decision by Roosevelt and Churchill, English work on atomic bombs was moved to the United States and Canada, where it was combined with the research already being conducted there by American and refugee European scientists. Work on the bomb project was driven forward by an overpowering fear that the Nazis would be the first to construct nuclear weapons.

In July, 1943, Robert Oppenheimer of the University of California was appointed director of a secret laboratory where atomic bombs would be built as soon as material for them became available. At the time of his appointment, Oppenheimer was 39 years old. He was a tall, thin man, with refined manners, and a somewhat ascetic appearance.

Oppenheimer was the son of a wealthy and cultured New York financier. He had graduated from Harvard with record grades, and had done post-graduate work in theoretical physics under Max Born at the University of Göttingen in Germany.

Robert Oppenheimer had then worked with E.O. Lawrence, who was separating the isotopes of uranium, using the Berkeley cyclotron, which had been converted to a mass spectrograph. After making a technical innovation which greatly reduced the cost of separation, Oppenheimer had been appointed the head of the theoretical group of the atomic bomb project. He proved to be a gifted leader. His charm was hypnotic; and under his leadership, "something got done, and done at astonishing speed", as Arthur Compton said later.

Oppenheimer proposed that all work on building atomic bombs should be assembled in a secret laboratory. This proposal was adopted; and because Oppenheimer had shown such gifts as a leader, he was made head of the secret laboratory.

At first, it was planned that this laboratory should be located near to the huge isotope separation plant at Oak Ridge, Tennessee. However, spies often were set on shore on the Atlantic coast of the United States by German submarines; and a number of spies were captured near to Oak Ridge. Therefore, Oppenheimer and General Leslie Groves (the military director of the project) looked for a more isolated site in the western part of the country.

Oppenheimer had boyhood memories of New Mexico, where he and his brother, Frank, had spent their vacations. He took General Groves to a boy's school, which he remembered, on a high plateau near the Los Alamos canyon. The mesa where the boy's school was located was the flat top of a mountain, 7,000 feet above sea level, overlooking the valley of the Rio Grande River.

It was a completely isolated place. Apart from the few buildings of the school, one saw only scattered aspens and fragrant pines, the red rock of the mesa, and the Jemez mountains on the horizon, standing out sharply in the dry, transparent air. Sixty miles separated Los Alamos from the nearest railway station, at Santa Fe, New Mexico.

Oppenheimer and Groves decided that this would be an excellent place for the secret laboratory which they were planning; and they told the headmaster that the school would have to be closed. It would be bought for government war work. The buildings of the school would accommodate the first scientists arriving at Los Alamos while other buildings were being constructed.

Within a year of the first visit to the lonely mesa by Oppenheimer and Groves, 3,500 people were working there; and in another year, the population of scientists and their families had grown to 6,000. More and more scientists received visits from the persuasive young director, Robert Oppenheimer; and more and more of them disappeared to the mysterious "Site Y", a place so secret that its location and name could not be mentioned, and knowledge of its mere existence was limited to very few people.

Many of the scientists who had fled from Hitler's Europe found themselves reunited with their friends at "Site Y". Fermi, Segrè, Rossi, Bethe, Peierls, Chadwick, Frisch, Szilard and Teller all were there. Even Niels Bohr arrived at Los Alamos, together with his son, Aage, who was also a physicist.

Bohr had remained in Denmark as long as possible, in order to protect his laboratory and his co-workers. However, in 1943, he heard that he had been marked by the Germans for arrest and deportation; and he

Fig. 16.4 *Oppenheimer with General Leslie Groves, military head of the Manhattan Project. Public domain, Wikimedia Commons*

escaped to Sweden in a small boat. In Sweden, he helped to rescue the Jewish population of Denmark from the Nazis; and finally he arrived at Los Alamos.

As time passed, many of the scientists at Los Alamos, including Niels Bohr, became deeply worried about the ethical aspects of work on the atomic bomb. When the project had first begun, everyone was sure that the Germans had a great lead in the development of nuclear weapons. They were convinced that the only way to save civilization from the threat of Nazi atomic bombs would be to have a counter-threat. In 1944, however, as the Allied invasion of Europe began, and no German atomic bombs appeared, this dogma seemed less certain.

In 1943, a special intelligence unit of the American Army had been established. Its purpose was to land with the first Allied troops invading Europe, and to obtain information about the German atomic bomb project. The code-name of the unit was *Aslos*, a literal Greek translation of the name of General Groves. The Dutch refugee physicist, Samuel Goudschmidt, was the scientific director of the *Aslos* mission.

When Strasbourg fell to the Allies, Goudschmidt found documents which made it clear that the Germans had not even come close to building atomic bombs. While walking with one of his military colleagues, Goudschmidt exclaimed with relief, "Isn't it wonderful? The Germans don't have atomic bombs! Now we won't have to use ours!"

He was shocked by the reply of his military colleague: "Of course you understand, Sam, that if we have such a weapon, we are going to use it." Goudschmidt's colleague unfortunately proved to have an accurate understanding of the psychology of military and political leaders.

The news that the Germans would not produce atomic bombs was classified as a secret. Nevertheless, it passed through the grapevine to the scientists working on the atomic bomb project in America; and it reversed their attitude to the project. Until then, they had been worried that Hitler would be the first to produce nuclear weapons. In 1944, they began to worry instead about what the American government might do if it came to possess such weapons.

At Los Alamos, Niels Bohr became the center of discussion and worry about the ethics of continued work on the bomb project. He was then 59 years old; and he was universally respected both for his pioneering work in atomic physics, and for his outstandingly good character.

Bohr was extremely worried because he foresaw a postwar nuclear arms race unless international control of atomic energy could be established. Consequently, as a spokesman for the younger atomic scientists, he approached both Roosevelt and Churchill to urge them to consider means by which international control might be established.

Roosevelt, too, was worried about the prospect of a postwar nuclear armaments race; and he was very sympathetic towards Bohr's proposals for international control. He suggested that Bohr travel to England and contact Churchill, to obtain his point of view.

Churchill was desperately busy, and basically unsympathetic towards Bohr's proposals; but on May 16, 1944, he agreed to a half-hour interview with the scientist. The meeting was a complete failure. Churchill and his scientific advisor, Lord Cherwell, spent most of the time talking with each other, so that Bohr had almost no time to present his ideas.

Although he could be very persuasive in long conversations, Bohr was unable to present his thoughts briefly. He wrote and spoke in a discursive style, similar to that of Henry James. Each of his long, convoluted sentences was heavily weighted with qualifications and dependent clauses. At one point in the conversation, Churchill turned to Lord Cherwell and asked: "What's he talking about, physics or politics?"

Bohr's low, almost whispering, way of speaking irritated Churchill. Furthermore, the two men were completely opposed in their views: Bohr was urging openness in approaching the Russians, with a view to establishing international control of nuclear weapons. Churchill, a defender of the old imperial order, was concerned mainly with maintaining British and American military supremacy.

After the interview, Churchill became worried that Bohr would give away "atomic secrets" to the Russians; and he even suggested that Bohr be arrested. However, Lord Cherwell explained to the Prime Minister that the possibility of making atomic bombs, as well as the basic means of doing so, had been common knowledge in the international scientific community ever since 1939.

After his disastrous interview with Churchill, Niels Bohr carefully prepared a memorandum to be presented to President Roosevelt. Realizing how much depended on its success or failure, Bohr wrote and rewrote the memorandum, sweating in the heat of Washington's summer weather. Aage Bohr, who acted as his father's secretary, typed the memorandum over and over, following his father's many changes of mind.

Finally, in July, 1944, Bohr's memorandum was presented to Roosevelt. It contains the following passages:

"...Quite apart from the question of how soon the weapon will be ready for use, and what role it will play in the present war, this situation raises a number of problems which call for urgent attention. Unless, indeed, some agreement about the control of the new and active materials can be obtained in due time, any temporary advantage, however great, may be outweighed by a perpetual menace to human society."

"Ever since the possibilities of releasing atomic energy on a vast scale came into sight, much thought has naturally been given to the question of control; but the further the exploration of the scientific problems is proceeding, the clearer it becomes that no kind of customary measures will suffice for this purpose, and that the terrifying prospect of a future competition between nations about a weapon of such formidable character can only be avoided by a universal agreement in true confidence..."

Roosevelt was sympathetic with the ideas expressed in this memorandum. In an interview with Bohr, he expressed his broad agreement with the idea of international control of atomic energy. Unfortunately, the President had only a few months left to live.

At the University of Chicago, worry and discussion were even more acute than at Los Alamos. The scientists at Chicago had better access to the news, and more time to think. A committee of seven was elected by the Chicago scientists to draft their views into a report on the social and political consequences of atomic energy. The chairman of the committee was the Nobel-laureate physicist James Franck, a man greatly respected for his integrity.

The Franck Report was submitted to the American Secretary of War in June, 1945; and it contains the following passages:

"In the past, science has been able to provide new methods of protection against new methods of aggression it made possible; but it cannot promise such effective protection against the destructive use of nuclear energy. This protection can only come from the political organization of the world. Among all the arguments calling for an efficient international organization for peace, the existence of nuclear weapons is the most compelling one..."

"If no efficient international agreement is achieved, the race for nuclear armaments will be on in earnest not later than the morning after our first demonstration of the existence of nuclear weapons. After this, it might take other nations three or four years to overcome our present head start..."

"It is not at all certain that American public opinion, if it could be enlightened as to the effect of atomic explosives, would approve of our own country being the first to introduce such an indiscriminate method for the wholesale destruction of civilian life.. The military advantages, and the saving of American lives, achieved by a sudden use of atomic bombs against Japan, may be outweighed by a wave of horror and revulsion sweeping over the rest of the world, and perhaps even dividing public opinion at home..."

"From this point of view, a demonstration of the new weapon might best be made, before the eyes of representatives of all the United Nations, on the desert, or on a barren island. The best possible atmosphere for.. an international agreement could be achieved if America could say to the world: 'You see what sort of weapon we had but did not use. We are ready to renounce its use in the future, if other nations join us in this renunciation, and join us in the establishment of an efficient control'."

"One thing is clear: Any international agreement on the prevention of nuclear armaments must be backed by actual and effective controls. No paper agreement can be sufficient, since neither this nor any other nation can stake its whole existence on trust in other nations' signatures."

The Franck report then goes on to outline the steps which would have to be taken in order to establish efficient international control of atomic energy. The report states that the most effective method would be for an international control board to restrict the mining of uranium ore. This would also prevent the use of atomic energy for generating electrical power; but the price would not be too high to pay in order to save humankind from the grave dangers of nuclear war.

Unfortunately, it was too late for the scientists to stop the machine which they themselves had set in motion. President Franklin Roosevelt might have stopped the use of the bomb; but in August, 1945, he was dead. On his desk, unread, lay letters from Albert Einstein and Leo Szilard - the same men who had written to Roosevelt six years previously, thus initiating the American atomic bomb project. In 1945, both Einstein and Szilard wrote again to Roosevelt, this time desperately urging him not to use nuclear weapons against Japan; but their letters arrived too late.

In Roosevelt's place was a new President, Harry Truman, who had been in office only a few weeks. He came from a small town in Missouri; and he was shocked to find himself suddenly thrust into a position of enormous power. He was overwhelmed with new responsibilities, and was cautiously feeling his way. Until Roosevelt's death he had known nothing whatever about the atomic bomb project; and he therefore had little chance to absorb its full meaning.

By contrast, General Leslie Groves, the military commander of the bomb project, was very sure of himself; and he was determined to use atomic bombs against Japan. General Groves had supervised the spending of two billion dollars of the American taxpayers' money. He was anxious to gain credit for winning the war, rather than to be blamed for the money's misuse.

Under these circumstances, it is understandable that Truman did nothing to stop the use of the atomic bomb. In General Groves' words, "Truman did not so much say 'yes', as not say 'no'. It would, indeed, have taken a lot of nerve to say 'no' at that time."

August 6

On August 6, 1945, at 8:15 in the morning, an atomic bomb was exploded in the air over Hiroshima. The force of the explosion was equivalent to twenty thousand tons of T.N.T.. Out of a city of two hundred and fifty thousand people, almost one hundred thousand were killed by the bomb; and another hundred thousand were hurt.

In some places, near the center of the city, people were completely vaporized, so that only their shadows on the pavement marked the places where they had been. Many people who were not killed by the blast or by burns from the explosion, were trapped under the wreckage of their houses. Unable to move, they were burned to death in the fire which followed.

Some accounts of the destruction of Hiroshima, written by children who survived it, have been collected by Professor Arata Osada. Among them is the following account, written by a boy named Hisato Ito. He was 11 years old when the atomic bomb was exploded over the city:

"On the morning of August 5th (we went) to Hiroshima to see my brother, who was at college there. My brother spent the night with us in a hotel... On the morning of the 6th, my mother was standing near the entrance, talking with the hotel proprietor before paying the bill, while I played with the cat. It was then that a violent flash of blue-white light swept in through the doorway."

"I regained consciousness after a little while, but everything was dark. I had been flung to the far end of the hall, and was lying under a pile of debris caused by the collapse of two floors of the hotel. Although I tried to crawl out of this, I could not move. The fine central pillar, of which the proprietor was so proud, lay flat in front of me. "

"I closed my eyes and was quite overcome, thinking that I was going to die, when I heard my mother calling my name. At the sound of her voice, I opened my eyes; and then I saw the flames creeping close to me. I called frantically to my mother, for I knew that I should be burnt alive if I did not escape at once. My mother pulled away some burning boards and saved me. I shall never forget how happy I felt at that moment - like a bird let out of a cage."

"Everything was so altered that I felt bewildered. As far as my eyes could see, almost all the houses were destroyed and on fire. People passed by, their bodies red, as if they had been peeled. Their cries were pitiful. Others were dead. It was impossible to go farther along the street on account of the bodies, the ruined houses, and the badly wounded who lay about moaning. I did not know what to do; and as I turned to the west, I saw that the flames were drawing nearer.."

"At the water's edge, opposite the old Sentai gardens, I suddenly realized that I had become separated from my mother. The people who had been burned were plunging into the river Kobashi, and then were crying out: 'It's hot! It's hot!' They were too weak to swim, and they drowned while crying for help."

Fig. 16.5 *It was like a scene from hell. Source: SGI International.*

In 1951, shortly after writing this account, Hisato Ito died of radiation sickness. His mother died soon afterward from the same cause.

When the news of the atomic bombing of Hiroshima and Nagasaki reached Albert Einstein, his sorrow and remorse were extreme. During the remainder of his life, he did his utmost to promote the cause of peace and to warn humanity against the dangers of nuclear warfare.

When Otto Hahn, the discoverer of fission, heard the news of the destruction of Hiroshima, he and nine other German atomic scientists were

Fig. 16.6 *Burned beyond recognition. Source: SGI International.*

being held prisoner at an English country house near Cambridge. Hahn became so depressed that his colleagues feared that he would take his own life.

Among the scientists who had worked at Chicago and Los Alamos, there

Fig. 16.7 *Memories of August 6. Source: SGI International.*

Fig. 16.8 *The effects lasted a lifetime. Source: SGI International.*

was relief that the war was over; but as descriptions of Hiroshima and Nagasaki became available, there were also sharp feelings of guilt. Many scientists who had worked on the bomb project made great efforts to persuade the governments of the United States, England and Russia to agree

Fig. 16.9 *After the bombing. Source: SGI International.*

to international control of atomic energy; but these efforts met with failure; and the nuclear arms race feared by Bohr developed with increasing momentum.

Suggestions for further reading

(1) Robert Jungk, *Brighter Than a Thousand Suns*, Penguin Books Ltd. (1964).

(2) Werner Braunbeck, *The Drama of the Atom*, Oliver and Boyd, Edinburgh (1958).

(3) Werner Heisenberg, *Physics and Beyond; Memories of a Life in Science*, George Allen and Unwin (1971).

(4) Emilio Segrè, *Enrico Fermi, Physicist*, The University of Chicago Press (1970).

(5) Laura Fermi, *Atoms in the Family*, The University of Chicago Press (1954).

(6) O.R. Frisch, *What Little I Remember*, Cambridge University Press (1979).

(7) James L. Henderson, *Hiroshima*, Longmans (1974).

(8) Arata Osada, *Children of the A-Bomb, The Testiment of Boys and*

Girls of Hiroshima, Putnam, New York (1963).
(9) Michehiko Hachiya, M.D., *Hiroshima Diary*, The University of North Carolina Press, Chapel Hill, N.C. (1955).
(10) Marion Yass, *Hiroshima*, G.P. Putnam's Sons, New York (1972).
(11) Robert Jungk, *Children of the Ashes*, Harcourt, Brace and World (1961).
(12) Burt Hirschfield, *A Cloud Over Hiroshima*, Baily Brothers and Swinfin Ltd. (1974).
(13) John Hersey, *Hiroshima*, Penguin Books Ltd. (1975).

Chapter 17

GENESPLICING

Genetics

Not only physicists, but also biologists, warned of the grave dangers of nuclear testing and nuclear warfare. During the postwar period, it became clear to the scientists that fall-out from nuclear explosions represented a danger to the genetic pool of humans and other living organisms.

During this period, there was a rapid development of genetic research, which culminated in an understanding of the molecular mechanism of heredity. It had been shown by Gregor Mendel that inherited characteristics, like the height of pea plants, were controlled by genes, which could be either dominant or recessive.

Mendel had crossed a strain of dwarf pea plants with a true-breeding tall variety, producing a generation of hybrids, all of which were tall. Next he had pollinated the hybrids with each other, and he had found that roughly one-quarter of the plants in the new generation were true-breeding tall plants, one quarter were true-breeding dwarfs, and one half were tall but not true-breeding. Mendel had deduced that the true-breeding dwarfs had recessive dwarf genes from both parents; and the true-breeding tall plants had dominant genes for tallness from both parents. Those plants which were tall, but not true-breeding, were hybrids, like the plants of the previous generation.

Mendel published his results in the *Transactions of the Brünn Natural History Society* in 1865, and no one noticed his paper[1]. At that time, Austria was being overrun by the Prussians, and people had other things to think about. Mendel was elected Abbot of his monastery; he grew too

[1] Mendel sent a copy of his paper to Darwin; but Darwin, whose German was weak, seems not to have read it.

Fig. 17.1 *Gregor Mendel (1822-1884). His extremely important discoveries went unnoticed for almost half a century. Public domain, Wikimedia Commons*

old and fat to bend over and cultivate his pea plants; his work on heredity was completely forgotten, and he died never knowing that he would one day be considered to be the founder of modern genetics.

In 1900 the Dutch botanist named Hugo de Vries, working on evening primroses, independently rediscovered Mendel's laws. Before publishing, he looked through the literature to see whether anyone else had worked on the subject, and to his amazement he found that Mendel had anticipated his great discovery by 35 years. De Vries could easily have published his own work without mentioning Mendel, but his honesty was such that he gave Mendel full credit and mentioned his own work only as a confirmation of Mendel's laws.

Astonishingly, the same story was twice repeated elsewhere in Europe during the same year. In 1900, two other botanists (Correns in Berlin and Tschermak in Vienna) independently rediscovered Mendel's laws, looked through the literature, found Mendel's 1865 paper, and gave him full credit for the discovery.

Besides rediscovering the Mendelian laws for the inheritance of dominant and recessive characteristics, de Vries made another very important discovery: He discovered genetic mutations - sudden unexplained changes of form which can be inherited by subsequent generations. In growing evening primroses, de Vries found that sometimes, but very rarely, a completely new variety would suddenly appear, and he found that the variation could be propagated to the following generations.

Actually, mutations had been observed before the time of de Vries. For example, a short-legged mutant sheep had suddenly appeared during the 18th century; and stock-breeders had taken advantage of this mutation to breed sheep that could not jump over walls. However, de Vries was the first scientist to study and describe mutations. He noticed that most mutations are harmful, but that a very few are beneficial, and those few tend in nature to be propagated to future generations.

After the rediscovery of Mendel's work by de Vries, many scientists began to suspect that chromosomes might be the carriers of genetic information. The word "chromosome" had been invented by the German physiologist, Walther Flemming, to describe the long, threadlike bodies which could be seen when cells were stained and examined through, the microscope during the process of division. It had been found that when an ordinary cell divides, the chromosomes also divide, so that each daughter cell has a full set of chromosomes.

The Belgian cytologist, Edouard van Benedin, had shown that in the formation of sperm and egg cells, the sperm and egg receive only half of the full number of chromosomes. It had been found that when the sperm of the father combines with the egg of the mother in sexual reproduction, the fertilized egg again has a full set of chromosomes, half coming from the mother and half from the father. This was so consistent with the genetic lottery studied by Mendel, de Vries and others, that it seemed almost certain that chromosomes were the carriers of genetic information.

The number of chromosomes was observed to be small (for example, each normal cell of a human has 46 chromosomes); and this made it obvious that each chromosome must contain thousands of genes. It seemed likely that all of the genes on a particular chromosome would stay together as they passed through the genetic lottery; and therefore certain characteristics should always be inherited together.

The sudden alteration or mutation of genes had been studied by the Dutch geneticist, Hugo de Vries. It was suspected that these genes (carriers of genetic information) were located on the chromosomes.

The Belgian cytologist, Edouard van Benedin, had shown that in the formation of sperm and egg cells, the sperm and egg receive only half of the full number of chromosomes. It had been found that when the sperm of the father combines with the egg of the mother in sexual reproduction, the fertilized egg again has a full set of chromosomes, half coming from the mother and half from the father. This was so like the genetic lottery studied by Mendel, de Vries and others, that it seemed almost certain that chromosomes were the carriers of genetic information.

The number of chromosomes was observed to be small (for example, each normal cell of a human has 46 chromosomes); and this made it obvious that each chromosome must contain thousands of genes. It seemed likely that all of the genes on a particular chromosome would stay together as they passed through the genetic lottery; and therefore certain characteristics should always be inherited together.

This problem had been taken up by Thomas Hunt Morgan, a professor of experimental zoology working at Colombia University. He had found it convenient to work with fruit flies, since they breed with lightning-like speed and since they have only four pairs of chromosomes.

Morgan had found that there was a tendency for all the genes on the same chromosome to be inherited together; but on rare occasions, there were "crosses", where apparently a pair of chromosomes broke at some point and exchanged segments. By studying these crosses statistically, Morgan and his "fly squad" were able to make maps of the fruit fly chromosomes showing the positions of the genes.

This work had been taken a step further by Hermann J. Muller, a member of Morgan's "fly squad", who exposed hundreds of fruit flies to X-rays. The result was a spectacular outbreak of man-made mutations in the next generation.

"They were a motley throng", recalled Muller. Some of the mutant flies had almost no wings, others bulging eyes, and still others brown, yellow or purple eyes; some had no bristles, and others curly bristles.

Muller's experiments indicated that mutations can be produced by radiation-induced physical damage; and he guessed that such damage alters the chemical structure of genes. His studies convinced him that exposing humans to too much radiation could lead to the genetic disintegration and extinction of our species. For this reason, Muller became a leader in the struggle to ban nuclear weapons, as did many other distinguished scientists, such as Linus Pauling, George Wald, Dorothy Crowfoot Hodgkin, Maurice Wilkins, and Sir Martin Ryle.

Fig. 17.2 *Thomas Hunt Morgan. This image is one of several created for the 1891 Johns Hopkins yearbook of 1891. Public domain, Wikimedia Commons*

The structure of DNA

Until 1944, most scientists had guessed that the genetic message was carried by the proteins of the chromosome. In 1944, however, O.T. Avery and his co-workers at the laboratory of the Rockefeller Institute in New York had performed a critical experiment, which proved that the material which carries genetic information is not protein, but deoxyribonucleic acid (DNA) - a giant chainlike molecule which had been isolated from cell nuclei by the Swiss chemist, Friedrich Miescher.

Avery had been studying two different strains of pneumococci, the bacteria which cause pneumonia. One of these strains, the S-type, had a smooth coat, while the other strain, the R-type, lacked an enzyme needed for the manufacture of a smooth carbohydrate coat. Hence, R-type pneumococci had a rough appearance under the microscope. Avery and his co-workers were able to show that an extract from heat-killed S-type pneumococci could convert the living R-type species permanently into S-type; and they also showed that this extract consisted of pure DNA.

In 1947, the Austrian-American biochemist, Erwin Chargaff, began to study the long, chainlike DNA molecules. It had already been shown by Levine and Todd that chains of DNA are built up of four bases: adenine (A), thymine (T), guanine (G) and cytosine (C), held together by a sugar-phosphate backbone. Chargaff discovered that in DNA from the nuclei of living cells, the amount of A always equals the amount of T; and the amount of G always equals the amount of C.

When Chargaff made this discovery, neither he nor anyone else understood its meaning. However, in 1953, the mystery was completely solved by Maurice Wilkins and Rosalind Franklin at Kings College, London, together with James Watson and Francis Crick at Cambridge University. By means of the Braggs' X-ray diffraction techniques, Wilkins and Franklin obtained crystallographic information about the structure of DNA. Using this information, together with Linus Pauling's model-building methods, Crick and Watson proposed a detailed structure for the giant DNA molecule.

The discovery of the molecular structure of DNA was an event of enormous importance for genetics, and for biology in general. The structure was a revelation! The giant, helical DNA molecule was like a twisted ladder: Two long, twisted sugar-phosphate backbones formed the outside of the ladder, while the rungs were formed by the base pairs, A, T, G and C.

The base adenine (A) could only be paired with thiamine (T), while guanine (G) fit only with cytosine (C). Each base pair was weakly joined

Fig. 17.3 *James Watson and Francis Crick at the Cavendish Laboratory with their model of the structure of DNA. Source: nitro.biosci.arizona.edu*

in the center by hydrogen bonds - in other words, there was a weak point in the center of each rung of the ladder - but the bases were strongly attached to the sugar-phosphate backbone. In their 1953 paper, Crick and Watson wrote:

"It has not escaped our notice that the specific pairing we have postulated suggests a possible copying mechanism for genetic material". Indeed, a sudden blaze of understanding illuminated the inner workings of heredity, and of life itself.

If the weak hydrogen bonds in the center of each rung were broken, the ladderlike DNA macromolecule could split down the center and divide into two single strands. Each single strand would then become a template for the formation of a new double-stranded molecule.

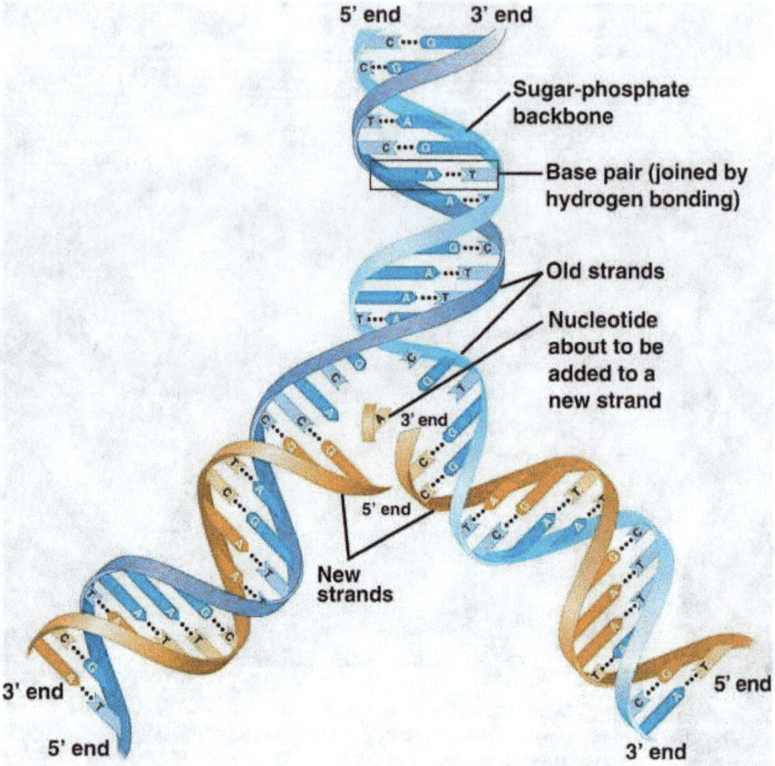

Fig. 17.4 *The mechanism by which DNA is copied. Source: bio1151.nicerweb.com*

Because of the specific pairing of the bases in the Watson-Crick model of DNA, the two strands had to be complementary. T had to be paired with A, and G with C. Therefore, if the sequence of bases on one strand was (for example) TTTGCTAAAGGTGAACCA... , then the other strand necessarily had to have the sequence AAACGATTTCCACTTGGT... The Watson-Crick model of DNA made it seem certain that all the genetic information needed for producing a new individual is coded into the long, thin, double-stranded DNA molecule of the cell nucleus, written in a four-letter language whose letters are the bases, adenine, thymine, guanine and cytosine.

The solution of the DNA structure in 1953 initiated a new kind of biology - molecular biology. This new discipline made use of recently-discovered physical techniques - X-ray diffraction, electron microscopy, electrophoresis, chromatography, ultracentrifugation, radioactive tracer techniques, autoradiography, electron spin resonance, nuclear magnetic resonance and ultraviolet spectroscopy. In the 1960's and 1970's, molecular biology became the most exciting and rapidly-growing branch of science.

Protein structure

In England, J.D. Bernal and Dorothy Crowfoot Hodgkin pioneered the application of X-ray diffraction methods to the study of complex biological molecules. In 1949, Mrs. Hodgkin determined the structure of penicillin; and in 1955, she followed this with the structure of vitamin B12.

In 1960, Max Perutz and John C. Kendrew obtained the structures of the blood proteins myoglobin and hemoglobin. This was an impressive achievement for the Cambridge crystallographers, since the hemoglobin molecule contains roughly 12,000 atoms.

The structure obtained by Perutz and Kendrew showed that hemoglobin is a long chain of amino acids, folded into a globular shape, like a small, crumpled ball of yarn. They found that the amino acids with an affinity for water were on the outside of the globular molecule; while the amino acids for which contact with water was energetically unfavorable were hidden on the inside. Perutz and Kendrew deduced that the conformation of the protein - the way in which the chain of amino acids folded into a 3-dimensional structure - was determined by the sequence of amino acids in the chain.

In 1966, D.C. Phillips and his co-workers at the Royal Institution in London found the crystallographic structure of the enzyme lysozyme (an egg-white protein which breaks down the cell walls of certain bacteria). Again, the structure showed a long chain of amino acids, folded into a roughly globular shape. The amino acids with hydrophilic groups were on the outside, in contact with water, while those with hydrophobic groups were on the inside. The structure of lysozyme exhibited clearly an active site, where sugar molecules of bacterial cell walls were drawn into a mouth-like opening and stressed by electrostatic forces, so that bonds between the sugars could easily be broken.

Meanwhile, at Cambridge University, Frederick Sanger developed methods for finding the exact sequence of amino acids in a protein chain. In 1945, he discovered a compound (2,4-dinitrofluorobenzene) which attaches itself preferentially to one end of a chain of amino acids. Sanger then broke down the chain into individual amino acids, and determined which of them was connected to his reagent. By applying this procedure many times to fragments of larger chains, Sanger was able to deduce the sequence of amino acids in complex proteins. In 1953, he published the sequence of insulin; and this led, in 1964, to the synthesis of insulin.

The picture of protein structure which began to emerge was as follows: A mammalian cell produces roughly 10,000 different proteins. All enzymes are proteins; and the majority of proteins are enzymes - that is, they catalyze reactions involving other biological molecules.

All proteins are built from chainlike polymers, whose monomeric subunits are the twenty amino acids (glycine, analine, valine, isoleucine, leucine, serine, threonine, proline, aspartic acid, glutamic acid, lysine, arginine, asparagine, glutamine, cysteine, methionine, tryptophan, phenylalanine, tyrosine and histidine). These monomers may be connected together into a polymer (called a polypeptide) in any order - hence the great number of possibilities. In such a polypeptide, the backbone is a chain of carbon and nitrogen atoms showing the pattern -C-C-N-C-C-N-C-C-N-...and so on. The -C-C-N- repeating unit is common to all amino acids. Their individuality is derived from differences in the side groups which are attached to the universal -C-C-N- group.

Some proteins, like hemoglobin, contain metal atoms, which may be oxidized or reduced as the protein performs its biological function. Other proteins, like lysozyme, contain no metal atoms, but instead owe their biological activity to an active site on the surface of the protein molecule.

In 1909, the English physician, Archibald Garrod, had proposed a one-gene-one-protein hypothesis. He believed that hereditary diseases are due to the absence of specific enzymes. According to Garrod's hypothesis, damage suffered by a gene results in the faulty synthesis of the corresponding enzyme; and loss of the enzyme ultimately results in the symptoms of the hereditary disease.

In the 1940's, Garrod's hypothesis was confirmed by experiments on the mold, *Neurospora*, performed at Stanford University by George Beadle and Edward Tatum. They demonstrated that mutant strains of the mold would grow normally, provided that specific extra nutrients were added to their diets. The need for these dietary supplements could in every case be traced to the lack of a specific enzyme in the mutant strains. Linus Pauling later extended these ideas to human genetics by showing that the hereditary disease, sickle-cell anemia, is due to a defect in the biosynthesis of hemoglobin.

RNA and ribosomes

Since DNA was known to carry the genetic message, coded into the sequence of the four nucleotide bases, A, T, G and C, and since proteins were known to be composed of specific sequences of the twenty amino acids, it was logical to suppose that the amino acid sequence in a protein was determined by the base sequence of DNA. The information somehow had to be read from the DNA and used in the biosynthesis of the protein.

It was known that, in addition to DNA, cells also contain a similar, but not quite identical, polynucleotide called ribonucleic acid (RNA). The sugar-phosphate backbone of RNA was known to differ slightly from that of DNA; and in RNA, the nucleotide thymine (T) was replaced by a chemically similar nucleotide, uracil (U). Furthermore, while DNA was found only in cell nuclei, RNA was found both in cell nuclei and in the cytoplasm of cells, where protein synthesis takes place. Evidence accumulated indicating that genetic information is first transcribed from DNA to RNA, and afterwards translated from RNA into the amino acid sequence of proteins.

At first, it was thought that RNA might act as a direct template, to which successive amino acids were attached. However, the appropriate chemical complementarity could not be found; and therefore, in 1955, Francis Crick proposed that amino acids are first bound to an adaptor molecule, which is afterward bound to RNA.

In 1956, George Emil Palade of the Rockefeller Institute used electron microscopy to study subcellular particles rich in RNA (ribosomes). Ribosomes were found to consist of two subunits - a smaller subunit, with a molecular weight one million times the weight of a hydrogen atom, and a larger subunit with twice this weight.

It was shown by means of radioactive tracers that a newly synthesized protein molecule is attached temporarily to a ribosome, but neither of the two subunits of the ribosome seemed to act as a template for protein synthesis. Instead, Palade and his coworkers found that genetic information is carried from DNA to the ribosome by a messenger RNA molecule (mRNA). Electron microscopy revealed that mRNA passes through the ribosome like a punched computer tape passing through a tape-reader. It was found that the adapter molecules, whose existence Crick had postulated, were smaller molecules of RNA; and these were given the name "transfer RNA" (tRNA). It was shown that, as an mRNA molecule passes through a ribosome, amino acids attached to complementary tRNA adaptor molecules are added to the growing protein chain.

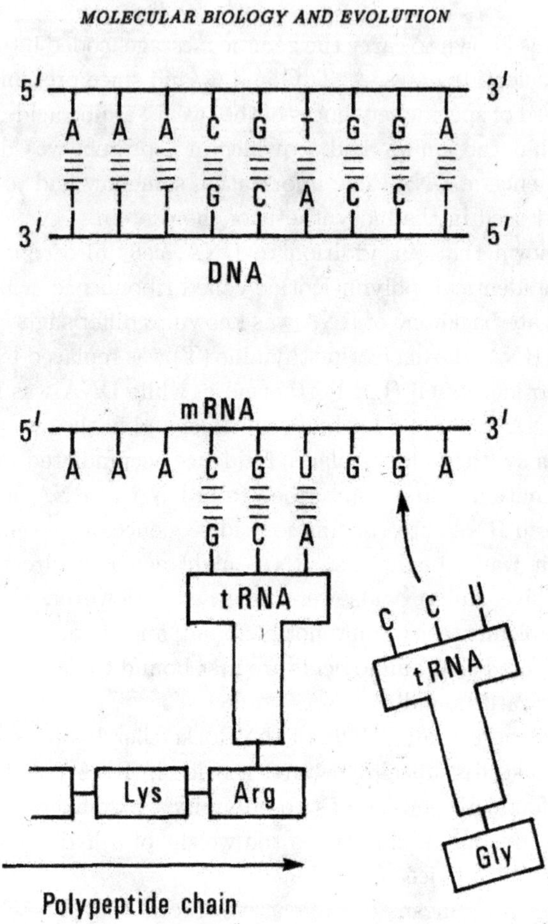

Fig. 17.5 *Information coded on DNA molecules in the cell nucleus is transcribed to mRNA molecules. The messenger RNA molecules in turn provide information for the amino acid sequence in protein synthesis. Drawing by Henning Vibeck.*

The relationship between DNA, RNA, the proteins and the smaller molecules of a cell was thus seen to be hierarchical: The cell's DNA controlled its proteins (through the agency of RNA); and the proteins controlled the synthesis and metabolism of the smaller molecules.

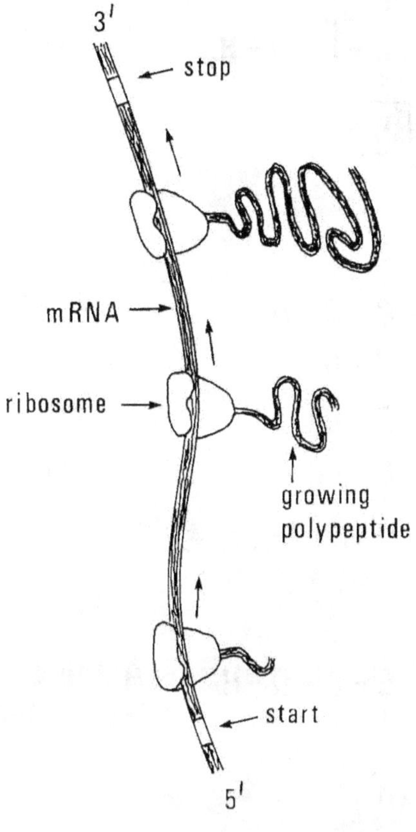

Fig. 17.6 *mRNA passes through the ribosome like a punched computer tape passing through a tape-reader. Drawing by Henning Vibeck.*

The genetic code

In 1955, Severo Ochoa, at New York University, isolated a bacterial enzyme (RNA polymerase) which was able join the nucleotides A, G, U and C so that they became an RNA strand. One year later, this feat was repeated for DNA by Arthur Kornberg.

With the help of Ochoa's enzyme, it was possible to make synthetic RNA molecules containing only a single nucleotide - for example, one could join uracil molecules into the ribonucleic acid chain, ...U-U-U-U-U-U-... In 1961, Marshall Nirenberg and Heinrich Matthaei used synthetic poly-U as

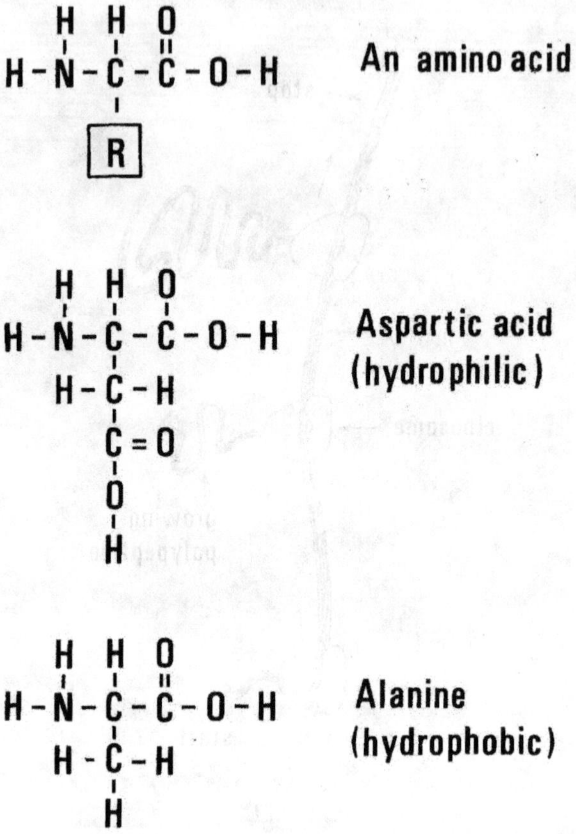

Fig. 17.7 *This figure shows aspartic acid, whose residue (R) is hydrophilic, contrasted with alanine, whose residue is hydrophobic. Drawing by Henning Vibeck.*

messenger RNA in protein synthesis; and they found that only polyphenylalanine was synthesized. In the same year, Sydney Brenner and Francis Crick reported a series of experiments on mutant strains of the bacteriophage, T4. The experiments of Brenner and Crick showed that whenever a mutation added or deleted either one or two base pairs, the proteins produced by the mutants were highly abnormal and non-functional. However, when the mutation added or subtracted three base pairs, the proteins often were functional. Brenner and Crick concluded that the genetic language has three-letter words (codons). With four different "letters", A, T, G and C, this gives sixty-four possible codons - more than enough to specify the twenty different amino acids.

Table 17.1 The genetic code

TTT=Phe	TCT=Ser	TAT=Tyr	TGT=Cys
TTC=Phe	TCC=Ser	TAC=Tyr	TGC=Cys
TTA=Leu	TCA=Ser	TAA=Ter	TGA=Ter
TTG=Leu	TGC=Ser	TAG=Ter	TGG=Trp
CTT=Leu	CCT=Pro	CAT=His	CGT=Arg
CTC=Leu	CCC=Pro	CAC=His	CGC=Arg
CTA=Leu	CCA=Pro	CAA=Gln	CGA=Arg
CTG=Leu	CGC=Pro	CAG=Gln	CGG=Arg
ATT=Ile	ACT=Thr	AAT=Asn	AGT=Ser
ATC=Ile	ACC=Thr	AAC=Asn	AGC=Ser
ATA=Ile	ACA=Thr	AAA=Lys	AGA=Arg
ATG=Met	AGC=Thr	AAG=Lys	AGG=Arg
GTT=Val	GCT=Ala	GAT=Asp	GGT=Gly
GTC=Val	GCC=Ala	GAC=Asp	GGC=Gly
GTA=Val	GCA=Ala	GAA=Glu	GGA=Gly
GTG=Val	GGC=Ala	GAG=Glu	GGG=Gly

In the light of the phage experiments of Brenner and Crick, Nirenberg and Matthaei concluded that the genetic code for phenylalanine is UUU in RNA and TTT in DNA. The remaining words in the genetic code were worked out by H. Gobind Khorana of the University of Wisconsin, who used other mRNA sequences (such as GUGUGU..., AAGAAGAAG... and GUUGUUGUU...) in protein synthesis. By 1966, the complete genetic code, specifying amino acids in terms of three-base sequences, was known. The code was found to be the same for all species studied, no matter how widely separated they were in form; and this showed that all life on earth belongs to the same family, as postulated by Darwin.

Genetic engineering

In 1970, Hamilton Smith of Johns Hopkins University observed that when the bacterium *Haemophilus influenzae* is attacked by a bacteriophage (a virus parasitic on bacteria), it can defend itself by breaking down the DNA of the phage. Following up this observation, he introduced DNA from the bacterium *E. coli* into *H. influenzae*. Again the foreign DNA was broken down.

Further investigation revealed that *H. influenzae* produced an enzyme, later named *Hin* dII, which cut a DNA strand only when it recognized a specific sequence of bases: The DNA was cut only if one strand contained the sequence GTPyPuAC, where Py stands for C or T, while Pu stands for A or G. The other strand, of course, contained the complementary sequence, CAPuPyTG. The enzyme *Hin* dII cut both strands in the middle of the six-base sequence.

Smith had, in fact, discovered the first of a class of bacterial enzymes which came to be called "restriction enzymes" or "restriction nucleases". Almost a hundred other restriction enzymes were subsequently discovered; and each was found to cut DNA at a specific base sequence. Smith's colleague, Daniel Nathans, used the restriction enzymes *Hin* dII and *Hin* dIII to produce the first "restriction map" of the DNA in a virus.

In 1971 and 1972, Paul Berg, and his co-workers Peter Lobban, Dale Kaiser and David Jackson at Stanford University, developed methods for adding cohesive ends to DNA fragments. Berg and his group used the calf thymus enzyme, terminal transferase, to add short, single-stranded polynucleotide segments to DNA fragments. For example, if they added the single-stranded segment AAAA to one fragment, and TTTT to another, then the two ends joined spontaneously when the fragments were incubated together. In this way Paul Berg and his group made the first recombinant DNA molecules.

The restriction enzyme *Eco* RI, isolated from the bacterium *E. coli*, was found to recognize the pattern, GAATTC, in one strand of a DNA molecule, and the complementary pattern, CTTAAG, in the other strand. Instead of cutting both strands in the middle of the six-base sequence, *Eco* RI was observed to cut both strands between G and A. Thus, each side of the cut was left with a "sticky end" - a four-base single-stranded segment, attached to the remainder of the double-stranded DNA molecule.

In 1972, Janet Mertz and Ron Davis, working at Stanford University, demonstrated that DNA strands cut with *Eco* RI could be rejoined by means of another enzyme - a DNA ligase. More importantly, when DNA strands from two different sources were cut with *Eco* RI, the sticky end of one fragment could form a spontaneous temporary bond with the sticky end of the other fragment. The bond could be made permanent by the addition of DNA ligase, even when the fragments came from different sources. Thus, DNA fragments from different organisms could be joined together.

Bacteria belong to a class of organisms (prokaryotes) whose cells do not have a nucleus. Instead, the DNA of the bacterial chromosome is arranged in a large loop. In the early 1950's, Joshua Lederberg had discovered that bacteria can exchange genetic information. He found that a frequently-exchanged gene, the F-factor (which conferred fertility), was not linked to other bacterial genes; and he deduced that the DNA of the F-factor was not physically a part of the main bacterial chromosome. In 1952, Lederberg coined the word "plasmid" to denote any extrachromosomal genetic system.

In 1959, it was discovered in Japan that genes for resistance to antibiotics can be exchanged between bacteria; and the name "R-factors" was given to these genes. Like the F-factors, the R-factors did not seem to be part of the main loop of bacterial DNA.

Because of the medical implications of this discovery, much attention was focused on the R-factors. It was found that they were plasmids, small loops of DNA existing inside the bacterial cell, but not attached to the bacterial chromosome. Further study showed that, in general, between one percent and three percent of bacterial genetic information is carried by plasmids, which can be exchanged freely even between different species of bacteria.

In the words of the microbiologist, Richard Novick, "Appreciation of the role of plasmids has produced a rather dramatic shift in biologists' thinking about genetics. The traditional view was that the genetic makeup of a species was about the same from one cell to another, and was constant over long periods of time. Now a significant proportion of genetic traits are known to be variable (present in some individual cells or strains, absent in others), labile (subject to frequent loss or gain) and mobile - all because those traits are associated with plasmids or other atypical genetic systems."

In 1973, Herbert Boyer, Stanley Cohen and their co-workers at Stanford University and the University of California carried out experiments in which they inserted foreign DNA segments, cut with *Eco* RI, into plasmids (also cut with *Eco* RI). They then resealed the plasmid loops with DNA ligase. Finally, bacteria were infected with the gene-spliced plasmids. The result was a new strain of bacteria, capable of producing an additional protein coded by the foreign DNA segment which had been spliced into the plasmids.

Cohen and Boyer used plasmids containing a gene for resistance to an antibiotic, so that a few gene-spliced bacteria could be selected from a large population by treating the culture with the antibiotic. The selected bacteria, containing both the antibiotic-resistance marker and the foreign DNA, could then be cloned on a large scale; and in this way a foreign gene could be "cloned". The gene-spliced bacteria were chimeras, containing genes from two different species.

The new recombinant DNA techniques of Berg, Cohen and Boyer had revolutionary implications: It became possible to produce many copies of a given DNA segment, so that its base sequence could be determined. With the help of direct DNA-sequencing methods developed by Frederick Sanger and Walter Gilbert, the new cloning techniques could be used for mapping and sequencing genes.

Since new bacterial strains could be created, containing genes from other species, it became possible to produce any protein by cloning the corresponding gene. Proteins of medical importance could be produced on a large scale. Thus, the way was open for the production of human insulin, interferon, serum albumin, clotting factors, vaccines, and protein hormones such as ACTH, human growth factor and leuteinizing hormone.

It also became possible to produce enzymes of industrial and agricultural importance by cloning gene-spliced bacteria. Since enzymes catalyze reactions involving smaller molecules, the production of these substrate molecules through gene-splicing also became possible.

It was soon discovered that the possibility of producing new, transgenic organisms was not limited to bacteria. Gene-splicing was also carried out on higher plants and animals as well as on fungi. It was found that the bacterium *Agrobacterium tumefaciens* contains a tumor-inducing (Ti) plasmid capable of entering plant cells and producing a crown gall. Genes spliced into the Ti plasmid frequently became incorporated in the plant chromosome, and afterwards were inherited in a stable, Mendelian fashion.

Transgenic animals were produced by introducing foreign DNA into embryo-derived stem cells (ES cells). The gene-spliced ES cells were then selected, cultured and introduced into a blastocyst, which afterwards was implanted in a foster-mother. The resulting chimeric animals were bred, and stable transgenic lines selected.

Thus, for the first time, humans had achieved direct control over the process of evolution. Selective breeding to produce new plant and animal varieties was not new - it was one of the oldest techniques of civilization. However, the degree and speed of intervention which recombinant DNA made possible was entirely new. In the 1970's it became possible to mix the genetic repetoires of different species: The genes of mice and men could be spliced together into new, man-made forms of life!

The Asilomar Conference

In the summer of 1971, Janet Mertz, who was then a student in Paul Berg's laboratory, gave a talk at Cold Spring Harbor. She discussed some proposed experiments applying recombinant techniques to the DNA of the tumor-inducing virus SV40.

This talk worried the cell biologist, Richard Pollack. He was working with SV40 and was already concerned about possible safety hazards in connection with the virus. Pollack telephoned to Berg, and asked whether it might not be dangerous to clone a gene capable of producing human cancer. As a result of this call, Berg decided not to clone genes from tumor-inducing viruses.

Additional concern over the safety of recombinant DNA experiments was expressed at the 1973 Gordon Conference on Nucleic Acids. The scientists attending the conference voted to send a letter to the President of the U.S. National Academy of Sciences:

"...We presently have the technical ability", the letter stated, "to join together, covalently, DNA molecules from diverse sources... This technique could be used, for example, to combine DNA from animal viruses with bacterial DNA... In this way, new kinds of hybrid plasmids or viruses, with biological activity of unpredictable nature, may eventually be created. These experiments offer exciting and interesting potential, both for advancing knowledge of fundamental biological processes, and for alleviation of human health problems."

"Certain such hybrid molecules may prove hazardous to laboratory workers and to the public. Although no hazard has yet been established, prudence suggests that the potential hazard be seriously considered."

"A majority of those attending the Conference voted to communicate their concern in this matter to you, and to the President of the Institute of Medicine... The conferees suggested that the Academies establish a study committee to consider this problem, and to recommend specific actions and guidelines."

As a result of this letter, the National Academy of Sciences set up a Committee on Recombinant DNA, chaired by Paul Berg. The Committee's report, published in July, 1974, contained the following passage:

"...There is serious concern that some of these artificial recombinant DNA molecules could prove biologically hazardous. One potential hazard in current experiments derives from the need to use a bacterium like *E. coli* to clone the recombinant DNA molecules and to amplify their number.

Strains of *E. coli* commonly reside in the human intestinal tract, and they are capable of exchanging genetic information with other types of bacteria, some of which are pathogenic to man. Thus, new DNA elements introduced into *E. coli* might possibly become widely disseminated among human, bacterial, plant, or animal populations, with unpredictable effects."

The Committee on Recombinant DNA recommended that scientists throughout the world should join in a voluntary postponement of two types of experiments: Type 1, introduction of antibiotic resistance factors into bacteria not presently carrying the R-factors; and Type 2, cloning of cancer-producing plasmids or viruses.

The Committee recommended caution in experiments linking DNA from animal cells to bacterial DNA, since animal-derived DNA can carry cancer-inducing base sequences. Finally, the Committee recommended that the National Institutes of Health establish a permanent advisory group to supervise experiments with recombinant DNA, and that an international meeting be held to discuss the biohazards of the new techniques.

In February, 1975, more than 100 leading molecular biologists from many parts of the world met at the Asilomar Conference Center near Monterey, California, to discuss safety guidelines for recombinant DNA research. There was an almost unanimous consensus at the meeting that, until more was known about the dangers, experiments involving cloning of DNA should make use of organisms and vectors incapable of living outside a laboratory environment.

The Asilomar Conference also recommended that a number of experiments be deferred. These included cloning of recombinant DNA derived from highly pathogenic organisms, or containing toxin genes, as well as large-scale experiments using recombinant DNA able to make products potentially harmful to man, animals or plants.

The Asilomar recommendations were communicated to a special committee appointed by the U.S. National Institutes of Health; and the committee drew up a set of guidelines for recombinant DNA research. The NIH Guidelines went into effect in 1976; and they remained in force until 1979. They were stricter than the Asilomar recommendations regarding cloning of DNA from cancer-producing viruses; and this was effectively forbidden by the NIH until 1979. (Of course, the NIH Guidelines were effective only for research conducted within the United States and funded by the U.S. government.)

In 1976, the first commercial genetic engineering company (Genentech) was founded. In 1980, the initial public offering of Genentech stock set a Wall Street record for the fastest increase of price per share. In 1981, another genetic engineering company (Cetus) set a Wall Street record for the largest amount of money raised in an initial public offering (125 million U.S. dollars). During the same years, Japan's Ministry of International Trade and Technology declared 1981 to be "The Year of Biotechnology"; and England, France and Germany all targeted biotechnology as an area for special development.

A number of genetic-engineering products reached the market in the early 1980's. These included rennin, animal growth hormones, foot and mouth vaccines, hog diarrhea vaccine, amino acids, antibiotics, anabolic steroids, pesticides, pesticide-resistant plants, cloned livestock, improved yeasts, cellulose-digesting bacteria, and a nitrogen-fixation enzyme.

Recently the United States and Japan have initiated large-scale programs whose aim is to map the entire human genome; and the European Economic Community is considering a similar program. The human genome project is expected to make possible prenatal diagnosis of many inherited diseases. For example, the gene for cystic fibrosis has been found; and DNA technology makes it possible to detect the disease prenatally.

The possibility of extensive genetic screening raises ethical problems which require both knowledge and thought on the part of the public. An expectant mother, in an early stage of pregnancy, often has an abortion if the foetus is found to carry a serious genetic defect. But with more knowledge, many more defects will be found. Where should the line be drawn between a serious defect and a minor one?

The cloning of genes for lethal toxins also needs serious thought and public discussion. From 1976 to 1982, this activity was prohibited in the United States under the NIH Guidelines. However, in April, 1982, the restriction was lifted, and by 1983, the toxins being cloned included several aflatoxins, lecithinase, cytochalasins, ochratoxins, sporidesmin, T-2 toxin, ricin and tremogen. Although international conventions exist under which chemical and biological weapons are prohibited, there is a danger that nations will be driven to produce and stockpile such weapons because of fear of what other nations might do.

Finally, the release of new, transgenic species into the environment requires thought and caution. Much benefit can come, for example, from the use of gene-spliced bacteria for nitrogen fixation or for cleaning up oil spills. However, once a gene-spliced microorganism has been released, it is virtually impossible to eradicate it; and thus the change produced by the release of a new organism is permanent. Permanent changes in the environment should not be made on the basis of short-term commercial considerations, nor indeed on the basis of short-term considerations of any kind; nor should such decisions be made unilaterally by single nations, since new organisms can easily cross political boundaries.

The rapid development of biotechnology has given humans enormous power over the fundamental mechanisms of life and evolution. But is society mature enough to use this power wisely and compassionately?

The polymerase chain reaction

One day in the early 1980's, an American molecular biologist, Kary Mullis, was driving to his mountain cabin with his girlfriend. The journey was a long one, and to pass the time, Kary Mullis turned over and over in his mind a problem which had been bothering him: He worked for a California biotechnology firm, and like many other molecular biologists he had been struggling to analyze very small quantities of DNA. Mullis realized that it would be desirable have a highly sensitive way of replicating a given DNA segment - a method much more sensitive than cloning. As he drove through the California mountains, he considered many ways of doing this, rejecting one method after the other as impracticable. Finally a solution came to him; and it seemed so simple that he could hardly believe that he was the first to think of it. He was so excited that he immediately pulled over to the side of the road and woke his sleeping girlfriend to tell her about his idea. Although his girlfriend was not entirely enthusiastic about being wakened from a comfortable sleep to be presented with a lecture on biochemistry, Kary Mullis had in fact invented a technique which was destined to revolutionize DNA technology: the polymerase chain reaction (PCR)[2].

The technique was as follows: Begin with a small sample of the genomic DNA to be analyzed. (The sample may be extremely small - only a few molecules.) Heat the sample to 95 °C to separate the double-stranded DNA molecule into single strands. Suppose that on the long DNA molecule there is a target segment which one wishes to amplify. If the target segment begins with a known sequence of bases on one strand, and ends with a known sequence on the complementary strand, then synthetic "primer" oligonucleotides[3] with these known beginning ending sequences are added

[2] The flash of insight didn't take long, but at least six months of hard work were needed before Mullis and his colleagues could convert the idea to reality.

[3] Short segments of single-stranded DNA.

in excess. The temperature is then lowered to 50-60 °C, and at the lowered temperature, the "start" primer attaches itself to one DNA strand at the beginning of the target segment, while the "stop" primer becomes attached to the complementary strand at the other end of the target segment. Polymerase (an enzyme which aids the formation of double-stranded DNA) is then added, together with a supply of nucleotides. On each of the original pieces of single-stranded DNA, a new complementary strand is generated with the help of the polymerase. Then the temperature is again raised to 95 °C, so that the double-stranded DNA separates into single strands, and the cycle is repeated.

In the early versions of the PCR technique, the polymerase was destroyed by the high temperature, and new polymerase had to be added for each cycle. However, it was discovered that polymerase from the bacterium *Thermus aquaticus* would withstand the high temperature. (Thermus aquaticus lives in hot springs.) This discovery greatly simplified the PCR technique. The temperature could merely be cycled between the high and low temperatures, and with each cycle, the population of the target segment doubled, concentrations of primers, deoxynucleotides and polymerase being continuously present.

After a few cycles of the PCR reaction, copies of copies begin to predominate over copies of the original genomic DNA. These copies of copies have a standard length, always beginning on one strand with the start primer, and ending on that strand with the complement of the stop primer.

Two main variants of the PCR technique are possible, depending on the length of the oligonucleotide primers: If, for example, trinucleotides are used as start and stop primers, they can be expected to match the genomic DNA at many points. In that case, after a number of PCR cycles, populations of many different segments will develop. Within each population, however, the length of the replicated segment will be standardized because of the predominance of copies of copies. When the resulting solution is placed on a damp piece of paper or a gel and subjected to the effects of an electric current (electrophoresis), the populations of different molecular weights become separated, each population appearing as a band. The bands are profiles of the original genomic DNA; and this variant of the PCR technique can be used in evolutionary studies to determine the degree of similarity of the genomic DNA of two species.

On the other hand, if the oligonucleotide primers contain as many as 20 nucleotides, they will be highly specific and will bind only to a particular target sequence of the genomic DNA. The result of the PCR reaction will

Fig. 17.8 *The Nobel-laureate biochemist, Melvin Calvin, famous for his studies of photosynthesis, later published an important book entitled "Chemical Evolution; Molecular Evolution Towards the Origin of Living Systems on Earth and Elsewhere". Public domain, Wikimedia Commons*

then be a single population, containing only the chosen target segment. The PCR reaction can be thought of as autocatalytic, and as we shall see in the next section, autocatylitic systems play an important role in modern theories of the origin of life.

Theories of chemical evolution towards the origin of life

The possibility of an era of chemical evolution prior to the origin of life entered the thoughts of Charles Darwin, but he considered the idea to be much too speculative to be included in his published papers and books. However, in February 1871, he wrote a letter to his close friend Sir Joseph Hooker containing the following words:

"It is often said that all the conditions for the first production of a living organism are now present, which could ever have been present. But if (and oh what a big if) we could conceive in some warm little pond with all sorts of ammonia and phosphoric salts, - light, heat, electricity etc. present, that a protein compound was chemically formed, ready to undergo still more complex changes, at the present day such matter would be instantly devoured, or absorbed, which would not have been the case before living creatures were formed."

The last letter which Darwin is known to have dictated and signed before his death in 1882 also shows that he was thinking about this problem:

"You have expressed quite correctly my views", Darwin wrote, "where you said that I had intentionally left the question of the Origin of Life uncanvassed as being altogether ultra vires in the present state of our knowledge, and that I dealt only with the manner of succession. I have met with no evidence that seems in the least trustworthy, in favor of so-called Spontaneous Generation. (However) I believe that I have somewhere said (but cannot find the passage) that the principle of continuity renders it probable that the principle of life will hereafter be shown to be a part, or consequence, of some general law.."

Modern researchers, picking up the problem where Darwin left it, have begun to throw a little light on the problem of chemical evolution towards the origin of life. In the 1930's J.B.S. Haldane in England and A.I. Oparin in Russia put forward theories of an era of chemical evolution prior to the appearance of living organisms.

In 1924 Oparin published a pamphlet on the origin of life. An expanded version of this pamphlet was translated into English and appeared in 1936 as a book entitled *The Origin of Life on Earth.* In this book Oparin pointed out that the time when life originated, conditions on earth were probably considerably different than they are at present: The atmosphere probably contained very little free oxygen, since free oxygen is produced by photosynthesis which did not yet exist. On the other hand, he argued, there were probably large amounts of methane and ammonia in the earth's primitive atmosphere[4]. Thus, before the origin of life, the earth probably had a reducing atmosphere rather than an oxidizing one. Oparin believed that energy-rich molecules could have been formed very slowly by the action of light from the sun. On the present-day earth, bacteria quickly consume energy-rich molecules, but before the origin of life, such molecules could have accumulated, since there were no living organisms to consume them. (This observation is similar to the remark made by Darwin in his 1871 letter to Hooker.)

The first experimental work in this field took place in 1950 in the laboratory of Melvin Calvin at the University of California, Berkeley. Calvin and his co-workers wished to determine experimentally whether the primitive atmosphere of the earth could have been converted into some of the molecules which are the building-blocks of living organisms. The energy needed to perform these conversions they imagined to be supplied by volcanism, radioactive decay, ultraviolet radiation, meteoric impacts, or by lightning strokes.

[4] It is now believed that the main constituents of the primordial atmosphere were carbon dioxide, water, nitrogen, and a little methane.

The earth is thought to be approximately 4.6 billion years old. At the time when Calvin and his co-workers were performing their experiments, the earth's primitive atmosphere was believed to have consisted primarily of hydrogen, water, ammonia, methane, and carbon monoxide, with a little carbon dioxide. A large quantity of hydrogen was believed to have been initially present in the primitive atmosphere, but it was thought to have been lost gradually over a period of time because the earth's gravitational attraction is too weak to effectively hold such a light and rapidly-moving molecule. However, Calvin and his group assumed sufficient hydrogen to be present to act as a reducing agent. In their 1950 experiments they subjected a mixture of hydrogen and carbon dioxide, with a catalytic amount of Fe^{2+}, to bombardment by fast particles from the Berkeley cyclotron. Their experiments resulted in a good yield of formic acid and a moderate yield of formaldehyde. (The fast particles from the cyclotron were designed to simulate an energy input from radioactive decay on the primitive earth.)

Two years later, Stanley Miller, working in the laboratory of Harold Urey at the University of Chicago, performed a much more refined experiment of the same type. In Miller's experiment, a mixture of the gases methane, ammonia, water and hydrogen was subjected to an energy input from an electric spark. Miller's apparatus was designed so that the gases were continuously circulated, passing first through the spark chamber, then through a water trap which removed the non-volatile water soluble products, and then back again through the spark chamber, and so on. The resulting products are shown as a function of time in Figure 3.5.

The Miller-Urey experiment produced many of the building-blocks of living organisms, including glycine, glycolic acid, sarcosine, alanine, lactic acid, N-methylalanine, β-alanine, succinic acid, aspartic acid, glutamic acid, iminodiacetic acid, iminoacetic-propionic acid, formic acid, acetic acid, propionic acid, urea and N-methyl urea[5]. Another major product was hydrogen cyanide, whose importance as an energy source in chemical evolution was later emphasized by Calvin.

The Miller-Urey experiment was repeated and extended by the Ceylonese-American biochemist Cyril Ponnamperuma and by the American expert in planetary atmospheres, Carl Sagan. They showed that when phosphorus is made available, then in addition to amino acids, the Miller-Urey experiment produces not only nucleic acids of the type that join

[5] The chemical reaction that led to the formation of the amino acids that Miller observed was undoubtedly the Strecker synthesis: $HCN + NH_3 + RC{=}O + H_2O \rightarrow RC(NH_2)COOH$.

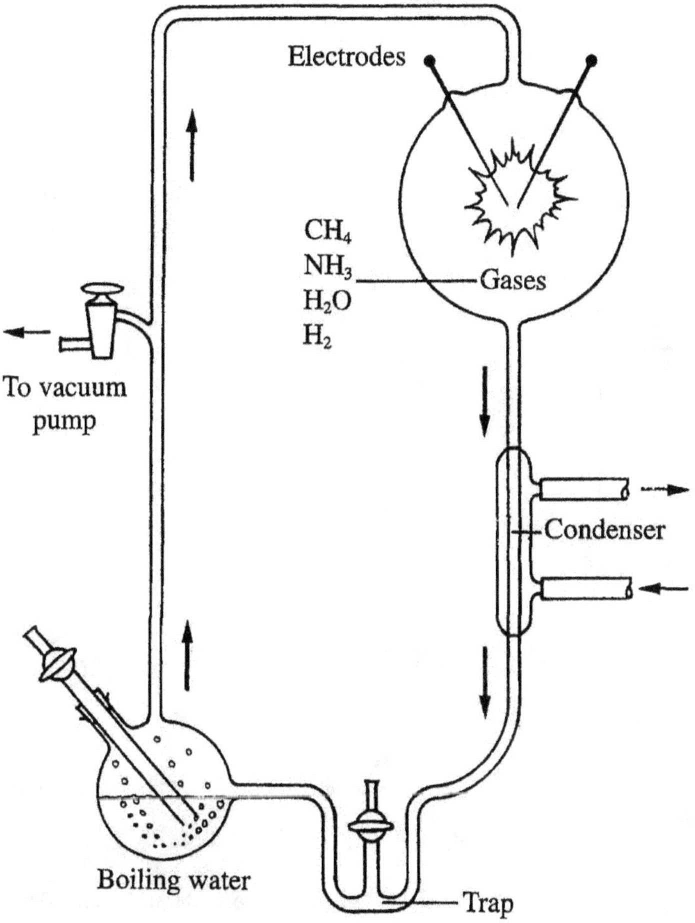

Fig. 17.9 *Miller's apparatus. Drawing by Henning Vibeck.*

together to form DNA, but also the energy-rich molecule ATP (adenosine triphosphate). ATP is extremely important in biochemistry, since it is a universal fuel which drives chemical reactions inside present-day living organisms.

Further variations on the Miller-Urey experiment were performed by Sydney Fox and his co-workers at the University of Miami. Fox and his group showed that amino acids can be synthesized from a primitive atmo-

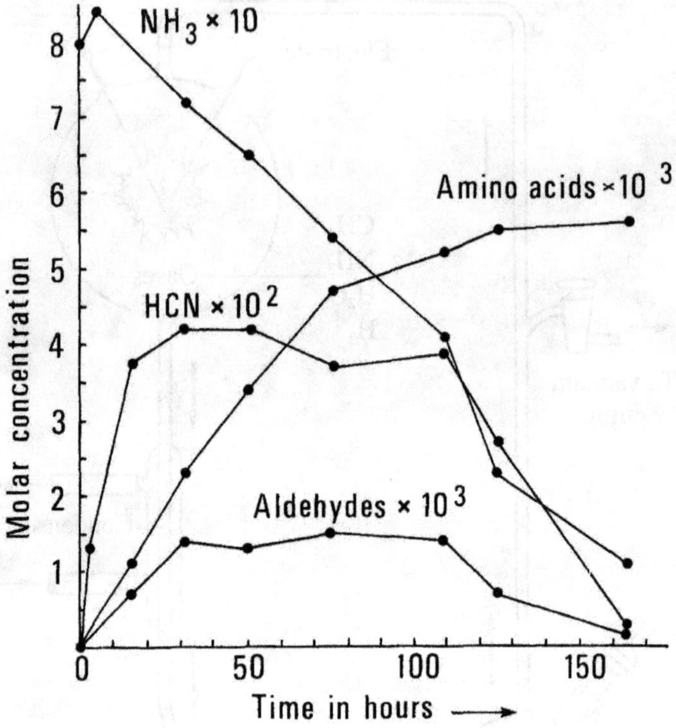

Fig. 17.10 *Products as a function of time in the Miller-Urey experiment. Drawing by Henning Vibeck.*

sphere by means of a thermal energy input, and that in the presence of phosphate esters, the amino acids can be thermally joined together to form polypeptides. However, some of the peptides produced in this way were cross linked, and hence not of biological interest.

In 1969, Melvin Calvin published an important book entitled *Chemical Evolution; Molecular Evolution Towards the Origin of Living Systems on Earth and Elsewhere.* In this book, Calvin reviewed the work of geochemists showing the presence in extremely ancient rock formations of molecules which we usually think of as being produced only by living organisms. He then discussed experiments of the Miller-Urey type - experiments simulating the first step in chemical evolution. According to Calvin, not only amino acids but also the bases adenine, thymine, guanine, cytosine and uracil, as well as various sugars, were probably present in the primitive ocean in moderate concentrations, produced from the primitive atmosphere by the available energy inputs, and not broken down because no organisms were present.

The next steps visualized by Calvin were dehydration reactions in which the building blocks were linked together into peptides, polynucleotides, lipids and porphyrins. Such dehydration reactions are in a thermodynamically uphill direction. In modern organisms, they are driven by a universally-used energy source, the high-energy phosphate bond of adenosine triphosphate (ATP). Searching for a substance present in the primitive ocean which could have driven the dehydrations, Calvin and his coworkers experimented with hydrogen cyanide (HC=N), and from the results of these experiments they concluded that the energy stored in the carbon-nitrogen triple bond of HC=N could indeed have driven the dehydration reactions necessary for polymerization of the fundamental building blocks. However, later work made it seem improbable that peptides could be produced from cyanide mixtures.

In Chemical Evolution, Calvin introduced the concept of autocatalysis as a mechanism for molecular selection, closely analogous to natural selection in biological evolution. Calvin proposed that there were a few molecules in the ancient oceans which could catalyze the breakdown of the energy-rich molecules present into simpler products. According to Calvin's hypothesis, in a very few of these reactions, the reaction itself produced more of the catalyst. In other words, in certain cases the catalyst not only broke down the energy-rich molecules into simpler products but also catalyzed their own synthesis. These autocatalysts, according to Calvin, were the first systems which might possibly be regarded as living organisms. They not only "ate" the energy-rich molecules but they also reproduced - i.e., they catalyzed the synthesis of molecules identical with themselves.

Autocatalysis leads to a sort of molecular natural selection, in which the precursor molecules and the energy-rich molecules play the role of "food", and the autocatalytic systems compete with each other for the food supply. In Calvin's picture of molecular evolution, the most efficient autocatalytic systems won this competition in a completely Darwinian way. These more efficient autocatalysts reproduced faster and competed more successfully for precursors and for energy-rich molecules. Any random change in the direction of greater efficiency was propagated by natural selection.

What were these early autocatalytic systems, the forerunners of life? Calvin proposed several independent lines of chemical evolution, which later, he argued, joined forces. He visualized the polynucleotides, the polypeptides, and the metallo-porphyrins as originally having independent lines of chemical evolution. Later, he argued, an accidental union of these independent autocatalysts showed itself to be a still more efficient autocatalytic system. He pointed out in his book that "autocatalysis" is perhaps too strong a word. One should perhaps speak instead of "reflexive catalysis" , where a molecule does not necessarily catalyze the synthesis of itself, but perhaps only the synthesis of a precursor. Like autocatalysis, reflexive catalysis is capable of exhibiting Darwinian selectivity.

The theoretical biologist, Stuart Kauffman, working at the Santa Fe Institute, has constructed computer models for the way in which the components of complex systems of reflexive catalysts may have been linked together. Kauffman's models exhibit a surprising tendency to produce orderly behavior even when the links are randomly programmed.

In 1967 and 1968, C. Woese, F.H.C. Crick and L.E. Orgel proposed that there may have been a period of chemical evolution involving RNA alone, prior to the era when DNA, RNA and proteins joined together to form complex self-reproducing systems. In the early 1980's, this picture of an "RNA world" was strengthened by the discovery (by Thomas R. Cech and Sydney Altman) of RNA molecules which have catalytic activity.

Today experiments aimed at throwing light on chemical evolution towards the origin of life are being performed in the laboratory of the Nobel Laureate geneticist Jack Sjostak at Harvard Medical School. The laboratory is trying to build a synthetic cellular system that undergoes Darwinian evolution.

In connection with autocatalytic systems, it is interesting to think of the polymerase chain reaction, which we discussed above. The target segment of DNA and the polymerase together form an autocatalytic system. The "food" molecules are the individual nucleotides in the solution. In the PCR system, a segment of DNA reproduces itself with an extremely high degree of fidelity. One can perhaps ask whether systems like the PCR system can have been among the forerunners of living organisms. The cyclic changes of temperature needed for the process could have been supplied by the cycling of water through a hydrothermal system. There is indeed evidence that hot springs and undersea hydrothermal vents may have played an important role in chemical evolution towards the origin of life. We will discuss this evidence in the next section.

Molecular evidence establishing family trees in evolution

Starting in the 1970's, the powerful sequencing techniques developed by Sanger and others began to be used to establish evolutionary trees. The evolutionary closeness or distance of two organisms could be estimated from the degree of similarity of the amino acid sequences of their proteins, and also by comparing the base sequences of their DNA and RNA. One of the first studies of this kind was made by R.E. Dickerson and his coworkers, who studied the amino acid sequences in Cytochrome C, a protein of very ancient origin which is involved in the "electron transfer chain" of respiratory metabolism. Some of the results of Dickerson's studies are shown in Figure 17.11.

Comparison of the base sequences of RNA and DNA from various species proved to be even more powerful tool for establishing evolutionary relationships. Figure 17.12 shows the universal phylogenetic tree established in this way by Iwabe, Woese and their coworkers.[6] In Figure 17.12, all presently living organisms are divided into three main kingdoms, Eukaryotes, Eubacteria, and Archaebacteria. Carl Woese, who proposed this classification on the basis of comparative sequencing, wished to call the three kingdoms "Eucarya, Bacteria and Archaea". However, the most widely accepted terms are the ones shown in capital letters on the figure. Before the comparative RNA sequencing work, which was performed on the ribosomes of various species, it had not been realized that there are two types of bacteria, so markedly different from each other that they must be classified as belonging to separate kingdoms. One example of the difference between archaebacteria and eubacteria is that the former have cell membranes which contain ether lipids, while the latter have ester lipids in their cell membranes. Of the three kingdoms, the eubacteria and the archaebacteria are "prokaryotes", that is to say, they are unicellular organisms having no cell nucleus. Most of the eukaryotes, whose cells contain a nucleus, are also unicellular, the exceptions being plants, fungi and animals.

One of the most interesting features of the phylogenetic tree shown in Figure 17.12 is that the deepest branches - the organisms with shortest pedigrees - are all hyperthermophiles, i.e. they live in extremely hot environments such as hot springs or undersea hydrothermal vents. The short-

[6] "Phylogeny" means "the evolutionary development of a species". "Ontogeny" means "the growth and development an individual, through various stages, for example, from fertilized egg to embryo, and so on." Ernst Haeckel, a 19th century follower of Darwin, observed that, in many cases, "ontogeny recapitulates phylogeny."

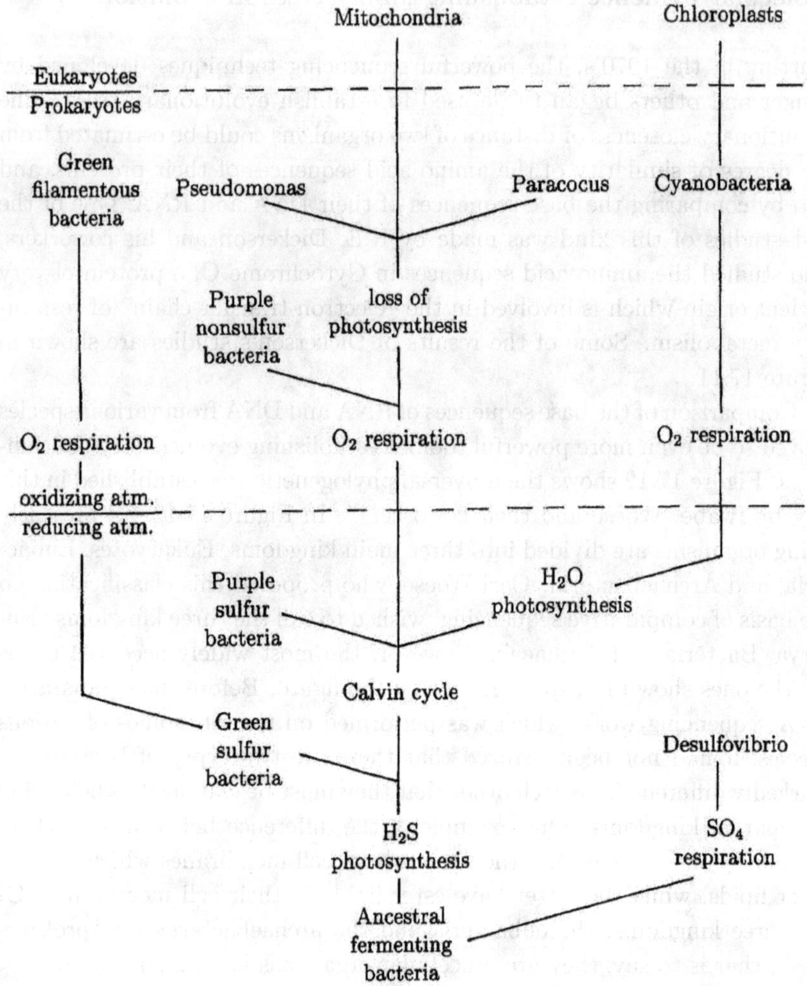

Fig. 17.11 *Evolutionary relationships established by Dickerson and coworkers by comparing the amino acid sequences of Cytochrome C from various species. Author's own diagram.*

est branches represent the most extreme hyperthermophiles. The group of archaebacteria indicated by (1) in the figure includes *Thermofilum*, *Thermoproteus*, *Pyrobaculum*, *Pyrodictium*, *Desulfurococcus*, and *Sulfolobus* - all hypothermophiles.[7] Among the eubacteria, the two shortest branches,

[7] Group (2) in Figure 17.12 includes *Methanothermus*, which is hyperthermophilic,

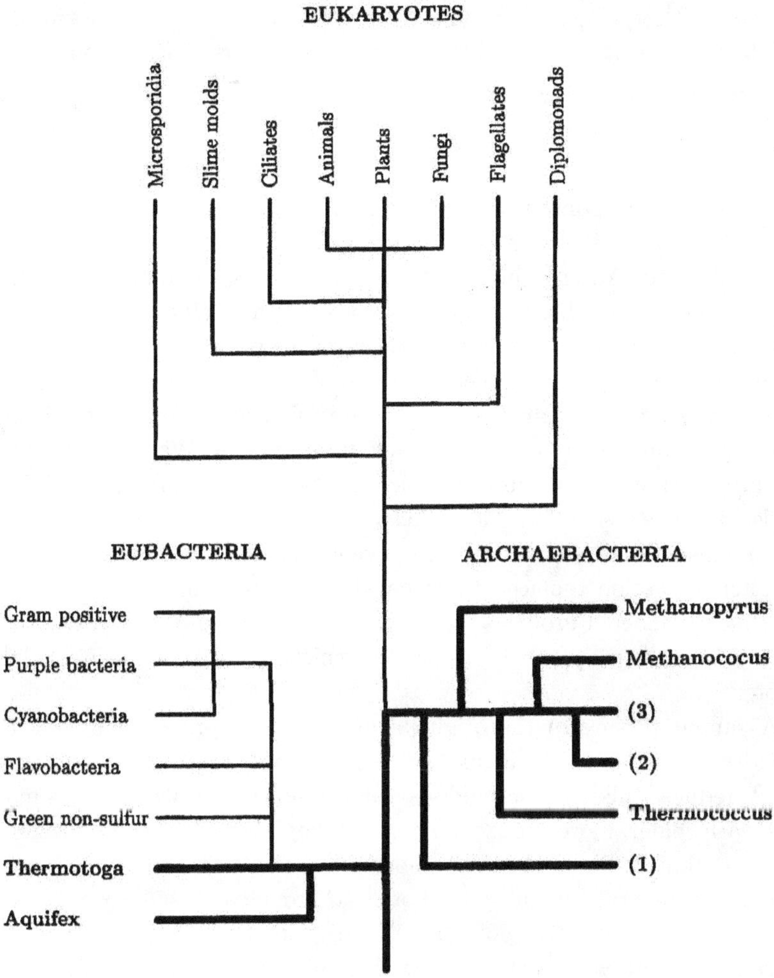

Fig. 17.12 *This figure shows the universal phylogenetic tree, established by the work of Woese, Iwabe et al. Hyperthermophiles are indicated by bold lines and by bold type. Author's own diagram.*

Aquifex and *Thermatoga* are both hyperthermophiles.[8]

---

and *Methanobacterium*, which is not. Group (3) includes *Archaeoglobus*, which is hyperthermophilic, and *Halococcus*, *Halobacterium*, *Methanoplanus*, *Methanospirilum*, and *Methanosarcina*, which are not.

[8] Thermophiles are a subset of the larger group of extremophiles.

The phylogenetic evidence for the existence of hyperthermophiles at a very early stage of evolution lends support to a proposal put forward in 1988 by the German biochemist Günter Wächterhäuser. He proposed that the reaction for pyrite formation,

$$FeS + H_2S \rightarrow FeS_2 + 2H + +2e^-$$

which takes place spontaneously at high temperatures, supplied the energy needed to drive the first stages of chemical evolution towards the origin of life. Wächterhäuser pointed out that the surface of the mineral pyrite (FeS_2) is positively charged, and he proposed that, since the immediate products of carbon-dioxide fixation are negatively charged, they would be attracted to the pyrite surface. Thus, in Wächterhäuser's model, pyrite formation not only supplied the reducing agent needed for carbon-dioxide fixation, but also the pyrite surface aided the process. Wächterhäuser further proposed an archaic autocatylitic carbon-dioxide fixation cycle, which he visualized as resembling the reductive citric acid cycle found in present-day organisms, but with all reducing agents replaced by FeS + H_2S, with thioester activation replaced by thioacid activation, and carbonyl groups replaced by thioenol groups. The interested reader can find the details of Wächterhäuser's proposals in his papers, which are listed at the end of this chapter.

A similar picture of the origin of life has been proposed by Michael J. Russell and Alan J. Hall in 1997. In this picture "...(i) life emerged as hot, reduced, alkaline, sulphide-bearing submarine seepage waters interfaced with colder, more oxidized, more acid, $Fe^{2+} >> Fe^{3+}$-bearing water at deep (*ca.* 4km) floors of the Hadean ocean *ca.* 4 Gyr ago; (ii) the difference in acidity, temperature and redox potential provided a gradient of pH (*ca.* 4 units), temperature (*ca.* 60°C) and redox potential (*ca.* 500 mV) at the interface of those waters that was sustainable over geological time-scales, providing the continuity of conditions conducive to organic chemical reactions needed for the origin of life..." [9]. Russell, Hall and their coworkers also emphasize the role that may have been played by spontaneously-formed 3-dimensional mineral chambers (bubbles). They visualize these as having prevented the reacting molecules from diffusing away, thus maintaining high concentrations.

[9] See W. Martin and M.J. Russell, *On the origins of cells: a hypothesis for the evolutionary transitions from abiotic geochemistry to chemoautotrophic prokaryotes, and from prokaryotes to nucleated cells*, Philos. Trans. R. Soc. Lond. B Biol. Sci., **358**, 59-85, (2003).

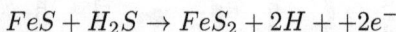

Table 17.2 Energy-yielding reactions of some litho-autotrophic hyperthermophiles. (After K.O. Setter)

Energy-yielding reaction	Genera
$4H_2+CO_2 \rightarrow CH_4+2H_2O$	Methanopyrus, Methanothermus, Methanococcus
$H_2+S^\circ \rightarrow H_2S$	Pyrodictium, Thermoproteus, Pyrobaculum, Acidianus, Stygiolobus
$4H_2+H_2SO_4 \rightarrow H_2S+4H_2O$	Archaeoglobus

Table 17.2 shows the energy-yielding reactions which drive the metabolisms of some organisms which are of very ancient evolutionary origin. All the reactions shown in the table make use of H_2, which could have been supplied by pyrite formation at the time when the organisms evolved. All these organisms are lithoautotrophic, a word which requires some explanation: A heterotrophic organism is one which lives by ingesting energy-rich organic molecules which are present in its environment. By contrast, an autotrophic organism ingests only inorganic molecules. The lithoautotrophs use energy from these inorganic molecules, while the metabolisms of photoautotrophs are driven by energy from sunlight.

Evidence from layered rock formations called "stromatolites", produced by colonies of photosynthetic bacteria, show that photoautotrophs (or phototrophs) appeared on earth at least 3.5 billion years ago. The geological record also supplies approximate dates for other events in evolution. For example, the date at which molecular oxygen started to become abundant in the earth's atmosphere is believed to have been 2.0 billion years ago, with equilibrium finally being established 1.5 billion years in the past. Multicellular organisms appeared very late on the evolutionary and geological time-scale - only 600 million years ago. By collecting such evidence, the Belgian cytologist Christian de Duve has constructed the phylogenetic tree shown in Figure 16.12, showing branching as a function of time. One very interesting feature of this tree is the arrow indicating the transfer of "endosymbionts" from the eubacteria to the eukaryotes. In the next section, we will look in more detail at this important event, which took place about 1.8 billion years ago.

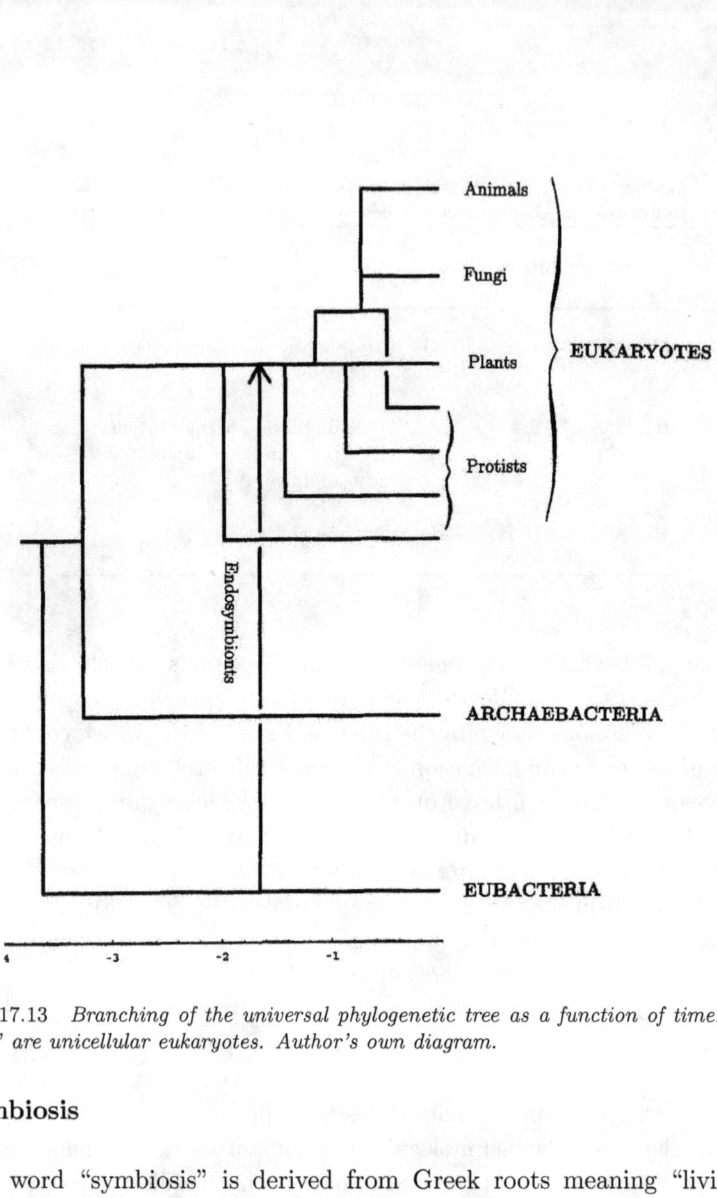

Fig. 17.13 *Branching of the universal phylogenetic tree as a function of time. "Protists" are unicellular eukaryotes. Author's own diagram.*

Symbiosis

The word "symbiosis" is derived from Greek roots meaning "living together". It was coined in 1877 by the German botanist Albert Bernard Frank. By that date, it had become clear that lichens are composite organisms involving a fungus and an alga; but there was controversy concerning whether the relationship was a parasitic one. Was the alga held captive and exploited by the fungus? Or did the alga and the fungus help each other, the former performing photosynthesis, and the latter leeching minerals from the lichen's environment? In introducing the word "symbiosis" (in German, "Symbiotismus"), Prank remarked that "We must bring all the cases where two different species live on or in one another under a comprehensive concept which does not consider the role which the two individuals play but is based on the mere coexistence, and for which the term symbiosis is to be recommended." Thus the concept of symbiosis, as defined by Frank, included all intimate relationships between two or more species, including parasitism at one extreme and "mutualism" at the other. However, as the word is used today, it usually refers to relationships which are mutually beneficial.

Charles Darwin himself had been acutely aware of close and mutually beneficial relationships between organisms of different species. For example, in his work on the fertilization of flowers,he had demonstrated the way in which insects and plants can become exquisitely adapted to each other's needs. However, T.H. Huxley, "Darwin's bulldog", emphasized competition as the predominant force in evolution. "The animal world is on about the same level as a gladiator's show", Huxley wrote in 1888, "The creatures are fairly well treated and set to fight - whereby the strongest, the swiftest and the cunningest live to fight another day. The spectator has no need to turn his thumbs down, as no quarter is given." The view of nature as a sort of "gladiator's contest" dominated the mainstream of evolutionary thought far into the 20th century; but there was also a growing body of opinion which held that symbiosis could be an extremely important mechanism for the generation of new species.

Among the examples of symbiosis studied by Frank were the nitrogen-fixing bacteria living in nodules on the roots of legumes, and the mycorrhizal fungi which live on the roots of forest trees such as oaks, beech and conifers. Frank believed that the mycorrhizal fungi aid in the absorption of nutrients. He distinguished between "ectotrophic" fungi, which form sheaths around the root fibers, and "endotrophic" fungi, which penetrate the root cells. Other examples of symbiosis studied in the 19th century included borderline cases between plants and animals, for example, paramecia, sponges, hydra, planarian worms and sea anemones, all of which frequently contain green bodies capable of performing photosynthesis.

Writing in 1897, the American lichenologist Albert Schneider prophesied that "future studies may demonstrate that.., plasmic bodies (within the eukaryote cell), such as chlorophyll granules, leucoplastids, chromoplastids, chromosomes, centrosomes, nucleoli, etc., are perhaps symbionts comparable to those in less highly specialized symbiosis. Reinke expresses the opinion that it is not wholly unreasonable to suppose that some highly skilled scientist of the future may succeed in cultivating chlorophyll-bodies in artificial media."

19th century cytologists such as Robert Altman, Andreas Schimper and A. Benda focused attention on the chlorophyll-bodies of plants, which Schimper named chloroplasts, and on another type of subcellular granule, present in large numbers in all plant and animal cells, which Benda named mitochondria, deriving the name from the Greek roots mitos (thread) and chrondos (granule). They observed that these bodies seemed to reproduce themselves within the cell in very much the manner that might be expected if they were independent organisms. Schimper suggested that chloroplasts are symbionts, and that green plants owe their origin to a union of a colorless unicellular organism with a smaller chlorophyll-containing species.

The role of symbiosis in evolution continued to be debated in the 20th century. Mitochondria were shown to be centers of respiratory metabolism; and it was discovered that both mitochondria and chloroplasts contain their own DNA. However, opponents of their symbiotic origin pointed out that mitochondria alone cannot synthesize all their own proteins: Some mitochondrial proteins require information from nuclear DNA. The debate was finally settled in the 1970's, when comparative sequencing of ribosomal RNA in the laboratories of Carl Woese, W. Ford Doolittle and Michael Gray showed conclusively that both chloroplasts and mitochondria were originally endosymbionts. The ribosomal RNA sequences showed that chloroplasts had their evolutionary root in the cyanobacteria, a species of eubacteria, while mitochondria were traced to a group of eubacteria called the alpha-proteobacteria. Thus the evolutionary arrow leading from the eubacteria to the eukaryotes can today be drawn with confidence, as in Figure 17.13.

Cyanobacteria are bluish photosynthetic bacteria which often become linked to one another so as to form long chains. They can be found today growing in large colonies on seacoasts in many parts of the world, for example in Baja California on the Mexican coast. The top layer of such colonies consists of the phototrophic cyanobacteria, while the organisms in underlying layers are heterotrophs living off the decaying remains of the cyanobacteria. In the course of time, these layered colonies can become fosilized, and they are the source of the layered rock formations called stromatolites (discussed above). Geological dating of ancient stromatolites has shown that cyanobacteria must have originated at least 3.5 billion years ago.

Cyanobacteria contain two photosystems, each making use of a different type of chlorophyll. Photosystem I, which is thought to have evolved first, uses the energy of light to draw electrons from inorganic compounds, and sometimes also from organic compounds (but never from water). Photosystem II, which evolved later, draws electrons from water. Hydrogen derived from the water is used to produce organic compounds from carbon-dioxide, and molecular oxygen is released into the atmosphere.

Photosystem II never appears alone. In all organisms which possess it, Photosystem II is coupled to Photosystem I, and together the two systems raise electrons to energy levels that are high enough to drive all the processes of metabolism. Dating of ancient stromatolites makes it probable that cyanobacteria began to release molecular oxygen into the earth's atmosphere at least 3.5 billion years ago; yet from other geological evidence we know that it was only 2 billion years ago that the concentration of molecular oxygen began to rise, equilibrium being reached 1.5 billion years ago. It is believed that ferrous iron, which at one time was very abundant, initially absorbed the photosynthetically produced oxygen. This resulted in the time-lag, as well as the ferrous-ferric mixture of iron which is found in the mineral magnetite.

When the concentrations of molecular oxygen began to rise in earnest, most of the unicellular microorganisms living at the time found themselves in deep trouble, faced with extinction, because for them oxygen was a deadly poison; and very many species undoubtedly perished. However, some of the archaebacteria retreated to isolated anaerobic niches where we find them today, while others found ways of detoxifying the poisonous oxygen. Among the eubacteria, the ancestors of the alpha-proteobacteria were particularly good at dealing with oxygen and even turning it to advantage: They developed the biochemical machinery needed for respiratory metabolism.

Meanwhile, during the period between 3.5 and 2.0 billion years before the present, an extremely important evolutionary development had taken place: Branching from the archaebacteria, a line of large[10] heterotrophic unicellular organisms had evolved. They lacked rigid cell walls, and they could surround smaller organisms with their flexible outer membrane, drawing the victims into their interiors to be digested. These new heterotrophs were the ancestors of present-day eukaryotes, and thus they were the ancestors of all multicellular organisms.

Not only are the cells of present-day eukaryotes very much larger than the cells of archaebacteria and eubacteria; their complexity is also astonishing. Every eukaryote cell contains numerous intricate structures: a nucleus,

[10] not large in an absolute sense, but large in relation to the prokaryotes

cytoskeleton, Golgi apparatus, endoplasmic reticulum, mitochondria, peroxisomes, chromosomes, the complex structures needed for mitotic cell division, and so on. Furthermore, the genomes of eykaryotes contain very much more information than those of prokaryotes. How did this huge and relatively sudden increase in complexity and information content take place? According to a growing body of opinion, symbiosis played an important role in this development.

The ancestors of the eukaryotes were in the habit of drawing the smaller prokaryotes into their interiors to be digested. It seems likely that in a few cases the swallowed prokaryotes resisted digestion, multiplied within the host, were transmitted to future generations when the host divided, and conferred an evolutionary advantage, so that the result was a symbiotic relationship. In particular, both mitochondria and chloroplasts have definitely been proved to have originated as endosymbionts. It is easy to understand how the photosynthetic abilities of the chloroplasts (derived from cyanobacteria) could have conferred an advantage to their hosts, and how mitochondria (derived from alpha-proteobacteria) could have helped their hosts to survive the oxygen crisis. The symbiotic origin of other subcellular organelles is less well understood and is currently under intense investigation.

If we stretch the definition of symbiosis a little, we can make the concept include cooperative relationships between organisms of the same species. For example, cyanobacteria join together to form long chains, and they live together in large colonies which later turn into stromatolites. Also, some eubacteria have a mechanism for sensing how many of their species are present, so that they know, like a wolf pack, when it is prudent to attack a larger organism. This mechanism, called "quorum sensing", has recently attracted much attention among medical researchers.

The cooperative behavior of a genus of unicellular eukaryotes called slime molds is particularly interesting because it gives us a glimpse of how multicellular organisms may have originated. The name of the slime molds is misleading, since they are not fungi, but heterotrophic protists similar to amoebae. Under ordinary circumstances, the individual cells wander about independently searching for food, which they draw into their interiors and digest, a process called "phagocytosis". However, when food is scarce, they send out a chemical signal of distress. Researchers have analyzed the molecule which expresses slime mold unhappiness, and they have found it to be cyclic adenosine monophosphate (cAMP). At this signal, the cells congregate and the mass of cells begins to crawl, leaving a slimy trail.

At it crawls, the community of cells gradually develops into a tall stalk, surmounted by a sphere - the "fruiting body". Inside the sphere, spores are produced by a sexual process. If a small animal, for example a mouse, passes by, the spores may adhere to its coat; and in this way they may be transported to another part of the forest where food is more plentiful.

Thus slime molds represent a sort of missing link between unicellular and multicellular or organisms. Normally the cells behave as individualists, wandering about independently, but when challenged by a shortage of food, the slime mold cells join together into an entity which closely resembles a multicellular organism. The cells even seem to exhibit altruism, since those forming the stalk have little chance of survival, and yet they are willing to perform their duty, holding up the sphere at the top so that the spores will survive and carry the genes of the community into the future. We should especially notice the fact that the cooperative behavior of the slime mold cells is coordinated by chemical signals.

Sponges are also close to the borderline which separates unicellular eukaryotes (protists) from multicellular organisms, but they are just on the other side of the border. Normally the sponge cells live together in a multicellular community, filtering food from water. However, if a living sponge is forced through a very fine cloth, it is possible to separate the cells from each other. The sponge cells can live independently for some time; but if many of them are left near to one another, they gradually join together and form themselves into a new sponge, guided by chemical signals. In a refinement of this experiment, one can take two living sponges of different species, separate the cells by passing the sponges through a fine cloth, and afterwards mix all the separated cells together. What happens next is amazing: The two types of sponge cells sort themselves out and become organized once more into two sponges - one of each species.

Slime molds and sponges hint at the genesis of multicellular organisms, whose evolution began approximately 600 million years ago. Looking at the slime molds and sponges, we can imagine how it happened. Some unicellular organisms must have experienced an enhanced probability of survival when they lived as colonies. Cooperative behavior and division of labor within the colonies were rewarded by the forces of natural selection, with the selective force acting on the entire colony of cells, rather than on the individual cell. This resulted in the formation of cellular societies and the evolution of mechanisms for cell differentiation. The division of labor within cellular societies (i.e., differentiation) came to be coordinated by chemical signals which affected the transcription of genetic information and the synthesis of proteins. Each cell within a society of cells possessed the entire genome characteristic of the colony, but once a cell had been assigned its specific role in the economy of the society, part of the information became blocked - that is, it was not expressed in the function of that particular cell. As multicellular organisms evolved, the chemical language of intercellular communication became very much more complex and refined. We will discuss the language of intercellular communication in more detail in a later section.

Geneticists have become increasingly aware that symbiosis has probably played a major role in the evolution of multicellular organisms. We mentioned above that, by means of genetic engineering techniques, transgenic plants and animals can be produced. In these chimeras, genetic material from a foreign species is incorporated into the chromosomes, so that it is inherited in a stable, Mendelian fashion. J.A. Shapiro, one of whose articles is referenced at the end of this chapter, believes that this process also occurs in nature, so that the conventional picture of evolutionary family trees needs to be corrected. Shapiro believes that instead of evolutionary trees, we should perhaps think of webs or networks.

For example, it is tempting to guess that symbiosis may have played a role in the development of the visual system of vertebrates. One of the archaebacteria, the purple *Halobacterium halobium* (recently renamed *Halobacterium salinarum*), is able to perform photosynthesis by means of a protein called bacterial rhodopsin, which transports hydrogen ions across the bacterial membrane. This protein is a near chemical relative of rhodopsin, which combines with a carotinoid to form the "visual purple" used in the vertebrate eye. It is tempting to think that the close similarity of the two molecules is not just a coincidence, and that vertebrate vision originated in a symbiotic relationship between the photosynthetic *Halobacterium* and an aquatic ancestor of the vertebrates, the host being able to sense when the *Halobacterium* was exposed to light and therefore transporting hydrogen ions across its cell membrane.

In this chapter, we have looked at the flow of energy and information in the origin and evolution of life on earth. We have seen how energy-rich molecules were needed to drive the first steps in the origin of life, and how during the evolutionary process, information was preserved, transmitted, and shared between increasingly complex organisms, the whole process being driven by an input of energy. In the next chapter, we will look closely at the relationships between energy and information.

Suggestions for further reading

(1) Richard Hutton, *Biorevolution, DNA and the Ethics of Man-Made Life*, The New American Library, New York (1968).
(2) Martin Ebon, *The Cloning of Man*, The New American Library, New York (1978).
(3) Sheldon Krimsky, *Genetic Alchemy: The Social History of the Recombinant DNA Controversy*, MIT Press, Cambridge Mass (1983).
(4) M. Lappé, *Germs That Won't Die*, Anchor/Doubleday, Garden City N.Y. (1982).
(5) M. Lappé, *Broken Code*, Sierra Club Books, San Francisco (1984).
(6) President's Commission for the Study of Ethical Problems in Medicine and Biomedical and Behavioral Research, *Splicing Life: The Social and Ethical Issues of Genetic Engineering with Human Beings*, U.S. Government Printing Office, Washington D.C. (1982).
(7) U.S. Congress, Office of Technology Assessment, *Impacts of Applied Genetics - Microorganisms, Plants and Animals*, U.S. Government Printing Office, Washington D.C. (1981).
(8) W.T. Reich (editor), *Encyclopedia of Bioethics*, The Free Press, New York (1978).
(9) Martin Brown (editor), *The Social Responsibility of the Scientist*, The Free Press, New York (1970).
(10) B. Zimmerman, *Biofuture*, Plenum Press, New York (1984).
(11) Alvin Toffler, *Future Shock*, Pan Books (1970).
(12) John Lear, *Recombinant DNA, Thc Untold Story*, Crown, New York (1978).
(13) B. Alberts, D. Bray, J. Lewis, M. Raff, K. Roberts and J.D. Watson, *Molecular Biology of the Cell*, Garland, New York (1983).
(14) H. Lodish, A. Berk, S.L. Zipursky, P. Matsudaira, D. Baltimore, and J. Darnell, *Molecular Cell Biology, 4th Edition*, W.H. Freeman, New York, (2000).
(15) Lily Kay, *Who Wrote the Book of Life? A History of the Genetic Code*, Stanford University Press, Stanford CA, (2000).
(16) Sahotra Sarkar (editor), *The Philosophy and History of Molecular Biology*, Kluwer Academic Publishers, Boston, (1996).
(17) James D. Watson et al. *Molecular Biology of the Gene, 4th Edition*, Benjamin-Cummings, (1988).
(18) J.S. Fruton, *Proteins, Enzymes, and Genes*, Yale University Press, New Haven, (1999).

(19) S.E. Lauria, *Life, the Unfinished Experiment*, Charles Scribner's Sons, New York (1973).
(20) A. Lwoff, *Biological Order*, MIT Press, Cambridge MA, (1962).
(21) James D. Watson, *The Double Helix*, Athenium, New York (1968).
(22) F. Crick, *The genetic code*, Scientific American, **202**, 66-74 (1962).
(23) F. Crick, *Central dogma of molecular biology*, Nature, **227**, 561-563 (1970).
(24) David Freifelder (editor), *Recombinant DNA, Readings from the Scientific American*, W.H. Freeman and Co. (1978).
(25) James D. Watson, John Tooze and David T. Kurtz, *Recombinant DNA, A Short Course*, W.H. Freeman, New York (1983).
(26) Richard Hutton, *Biorevolution, DNA and the Ethics of Man-Made Life*, The New American Library, New York (1968).
(27) Martin Ebon, *The Cloning of Man*, The New American Library, New York (1978).
(28) Sheldon Krimsky, *Genetic Alchemy: The Social History of the Recombinant DNA Controversy*, MIT Press, Cambridge Mass (1983).
(29) M. Lappe, *Germs That Won't Die*, Anchor/Doubleday, Garden City N.Y. (1982).
(30) M. Lappe,*Broken Code*, Sierra Club Books, San Francisco (1984).
(31) President's Commission for the Study of Ethical Problems in Medicine and Biomedical and Behavioral Research, *Splicing Life: The Social and Ethical Issues of Genetic Engineering with Human Beings*, U.S. Government Printing Office, Washington D.C. (1982).
(32) U.S. Congress, Office of Technology Assessment, *Impacts of Applied Genetics - Microorganisms, Plants and Animals*, U.S. Government Printing Office, Washington D.C. (1981).
(33) W.T. Reich (editor), *Encyclopedia of Bioethics*, The Free Press, New York (1978).
(34) Martin Brown (editor), *The Social Responsibility of the Scientist*, The Free Press, New York (1970).
(35) B. Zimmerman, *Biofuture*, Plenum Press, New York (1984).
(36) John Lear, *Recombinant DNA, The Untold Story*, Crown, New York (1978).
(37) B. Alberts, D. Bray, J. Lewis, M. Raff, K. Roberts and J.D. Watson, *Molecular Biology of the Cell*, Garland, New York (1983).
(38) C. Woese, *The Genetic Code; The Molecular Basis for Genetic Expression*, Harper and Row, New York, (1967).

(39) F.H.C. Crick, *The Origin of the Genetic Code*, J. Mol. Biol. **38**, 367-379 (1968).
(40) M.W. Niernberg, *The genetic code: II*, Scientific American, **208**, 80-94 (1962).
(41) L.E. Orgel, *Evolution of the Genetic Apparatus*, J. Mol. Biol. **38**, 381-393 (1968).
(42) Melvin Calvin, *Chemical Evolution Towards the Origin of Life, on Earth and Elsewhere*, Oxford University Press (1969).
(43) R. Shapiro, *Origins: A Skeptic's Guide to the Origin of Life*, Summit Books, New York, (1986).
(44) J. William Schopf, *Earth's earliest biosphere: its origin and evolution*, Princeton University Press, Princeton, N.J., (1983).
(45) J. William Schopf (editor), *Major Events in the History of Life*, Jones and Bartlet, Boston, (1992).
(46) Robert Rosen, *Life itself: a comprehensive inquiry into the nature, origin and fabrication of life*, Colombia University Press, (1991).
(47) R.F. Gesteland, T.R Cech, and J.F. Atkins (editors), *The RNA World, 2nd Edition*, Cold Spring Harbor Laboratory Press, Cold Spring Harbor, New York, (1999).
(48) C. de Duve, Blueprint of a Cell, Niel Patterson Publishers, Burlington N.C., (1991).
(49) C. de Duve, *Vital Dust; Life as a Cosmic Imperative*, Basic Books, New York, (1995).
(50) F. Dyson, *Origins of Life*, Cambridge University Press, (1985).
(51) S.A. Kaufman, *Antichaos and adaption*, Scientific American, 265, 78-84, (1991).
(52) S.A. Kauffman, *The Origins of Order*, Oxford University Press, (1993).
(53) F.J. Varela and J.-P. Dupuy, *Understanding Origins: Contemporary Views on the Origin of Life, Mind and Society*, Kluwer, Dordrecht, (1992).
(54) Stefan Bengtson (editor) *Early Life on Earth; Nobel Symposium No. 84*, Colombia University Press, New York, (1994).
(55) Herrick Baltscheffsky, *Origin and Evolution of Biological Energy Conversion*, VCH Publishers, New York, (1996).
(56) J. Chilea-Flores, T. Owen and F. Raulin (editors), *First Steps in the Origin of Life in the Universe*, Kluwer, Dordrecht, (2001).
(57) R.E. Dickerson, Nature **283**, 210-212 (1980).
(58) R.E. Dickerson, Scientific American **242**, 136-153 (1980).

(59) C.R. Woese, *Archaebacteria*, Scientific American **244**, 98-122 (1981).
(60) N. Iwabe, K. Kuma, M. Hasegawa, S. Osawa and T. Miyata, *Evolutionary relationships of archaebacteria, eubacteria, and eukaryotes inferred phylogenetic trees of duplicated genes*, Proc. Nat. Acad. Sci. USA **86**, 9355-9359 (1989).
(61) C.R. Woese, O. Kundler, and M.L. Wheelis, *Towards a Natural System of Organisms: Proposal for the Domains Archaea, Bacteria and Eucaria*, Proc. Nat. Acad. Sci. USA **87**, 4576-4579 (1990).
(62) W. Ford Doolittle, Phylogenetic Classification and the Universal Tree, Science, **284**, (1999).
(63) G. Wächterhäuser, *Pyrite formation, the first energy source for life: A hypothesis*, Systematic and Applied Microbiology **10**, 207-210 (1988).
(64) G. Wächterhäuser, *Before enzymes and templates: Theory of surface metabolism*, Microbiological Reviews, **52**, 452-484 (1988).
(65) G. Wächterhäuser, Evolution of the first metabolic cycles, Proc. Nat. Acad. Sci. USA **87**, 200-204 (1990).
(66) G. Wächterhäuser, *Groundworks for an evolutionary biochemistry the iron-sulfur world*, Progress in Biophysics and Molecular Biology **58**, 85-210 (1992).
(67) M.J. Russell and A.J. Hall, *The emergence of life from iron monosulphide bubbles at a submarine hydrothermal redox and pH front* J. Geol. Soc. Lond. **154**, 377-402, (1997).
(68) L.H. Caporale (editor), *Molecular Strategies in Biological Evolution*, Ann. N.Y. Acad. Sci., May 18, (1999).
(69) W. Martin and M.J. Russell, *On the origins of cells: a hypothesis for the evolutionary transitions from abiotic geochemistry to chemoautotrophic prokaryotes, and from prokaryotes to nucleated cells*, Philos. Trans. R. Soc. Lond. B Biol. Sci., **358**, 59-85, (2003).
(70) Werner Arber, *Elements in Microbal Evolution*, J. Mol. Evol. **33, 4** (1991).
(71) Michael Gray, *The Bacterial Ancestry of Plastids and Mitochondria*, BioScience, **33**, 693-699 (1983).
(72) Michael Grey, *The Endosymbiont Hypothesis Revisited*, International Review of Cytology, **141**, 233-257 (1992).
(73) Lynn Margulis and Dorian Sagan, *Microcosmos: Four Billion Years of Evolution from Our Microbal Ancestors*, Allan and Unwin, London, (1987).
(74) Lynn Margulis and Rene Fester, eds., *Symbiosis as as Source of Evolutionary Innovation: Speciation and Morphogenesis*, MIT Press, (1991).

(75) Charles Mann, *Lynn Margulis: Science's Unruly Earth Mother*, Science, **252**, 19 April, (1991).

(76) Jan Sapp, *Evolution by Association; A History of Symbiosis*, Oxford University Press, (1994).

(77) J.A. Shapiro, *Natural genetic engineering in evolution*, Genetics, **86**, 99-111 (1992).

(78) E.M. De Robertis et al., *Homeobox genes and the vertebrate body plan*, Scientific American, July, (1990).

(79) J.S. Schrum, T.F. Zhu and J.W. Szostak, *The origins of cellular life*, Cold Spring Harb. Perspect. Biol., May 19 (2010).

(80) I. Budin and J.W. Szostak, *Expanding Roles for Diverse Physical Phenomena During the Origin of Life*, Annu. Rev. Biophys., **39**, 245-263, (2010).

Chapter 18

ARTIFICIAL INTELLIGENCE

The first computers

The dramatic development of molecular biology during the period following World War II would have been impossible without X-ray crystallography; and the application of X-ray crystallography to large biological molecules would have been impossible without another equally dramatic postwar development - the advent of high-speed electronic digital computers. The first programmable universal computers were completed in the middle 1940's; but they had their roots in the much earlier ideas of Blaise Pascal (1623-1662), Gottfried Wilhelm Leibniz (1646-1716), Joseph Marie Jacquard (1752-1834) and Charles Babbage (1791-1871).

In 1642, the distinguished French mathematician and philosopher, Blaise Pascal, completed a working model of a machine for adding and subtracting. According to tradition, the idea for his "calculating box" came to Pascal when, as a young man of 17, he sat thinking of ways to help his father (who was a tax collector). In describing his machine, Pascal wrote:

"I submit to the public a small machine of my own invention, by means of which you alone may, without any effort, perform all the operations of arithmetic, and may be relieved of the work which has often times fatigued your spirit when you have worked with the counters or with the pen."

Pascal's machine, which worked by means of toothed wheels, was much improved by Leibniz, who constructed a mechanical calculator which, besides adding and subtracting, could also multiply and divide. His first machine was completed in 1671; and Leibniz' description of it, written in Latin, is preserved in the Royal Library at Hanover:

"There are two parts of the machine, one designed for addition (and subtraction), and the other designed for multiplication (and division); and

Fig. 18.1 *Blaise Pascal (1623-1662), who invented the first calculating machine. Public domain, Wikimedia Commons.*

they should fit together. The adding (and subtracting) machine coincides completely with the calculating box of Pascal. Something, however, must be added for the sake of multiplication..."

"The wheels which represent the multiplicand are all of the same size, equal to that of the wheels of addition, and are also provided with ten teeth which, however, are movable so that at one time there should protrude 5, at another 6 teeth, etc., according to whether the multiplicand is to be represented five times or six times, etc."

"For example, the multiplicand 365 consists of three digits, 3, 6, and 5. Hence the same number of wheels is to be used. On these wheels, the multiplicand will be set if from the right wheel there protrude 5 teeth, from the middle wheel 6, and from the left wheel 3."

By 1810, calculating machines based on Leibniz' design were being manufactured commercially; and mechanical calculators of a similar design could be found in laboratories and offices until the 1960's.

The idea of a programmable universal computer is due to the English mathematician, Charles Babbage, who was the Lucasian Professor of Mathematics at Cambridge University. (In the 17th century, Isaac Newton held this post, and in the 20th century, P.A.M. Dirac also held it.)

In 1812, Babbage conceived the idea of constructing a machine which could automatically produce tables of functions, provided that the functions could be approximated by polynomials. He constructed a small machine, which was able to calculate tables of quadratic functions to eight decimal places; and in 1832 he demonstrated this machine to the Royal Society and to representatives of the British government.

The demonstration was so successful that Babbage secured financial support for the construction of a large machine which would tabulate sixth-order polynomials to twenty decimal places. The large machine was never completed, and twenty years later, after having spent seventeen thousand pounds on the project, the British government withdrew its support. The reason why Babbage's large machine never was finished can be understood from the following account by Lord Moulton of a visit to the mathematician's laboratory:

"One of the sad memories of my life is a visit to the celebrated mathematician and inventor, Mr. Babbage. He was far advanced in age, but his mind was still as vigorous as ever. He took me through his workrooms."

"In the first room I saw the parts of the original Calculating Machine, which had been shown in an incomplete state many years before, and had even been put to some use. I asked him about its present form. 'I have not finished it, because in working at it, I came on the idea of my Analytical Machine, which would do all that it was capable of doing, and much more. Indeed, the idea was so much simpler that it would have taken more work to complete the Calculating Machine than to design and construct the other in its entirety; so I turned my attention to the Analytical Machine.'"

"After a few minutes talk, we went into the next workroom, where he showed me the working of the elements of the Analytical Machine. I asked if I could see it. 'I have never completed it,' he said, 'because I hit upon the idea of doing the same thing by a different and far more effective method, and this rendered it useless to proceed on the old lines.'"

"Then we went into a third room. There lay scattered bits of mechanism, but I saw no trace of any working machine. Very cautiously I approached

Fig. 18.2 *19th century engraving of Charles Babbage. Public domain, Wikimedia Commons.*

the subject, and received the dreaded answer: 'It is not constructed yet, but I am working at it, and will take less time to construct it altogether than it would have taken to complete the Analytical Machine from the stage in which I left it.' I took leave of the old man with a heavy heart."

Babbage's first calculating machine was a special-purpose mechanical computer, designed to tabulate polynomial functions; and he abandoned this design because he had hit on the idea of a universal programmable computer. Several years earlier, the French inventor, Joseph Marie Jacquard, had constructed an automatic loom in which punched cards were used to control the warp threads. Inspired by Jacquard's invention, Babbage planned to use punched cards to program his universal computer.

(Jacquard's looms could be programmed to weave extremely complex patterns: A portrait of the inventor, woven on one of his looms in Lyons, hung in Babbage's drawing room.)

One of Babbage's frequent visitors was Augusta Ada, Countess of Lovelace (1815-1852), the daughter of Lord and Lady Byron. She was a mathematician of considerable ability, and it is through her lucid descriptions that we know how Babbage's never-completed Analytical Machine was to have worked.

The next step towards modern computers was taken by Hermann Hollerith, a statistician working for the United States Bureau of the Census. He invented electromechanical machines for reading and sorting data punched onto cards. Hollerith's machines were used to analyse the data from the 1890 United States Census; and similar machines began to be manufactured and used in business and administration.

In 1937, Howard Aiken, of Harvard University, became interested in combining Babbage's ideas with some of the techniques which had developed from Hollerith's punched card machines. He approached the International Business Machine Corporation, the largest manufacturer of punched card equipment, with a proposal for the construction of a large, automatic, programmable calculating machine.

Aiken's machine, the Automatic Sequence Controlled Calculator (ASCC), was completed in 1944 and presented to Harvard University. Based on geared wheels, in the Pascal-Leibniz-Babbage tradition, ASCC had more than three quarters of a million parts and used 500 miles of wire. ASCC was unbelievably slow by modern standards - it took three-tenths of a second to perform an addition - but it was one of the first programmable general-purpose digital computers ever completed. It remained in continuous use, day and night, for fifteen years.

In the ASCC, binary numbers were represented by relays, which could be either on or off. The on position represented 1, while the off position represented 0, these being the only two digits required to represent numbers in the binary (base 2) system. Electromechanical calculators similar to ASCC were developed independently by Konrad Zuse in Germany and by George R. Stibitz at Bell Telephone Laboratories.

Fig. 18.3 *One of Babbage's frequent visitors was Agusta Ada, Countess of Lovelace, seen here in a painting by Margaret Carpenter. She was an excellent mathematician, and it is through her lucid writings that we know how Babbage's universal calculating machine was to have worked. The programming language ADA is named after her. Public domain, Wikimedia Commons.*

Meanwhile, at Iowa State University, the physicist John V. Atanasoff and his student, Clifford E. Berry, had developed a special-purpose electronic digital computer designed to solve large sets of simultaneous equations. The Atanasoff-Berry Computer (ABC) was completed in 1943. It used capacitors as a memory device; but since they gradually lost their charge, Atanasoff included a device for periodically "jogging" the memory (i.e. recharging the capacitors). Because of a relatively minor fault with the input-output system, ABC was never used for practical computational problems; and Atanasoff and Berry had to abandon it to work on research related to the war effort.

Like ASCC, ABC represented numbers in binary notation. Although it was a special-purpose machine, ABC represented a milestone in computing: It was the first electronic digital computer. (Analogue computers, such as the Differential Analyser designed by Vannevar Bush at M.I.T., have a separate history, and we will not discuss them here.)

In 1943, the electronic digital computer, Colossus, was completed in England by a group inspired by the mathematicians A.M. Turing, M.H.A. Newman. Colossus was the first large-scale electronic computer. It was used to break the German Enigma code; and it thus affected the course of World War II.

In 1946, ENIAC (Electronic Numerical Integrator and Calculator) became operational. This general-purpose computer, designed by J.P. Eckert and J.W. Mauchley of the University of Pennsylvania, contained 18,000 vacuum tubes, one or another of which was often out of order. However, during the periods when all its vacuum tubes were working, an electronic computer like Colossus or ENIAC could shoot ahead of an electromechanical machine (such as ASCC) like a hare outdistancing a tortoise.

Cybernetics

The word "Cybernetics", was coined by the American mathematician Norbert Wiener (1894-1964) and his colleagues, who defined it as "the entire field of control and communication theory, whether in the machine or in the animal". Wiener derived the word from the Greek term for "steersman".

Norbert Wiener began life as a child prodigy: He entered Tufts University at the age of 11 and received his Ph.D. from Harvard at 19. He later became a professor of mathematics at the Massachusetts Institute of Technology. In 1940, with war on the horizon, Wiener sent a memorandum to Vannevar Bush, another MIT professor who had done pioneering work with analogue computers, and had afterwards become the chairman of the U.S. National Defense Research Committee. Wiener's memorandum urged the American government to support the design and construction of electronic digital computers, which would make use of binary numbers, vacuum tubes, and rapid memories. In such machines, the memorandum emphasized, no human intervention should be required except when data was to be read into or out of the machine.

Like Leo Szilard, John von Neumann, Claude Shannon and Erwin Schrödinger, Norbert Wiener was aware of the relation between information and entropy. In his 1948 book Cybernetics he wrote: "...we had to develop a statistical theory of the amount of information, in which the unit amount of information was that transmitted by a single decision between equally probable alternatives. This idea occurred at about the same time to several writers, among them the statistician R.A. Fisher, Dr. Shannon of Bell Telephone Laboratories, and the author. Fisher's motive in studying this subject is to be found in classical statistical theory; that of Shannon in the problem of coding information; and that of the author in the problem of noise and message in electrical filters... The notion of the amount of information attaches itself very naturally to a classical notion in statistical mechanics: that of entropy. Just as the amount of information in a system is a measure of its degree of organization, so the entropy of a system is a measure of its degree of disorganization; and the one is simply the negative of the other."

During World War II, Norbert Wiener developed automatic systems for control of anti-aircraft guns. His systems made use of feedback loops closely analogous to those with which animals coordinate their movements. In the early 1940's, he was invited to attend a series of monthly dinner parties organized by Arturo Rosenblueth, a professor of physiology at Harvard University. The purpose of these dinners was to promote discussions and collaborations between scientists belonging to different disciplines. The discussions which took place at these dinners made both Wiener and Rosenblueth aware of the relatedness of a set of problems that included homeostasis and feedback in biology, communication and control mechanisms in neurophysiology, social communication among animals (or humans), and control and communication involving machines.

Wiener and Rosenblueth therefore tried to bring together workers in the relevant fields to try to develop common terminology and methods. Among the many people whom they contacted were the anthropologists Gregory Bateson and Margaret Mead, Howard Aiken (the designer of the Automatic Sequence Controlled Calculator), and the mathematician John von Neumann. The Josiah Macy Jr. Foundation sponsored a series of ten yearly meetings, which continued until 1949 and which established cybernetics as a new research discipline. It united areas of mathematics, engineering, biology, and sociology which had previously been considered unrelated. Among the most important participants (in addition to Wiener, Rosenblueth, Bateson, Mead, and von Neumann) were Heinz von Foerster, Kurt Lewin, Warren McCulloch and Walter Pitts. The Macy conferences were small and informal, with an emphasis on discussion as opposed to the presentation of formal papers. A stenographic record of the last five conferences has been published, edited by von Foerster. Transcripts of the discussions give a vivid picture of the enthusiastic and creative atmosphere of the meetings. The participants at the Macy Conferences perceived Cybernetics as a much-needed bridge between the natural sciences and the humanities. Hence their enthusiasm. Weiner's feedback loops and von Neumann's theory of games were used by anthropologists Mead and Bateson to explain many aspects of human behavior.

Microelectronics

During the summer of 1946, a course on "The Theory and Techniques of Electronic Digital Computers" was given at the University of Pennsylvania. The ideas put forward in this course had been worked out by a group of mathematicians and engineers headed by J.P. Eckert, J.W. Mauchley and John von Neumann, and these ideas very much influenced all subsequent computer design.

The problem of unreliable vacuum tubes was solved in 1948 by John Bardeen, William Shockley and Walter Brattain of the Bell Telephone Laboratories. Application of quantum theory to solids had led to an understanding of the electronic properties of crystals. Like atoms, crystals were found to have allowed and forbidden energy levels.

The allowed energy levels for an electron in a crystal were known to form bands; i.e., some energy ranges with a quasi-continuum of allowed states (allowed bands), and other energy ranges with none (forbidden bands). The lowest allowed bands were occupied by electrons, while higher bands were empty. The highest filled band was called the valence band, and the lowest empty band was called the conduction band.

According to quantum theory, whenever the valence band of a crystal is only partly filled, the crystal is a conductor of electricity; but if the valence band is completely filled with electrons, the crystal is an electrical insulator. (A completely filled band is analogous to a room so packed with people that none of them can move.)

In addition to explaining conductors and insulators, quantum theory yielded an understanding of semiconductors - crystals where the valence band is completely filled with electrons, but where the energy gap between the conduction band and the valence band is relatively small. For example, crystals of the elements silicon and germanium are semiconductors. For such a crystal, thermal energy is sometimes enough to lift an electron from the valence band to the conduction band.

Bardeen, Shockley and Brattain found ways to control the conductivity of germanium crystals by injecting electrons into the conduction band, or alternatively by removing electrons from the valence band. They could do this by forming junctions between crystals "doped" with appropriate impurities, and by injecting electrons with a special electrode. The semiconducting crystals whose conductivity was controlled in this way could be used as electronic valves, in place of vacuum tubes.

By the 1960's, replacement of vacuum tubes by transistors in electronic computers had led not only to an enormous increase in reliability and a great reduction in cost, but also to an enormous increase in speed. It was found that the limiting factor in computer speed was the time needed for an electrical signal to propagate from one part of the central processing unit to another. Since electrical impulses propagate with the speed of light, this time is extremely small; but nevertheless, it is the limiting factor in the speed of electronic computers.

In order to reduce the propagation time, computer designers tried to make the central processing units very small; and the result was the development of integrated circuits and microelectronics. (Another motive for miniaturization of electronics came from the requirements of space exploration.)

Integrated circuits were developed, in which single circuit elements were not manufactured separately, but instead the whole circuit was made at one time. An integrated circuit is a multilayer sandwich-like structure, with conducting, resisting and insulating layers interspersed with layers of germanium or silicon, "doped" with appropriate impurities. At the start of the manufacturing process, an engineer makes a large drawing of each layer. For example, the drawing of a conducting layer would contain pathways which fill the role played by wires in a conventional circuit, while the remainder of the layer would consist of areas destined to be etched away by acid.

The next step is to reduce the size of the drawing and to multiply it photographically. The pattern of the layer is thus repeated many times, like the design on a piece of wallpaper. The multiplied and reduced drawing is then focused through a reversed microscope onto the surface to be etched.

Successive layers are built up by evaporating or depositing thin films of the appropriate substances onto the surface of a silicon or germanium wafer. If the layer being made is to be conducting, the surface might consist of an extremely thin layer of copper, covered with a photosensitive layer called a "photoresist". On those portions of the surface receiving light from the pattern, the photoresist becomes insoluble, while on those areas not receiving light, the photoresist can be washed away.

The surface is then etched with acid, which removes the copper from those areas not protected by photoresist. Each successive layer of a wafer is made in this way, and finally the wafer is cut into tiny "chips", each of which corresponds to one unit of the wallpaper-like pattern. Although the area of a chip may be much smaller than a square centimeter, the chip can contain an extremely complex circuit.

In 1965, only four years after the first integrated circuits had been produced, Dr. Gordon E. Moore, one of the founders of Intel, made a famous prediction which has come to be known as "Moore's Law". He predicted that the number of transistors per integrated circuit would double every two years, and that this trend would continue through 1975. In fact, the general trend predicted by Moore has continued for a much longer time. Although the number of transistors per unit area has not continued to double every two years, the logic density (bits per unit area) has done so, and thus a modified version of Moore's law still holds today. How much longer the trend can continue remains to be seen. Physical limits to miniaturization of transistors of the present type will soon be reached; but there is hope that further miniaturization can be achieved through "quantum dot" technology, molecular switches, and autoassembly, as will be discussed in Chapter 8.

A typical programmable minicomputer or "microprocessor", manufactured in the 1970's, could have 30,000 circuit elements, all of which were contained on a single chip. By 1989, more than a million transistors were being placed on a single chip; and by 2000, the number reached 42,000,000.

As a result of miniaturization and parallelization, the speed of computers rose exponentially. In 1960, the fastest computers could perform a hundred thousand elementary operations in a second. By 1970, the fastest computers took less than a second to perform a million such operations. In 1987, a massively parallel computer, with 566 parallel processors, called GFll was designed to perform 11 billion floating-point operations per second (flops). By 2002 the fastest computer performed 40 at teraflops, making use of 5120 parallel CPU's.

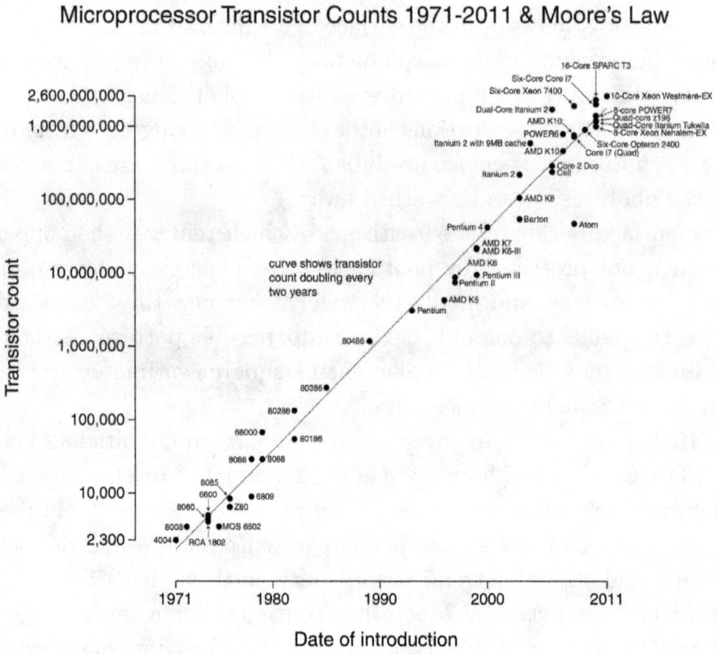

Fig. 18.4 *In 1965, George E. Moore, one of the co-founders of Intel, predicted that the number of transistors that could be placed on an integrated circuit would double every two years, and that this trend would continue until 1975. In fact, as is shown by the figure, the trend has continued much longer than that. In 2011, the number of transistors per chip reached 2.6 billion. (After Wgsimon, Wikimedia Commons)*

Computer disk storage has also undergone a remarkable development. In 1987, the magnetic disk storage being produced could store 20 million bits of information per square inch; and even higher densities could be achieved by optical storage devices. Storage density has until followed a law similar to Moore's law.

In the 1970's and 1980's, computer networks were set up linking machines in various parts of the world. It became possible (for example) for a scientist in Europe to perform a calculation interactively on a computer in the United States just as though the distant machine were in the same room; and two or more computers could be linked for performing large calculations. It also became possible to exchange programs, data, letters and manuscripts very rapidly through the computer networks.

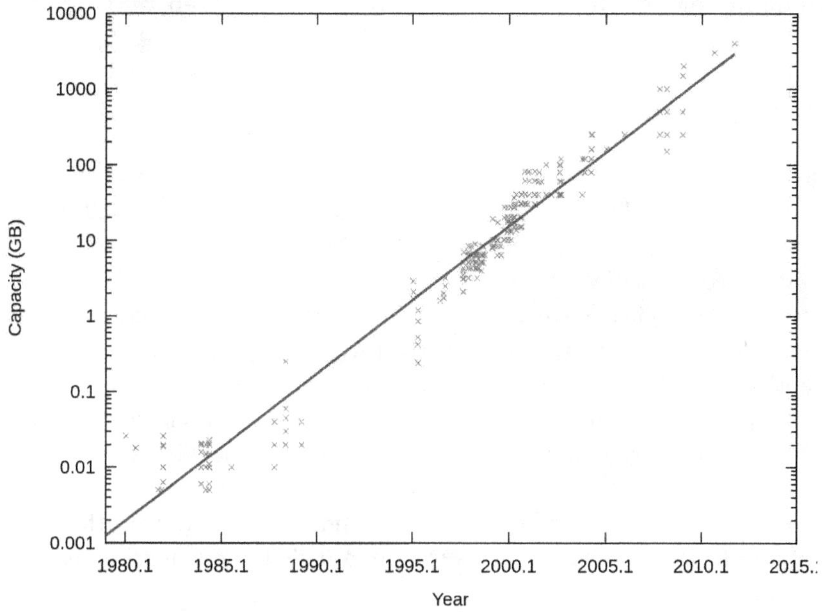

Fig. 18.5 *A logarithmic plot of the increase in PC hard-drive capacity in gigabytes. An extrapolation of the rate of increase predicts that the individual capacity of a commercially available PC will reach 10,000 gigabytes by 2015, i.e. 10,000,000,000,000 bytes. (After Hankwang and Rentar, Wikimedia Commons)*

The exchange of large quantities of information through computer networks was made easier by the introduction of fiber optics cables. By 1986, 250,000 miles of such cables had been installed in the United States. If a ray of light, propagating in a medium with a large refractive index, strikes the surface of the medium at a grazing angle, then the ray undergoes total internal reflection. This phenomenon is utilized in fiber optics: A light signal can propagate through a long, hairlike glass fiber, following the bends of the fiber without losing intensity because of total internal reflection. However, before fiber optics could be used for information transmission over long distances, a technological breakthrough in glass manufacture was needed, since the clearest glass available in 1940 was opaque in lengths more than 10 m. Through studies of the microscopic properties of glasses, the problem of absorption was overcome. By 1987, devices were being manufactured commercially that were capable of transmitting information through fiber-optic cables at the rate of 1.7 billion bits per second.

The history of the Internet and World Wide Web

The history of the Internet began in 1961, when Leonard Kleinrock, a student at MIT, submitted a proposal for Ph.D. thesis entitled "Information Flow in Large Communication Nets". In his statement of the problem, Kleinrock wrote: "The nets under consideration consist of nodes, connected to each other by links. The nodes receive, sort, store, and transmit messages that enter and leave via the links. The links consist of one-way channels, with fixed capacities. Among the typical systems which fit this description are the Post Office System, telegraph systems, and satellite communication systems." Kleinrock's theoretical treatment of package switching systems anticipated the construction of computer networks which would function on a principle analogous to a post office rather than a telephone exchange: In a telephone system, there is a direct connection between the sender and receiver of information. But in a package switching system, there is no such connection - only the addresses of the sender and receiver on the package of information, which makes its way from node to node until it reaches its destination.

Further contributions to the concept of package switching systems and distributed communications networks were made by J.C.R. Licklider and W. Clark of MIT in 1962, and by Paul Baran of the RAND corporation in 1964. Licklider visualized what he called a "Galactic Network", a globally interconnected network of computers which would allow social interactions and interchange of data and software throughout the world. The distributed computer communication network proposed by Baran was motivated by the desire to have a communication system that could survive a nuclear war. The Cold War had also provoked the foundation (in 1957) of the Advanced Research Projects Agency (ARPA) by the U.S. government as a response to the successful Russian satellite "Sputnik".

In 1969, a 4-node network was tested by ARPA. It connected computers at the University of California divisions at Los Angeles and Santa Barbara with computers at the Stanford Research Institute and the University of Utah. Describing this event, Leonard Kleinrock said in an interview: "We set up a telephone connection between us and the guys at SRI. We typed the L and we asked on the phone 'Do you see the L?' 'Yes we see the L', came the response. We typed the 0 and we asked 'Do you see the 0?' 'Yes we see the O.' Then we typed the G and the system crashed." The ARPANET (with 40 nodes) performed much better in 1972 at the Washington Hilton Hotel where the participants at a Conference on Computer Communications were invited to test it.

Although the creators of ARPANET visualized it as being used for long-distance computations involving several computers, they soon discovered that social interactions over the Internet would become equally important if not more so. An electronic mail system was introduced in the early 1970's, and in 1976 Queen Elizabeth II of the United Kingdom became one of the increasing number of e-mail users.

In September, 1973, Robert F. Kahn and Vinton Cerf presented the basic ideas of the Internet at a meeting of the International Network Working Group at the University Sussex in Brighton, England. Among these principles was the rule that the networks to be connected should not be changed internally. Another rule was that if a packet did not arrive at its destination, it would be retransmitted from its original source. No information was to be retained by the gateways used to connect networks; and finally there was to be no global control of the Internet at the operations level.

Computer networks devoted to academic applications were introduced in the 1970's and 1980's, both in England, the United States and Japan. The Joint Academic Network (JANET) in the U.K. had its counterpart in the National Science Foundation's network (NSFNET) in America and Japan's JUNET (Japan Unix Network). Internet traffic is approximately doubling each year,[1] and it is about to overtake voice communication in the volume of information transferred.

In March, 2011, there were more than two billion Internet users in the world. In North America they amounted to 78.3 % of the total population, in Europe 58.3 % and worldwide, 30.2 %. Another index that can give us an impression of the rate of growth of digital data generation and exchange is the "digital universe", which is defined to be the total volume of digital information that human information technology creates and duplicates in a year. In 2011 the digital universe reached 1.2 zettabytes, and it is projected to quadruple by 2015. A zettabyte is 10^{21} bytes, an almost unimaginable number, equivalent to the information contained in a thousand trillion books, enough books to make a pile that would stretch twenty billion kilometers.

[1]In the period 1995-1996, the rate of increase was even faster - a doubling every four months

Table 18.1 Historical total world Internet traffic (after Cisco Visual Networking Index Forecast). 1 terrabyte = 1,000,000,000,000 bytes

year	terabytes per month
1990	1
1991	2
1992	4
1993	10
1994	20
1995	170
1996	1,800
1997	5,000
1998	11,000
1999	26,000
2000	75,000
2001	175,000
2002	358,000
2003	681,000
2004	1,267,000
2005	2,055,000
2006	3,339,000
2007	5,219,000
2008	7,639,000
2009	10,676,000
2010	14,984,000

Self-reinforcing information accumulation

Humans have been living on the earth for roughly two million years (more or less, depending on where one draws the line between our human and prehuman ancestors). During almost all of this time, our ancestors lived by hunting and food-gathering. They were not at all numerous, and did not stand out conspicuously from other animals. Then, suddenly, during the brief space of ten thousand years, our species exploded in numbers from a few million to seven billion, populating all parts of the earth, and even setting foot on the moon. This population explosion, which is still going on, has been the result of dramatic cultural changes. Genetically we are almost identical with our hunter-gatherer ancestors, who lived ten thousand years ago, but cultural evolution has changed our way of life beyond recognition.

Beginning with the development of speech, human cultural evolution began to accelerate. It started to move faster with the agricultural revolution, and faster still with the invention of writing and printing. Finally, modern science has accelerated the rate of social and cultural change to a completely unprecedented speed.

The growth of modern science is accelerating because knowledge feeds on itself. A new idea or a new development may lead to several other innovations, which can in turn start an avalanche of change. For example, the quantum theory of atomic structure led to the invention of transistors, which made high-speed digital computers possible. Computers have not only produced further developments in quantum theory; they have also revolutionized many other fields.

The self-reinforcing accumulation of knowledge - the information explosion - which characterizes modern human society is reflected not only in an explosively-growing global population, but also in the number of scientific articles published, which doubles roughly every ten years. Another example is Moore's law - the doubling of the information density of integrated circuits every two years. Yet another example is the explosive growth of Internet traffic shown in Table 17.1.

The Internet itself is the culmination of a trend towards increasing societal information exchange - the formation of a collective human consciousness. This collective consciousness preserves the observations of millions of eyes, the experiments of millions of hands, the thoughts of millions of brains; and it does not die when the individual dies.

Automation

During the last three decades, the cost of computing has decreased exponentially by between twenty and thirty percent per year. Meanwhile, the computer industry has grown exponentially by twenty percent per year (faster than any other industry). The astonishing speed of this development has been matched by the speed with which computers have become part of the fabric of science, engineering, industry, commerce, communications, transport, publishing, education and daily life in the industrialized parts of the world.

The speed, power and accuracy of computers has revolutionized many branches of science. For example, before the era of computers, the determination of a simple molecular structure by the analysis of X-ray diffraction data often took years of laborious calculation; and complicated structures were completely out of reach. In 1949, however, Dorothy Crowfoot Hodgkin used an electronic computer to work out the structure of penicillin from X-ray data. This was the first application of a computer to a biochemical problem; and it was followed by the analysis of progressively larger and more complex structures.

Proteins, DNA, and finally even the detailed structures of viruses were studied through the application of computers in crystallography. The enormous amount of data needed for such studies was gathered automatically by computer-controlled diffractometers; and the final results were stored in magnetic-tape data banks, available to users through computer networks.

The application of quantum theory to chemical problems is another field of science which owes its development to computers. When Erwin Schrödinger wrote down his wave equation in 1926, it became possible, in principle, to calculate most of the physical and chemical properties of matter. However, the solutions to the Schrödinger equation for many-particle systems can only be found approximately; and before the advent of computers, even approximate solutions could not be found, except for the simplest systems.

When high-speed electronic digital computers became widely available in the 1960's, it suddenly became possible to obtain solutions to the Schrödinger equation for systems of chemical and even biochemical interest. Quantum chemistry (pioneered by such men as J.C. Slater, R.S. Mullikin, D.R. Hartree, V. Fock, J.H. Van Vleck, L. Pauling, E.B. Wilson, P.O. Löwdin, E. Clementi, C.J. Ballhausen and others) developed into a rapidly-growing field, as did solid state physics. Through the use of computers, it became possible to design new materials with desired chemical, mechanical, electrical or magnetic properties. Applying computers to the analysis of reactive scattering experiments, D. Herschbach, J. Polanyi and Y. Lee were able to achieve an understanding of the dynamics of chemical reactions.

The successes of quantum chemistry led Albert Szent-Györgyi, A. and B. Pullman, H. Scheraga and others to pioneer the fields of quantum biochemistry and molecular dynamics. Computer programs for drug design were developed, as well as molecular-dynamics programs which allowed the conformations of proteins to be calculated from a knowledge of their amino acid sequences. Studies in quantum biochemistry have yielded insights into the mechanisms of enzyme action, photosynthesis, active transport of ions across membranes, and other biochemical processes.

In medicine, computers began to be used for monitoring the vital signs of critically ill patients, for organizing the information flow within hospitals, for storing patients' records, for literature searches, and even for differential diagnosis of diseases.

The University of Pennsylvania has developed a diagnostic program called INTERNIST-1, with a knowledge of 577 diseases and their interrelations, as well as 4,100 signs, symptoms and patient characteristics. This program was shown to perform almost as well as an academic physician in diagnosing difficult cases. QMR (Quick Medical Reference), a microcomputer adaptation of INTERNIST-1, incorporates the diagnostic functions of the earlier program, and also offers an electronic textbook mode.

Beginning in the 1960's, computers played an increasingly important role in engineering and industry. For example, in the 1960's, Rolls Royce Ltd. began to use computers not only to design the optimal shape of turbine blades for aircraft engines, but also to control the precision milling machines which made the blades. In this type of computer-assisted design and manufacture, no drawings were required. Furthermore, it became possible for an industry requiring a part from a subcontractor to send the machine-control instructions for its fabrication through the computer network to the subcontractor, instead of sending drawings of the part.

In addition to computer-controlled machine tools, robots were also introduced. They were often used for hazardous or monotonous jobs, such as spray-painting automobiles; and they could be programmed by going through the job once manually in the programming mode. By 1987, the population of robots in the United States was between 5,000 and 7,000, while in Japan, the Industrial Robot Association reported a robot population of 80,000.

Chemical industries began to use sophisticated computer programs to control and to optimize the operations of their plants. In such control systems, sensors reported current temperatures, pressures, flow rates, etc. to the computer, which then employed a mathematical model of the plant to calculate the adjustments needed to achieve optimum operating conditions.

Not only industry, but also commerce, felt the effects of computerization during the postwar period. Commerce is an information-intensive activity; and in fact some of the crucial steps in the development of information-handling technology developed because of the demands of commerce: The first writing evolved from records of commercial transactions kept on clay tablets in the Middle East; and automatic business machines, using punched cards, paved the way for the development of the first programmable computers.

Computerization has affected wholesaling, warehousing, retailing, banking, stockmarket transactions, transportation of goods - in fact, all aspects of commerce. In wholesaling, electronic data is exchanged between companies by means of computer networks, allowing order-processing to be handled automatically; and similarly, electronic data on prices is transmitted to buyers.

The key to automatic order-processing in wholesaling was standardization. In the United States, the Food Marketing Institute, the Grocery Manufacturers of America, and several other trade organizations, established the Uniform Communications System (UCS) for the grocery industry. This system specifies a standard format for data on products, prices and orders.

Automatic warehouse systems were designed as early as 1958. In such systems, the goods to be stored are placed on pallets (portable platforms), which are stacked automatically in aisles of storage cubicles. A computer records the position of each item for later automatic retrieval.

In retailing, just as in wholesaling, standardization proved to be the key requirement for automation. Items sold in supermarkets in most industrialized countries are now labeled with a standard system of machine-readable thick and thin bars known as the Universal Product Code (UPC). The left-hand digits of the code specify the manufacturer or packer of the item, while the right-hand set of digits specify the nature of the item. A final digit is included as a check, to make sure that the others were read correctly. This last digit (called a modulo check digit) is the smallest number which yields a multiple of ten when added to the sum of the previous digits.

When a customer goes through a check-out line, the clerk passes the purchased items over a laser beam and photocell, thus reading the UPC code into a small embedded computer or microprocessor at the checkout counter, which adds the items to the customer's bill. The microprocessor also sends the information to a central computer and inventory data base. When stocks of an item become low, the central computer generates a replacement order. The financial book-keeping for the retailing operation is also carried out automatically by the central computer.

In many places, a customer passing through the checkout counter of a supermarket is able to pay for his or her purchases by means of a plastic card with a magnetic, machine-readable identification number. The amount of the purchase is then transmitted through a computer network and deducted automatically from the customer's bank account. If the customer pays by check, the supermarket clerk may use a special terminal to determine whether a check written by the customer has ever "bounced".

Most checks are identified by a set of numbers written in the Magnetic-Ink Character Recognition (MICR) system. In 1958, standards for the MICR system were established, and by 1963, 85 percent of all checks written in the United States were identified by MICR numbers. By 1968, almost all banks had adopted this system; and thus the administration of checking accounts was automated, as well as the complicated process by which a check, deposited anywhere in the world, returns to the payors bank.

Container ships were introduced in the late 1950's, and since that time, container systems have increased cargo-handling speeds in ports by at least an order of magnitude. Computer networks contributed greatly to the growth of the container system of transportation by keeping track of the position, ownership and contents of the containers.

In transportation, just as in wholesaling and retailing, standardization proved to be a necessary requirement for automation. Containers of a standard size and shape could be loaded and unloaded at ports by specialized tractors and cranes which required only a very small staff of operators. Standard formats for computerized manifests, control documents, and documents for billing and payment, were instituted by the Transportation Data Coordinating Committee, a non-profit organization supported by dues from shipping firms.

In the industrialized parts of the world, almost every type of work has been made more efficient by computerization and automation. Even artists, musicians, architects and authors find themselves making increasing use of computers: Advanced computing systems, using specialized graphics chips, speed the work of architects and film animators. The author's traditional typewriter has been replaced by a word-processor, the composer's piano by a music synthesizer.

In the Industrial Revolution of the 18th and 19th centuries, muscles were replaced by machines. Computerization represents a Second Industrial Revolution: Machines have begun to perform not only tasks which once required human muscles, but also tasks which formerly required human intelligence.

In industrial societies, the mechanization of agriculture has very much reduced the fraction of the population living on farms. For example, in the United States, between 1820 and 1980, the fraction of workers engaged in agriculture fell from 72% to 3.1%. There are signs that computerization and automation will similarly reduce the number of workers needed in industry and commerce.

Computerization is so recent that, at present, we can only see the beginnings of its impact; but when the Second Industrial Revolution is complete, how will it affect society? When our children finish their education, will they face technological unemployment?

As we saw in an previous chapter, the initial stages of the First Industrial Revolution produced much suffering, because labor was regarded as a commodity to be bought and sold according to the laws of supply and demand, with almost no consideration for the needs of the workers. Will we repeat this mistake? Or will society learn from its earlier experience, and use the technology of automation to achieve widely-shared human happiness?

The Nobel-laureate economist, Wassily W. Leontief, has made the following comment on the problem of technological unemployment:

"Adam and Eve enjoyed, before they were expelled from Paradise, a high standard of living without working. After their expulsion, they and their successors were condemned to eke out a miserable existence, working from dawn to dusk. The history of technological progress over the last 200 years is essentially the story of the human species working its way slowly and steadily back into Paradise. What would happen, however, if we suddenly found ourselves in it? With all goods and services provided without work, no one would be gainfully employed. Being unemployed means receiving no wages. As a result, until appropriate new income policies were formulated to fit the changed technological conditions, everyone would starve in Paradise."

To say the same thing in a slightly different way: consider what will happen when a factory which now employs a thousand workers introduces microprocessor-controlled industrial robots and reduces its work force to only fifty. What will the nine hundred and fifty redundant workers do? They will not be able to find jobs elsewhere in industry, commerce or agriculture, because all over the economic landscape, the scene will be the same.

There will still be much socially useful work to be done - for example, taking care of elderly people, beautifying the cities, starting youth centers, planting forests, cleaning up pollution, building schools in developing countries, and so on. These socially beneficial goals are not commercially "profitable". They are rather the sort of projects which governments sometimes support if they have the funds for it. However, the money needed to usefully employ the nine hundred and fifty workers will not be in the hands of the government. It will be in the hands of the factory owner who has just automated his production line.

In order to make the economic system function again, either the factory owner will have to be persuaded to support socially beneficial but commercially unprofitable projects, or else an appreciable fraction of his profits will have to be transferred to the government, which will then be able to constructively re-employ the redundant workers.

The future problems of automation and technological unemployment may force us to rethink some of our economic ideas. It is possible that helping young people to make a smooth transition from education to secure jobs will become one of the important responsibilities of governments, even in countries whose economies are based on free enterprise. If such a change does take place in the future, while at the same time socialistic countries are adopting a few of the better features of free enterprise, then one can hope that the world will become less sharply divided by contrasting economic systems.

Neural networks

If civilization survives, future historians may regard the invention of computers as an even more important step in cultural evolution than the invention of printing or the invention of writing. Exploration of the possibilities of artificial intelligence has only barely begun. In part, the future development of computers will depend on more sophisticated programs (software), and in part on new types of computer architecture (hardware).

Physiologists have begun to make use of insights derived from computer design in their efforts to understand the mechanism of the brain; and computer designers are beginning to construct computers modeled after neural networks. We may soon see the development of computers capable of learning complex ideas, generalization, value judgements, artistic creativity, and much else that was once thought to be uniquely characteristic of the human mind. Efforts to design such computers will undoubtedly give us a better understanding of the way in which the brain performs its astonishing functions.

Much of our understanding of the nervous systems of higher animals is due to the Spanish microscopist, Ramón y Cajal, and to the English physiologists, Alan Hodgkin and Andrew Huxley. Cajal's work, which has been confirmed and elaborated by modern electron microscopy, showed that the central nervous system is a network of nerve cells (neurons) and threadlike fibers growing from them. Each neuron has many input fibers (dendrites), and one output fiber (the axon), which may have several branches.

In 1952, working with the giant axon of the squid (which can be as large as a millimeter in diameter), Hodgkin and Huxley showed that nerve fibers are like long tubes. Inside the tube is a fluid which contains potassium and sodium ions. In a resting nerve, the concentration of potassium inside is higher than it is in the normal body fluids outside, and the concentration of sodium is lower. These abnormal concentrations are maintained by a "pump", which uses metabolic energy to bring potassium ions into the nerve and to expel sodium ions.

The tubelike membrane surrounding the nerve fiber is more permeable to sodium than to potassium; and the positively-charged sodium ions tend to leak back into the resting nerve, producing a small difference in electrical potential between the inside and outside. This electrical potential helps to hold the molecules of the nerve membrane in an orderly layer, so that the membrane's permeability to ions is low.

Hodgkin and Huxley showed that when a nerve cell "fires", the whole situation changes dramatically. Potassium ions begin to flow out of the nerve, destroying the electrical potential which maintained order in the membrane. A wave of depolarization passes along the nerve. Like a row of dominos falling, the disturbance propagates from one section to the next: Potassium ions flow out, the order-maintaining electrical potential disappears, the next small section of the nerve membrane becomes permeable, and so on. Thus, Hodgkin and Huxley showed that when a nerve cell fires, a quick pulse-like electrical and chemical disturbance is transmitted along the fiber.

The fibers of nerve cells can be very long, but finally the signal reaches a junction where one nerve cell is joined to another, or where a nerve is joined to a muscle. The junction is called a "synapse". At the synapse, chemical transmitters are released which may cause the next nerve cell to fire, or which may inhibit it from firing, depending on the type of synapse. The chemical transmitters released by nerve impulses were first studied by Sir Henry Dale, Sir John Eccles and Otto Loewi, who found that they can also trigger muscle contraction. (Among the substances believed to be exitatory transmitters are acetylcholine, noradrenalin, norepinephrine, serotonin, dopamine and glutamate, while gamma-amino-butyric acid is believed to be an inhibitory transmitter.)

Once a nerve cell fires, a signal will certainly go out along its axon. However, when the signal comes to a synapse, where the axon makes contact with the dendrite of another cell, it is not at all certain that the next nerve cell will fire. Whether it does so or not depends on many things: It depends on the frequency of the pulses arriving along the axon. (The transmitter substances are constantly being broken down.) It depends on the type of transmitter substance. (Some of them inhibit the firing of the next cell.) And finally, the firing of the next neuron depends on the way in which the synapse has been modified by its previous history and by the concentration of various chemicals in the blood.

The variety and plasticity of synapses, and the complex, branching interconnections of dendrites and axons, help to account for the subtlety of the nervous system, as well as its sensitivity to various chemicals in the blood. Some neurons (called "and" cells) fire only when all their input dendrites are excited. Other neurons (called "or" cells) fire when any one of the dendrites is excited. Still other neurons (called "inhibited" cells) fire when certain dedrites are excited only if other inhibiting dendrites are not excited. Interestingly, "and" circuits, "or" circuits and "inhibited" circuits have played a fundamental role in computer design ever since the the beginning of electronic computers.

In the 1960's, the English neuroanatomist J.Z. Young proposed a model of the visual cortex of the octopus brain. In Young's model, the arrangement of "and", "or" and "inhibited" cells performs the function of pattern abstraction. The model is based both on learning experiments with the octopus, and on microscopic studies of the octopus brain.

According to Young's model, the visual pattern received by the retina of the octopus eye is mapped in a direct way onto the outer layer of neurons in the animal's visual cortex. The image on the retina forms a picture on the cortex, just as though it were projected onto a screen. However, the arrangement of "and", "or" and "inhibited" cells in the cortex is such that as the signals from the retina are propagated inward to more deeply-lying layers, certain deep cortical cells will fire only in response to a particular pattern on the retina.

In Young's model, the signal then comes to a branch, where it can either stimulate the octopus to attack or to retreat. There is a bias towards the attack pathway; and therefore, the first time an octopus is presented with an object of any shape, it tends to attack it. However, if the experimenter administers an electric shock to the animal, synapses in the attack pathway are modified, and the attack pathway is blocked.

When the octopus later is presented with an object of the same shape, the signal comes through in exactly the same way as before. However, this time when it reaches the attack-retreat branch, the attack pathway is blocked, and the signal causes the animal to retreat. The octopus has learned!

It is possible the computers of the future will have pattern-recognition and learning abilities derived from architecture inspired by our understanding of the synapse, by Young's model, or by other biological models. However, pattern recognition and learning can also be achieved by programming, using computers of conventional architecture. Programs already exist which allow computers to understand both handwriting and human speech; and a recent chess-playing program was able to learn by studying a large number of championship games. Having optimized its parameters by means of this learning experience, the chess-playing program was able to win against grand masters!

In 1943, W. McCulloch and W. Pitts published a paper entitled *A Logical Calculus of the Ideas Immanent in Nervous Activity.* In this pioneering paper, they proposed the idea of a Threshold Logic Unit (TLU), which they visualized not only as a model of the way in which neurons function in the brain but also as a possible subunit for artificial systems which might be constructed to perform learning and pattern-recognition tasks. Problems involving learning, generalization, pattern recognition and noisy data are easily handled by the brains of humans and animals, but computers of the conventional von Neumann type find such tasks especially difficult.

Conventional computers consist of a memory and one or more central processing units (CPUs). Data and instructions are repeatedly transferred from the memory to the CPUs, where the data is processed and returned to the memory. The repeated performance of many such cycles requires a long and detailed program, as well as high-quality data. Thus conventional computers, despite their great speed and power, lack the robustness, intuition, learning powers and powers of generalization which characterize biological neural networks. In the 1950's, following the suggestions of McCulloch and Pitts, and inspired by the growing knowledge of brain structure and function which was being gathered by histologists and neurophysiologists, computer scientists began to construct artificial neural networks - massively parallel arrays of TLU's.

The analogy between a TLU and a neuron can be seen by comparing Figure 17.6, which shows a neuron, with Figure 17.7, which shows a TLU. A neuron is a specialized cell consisting of a cell body (*soma*) from which an extremely long, tubelike fiber called an *axon* grows. The axon is analogous to the output channel of a TLU. From the soma, a number of slightly shorter, rootlike extensions called *dendrites* also grow. The dendrites are analogous to the input channels of a TLU.

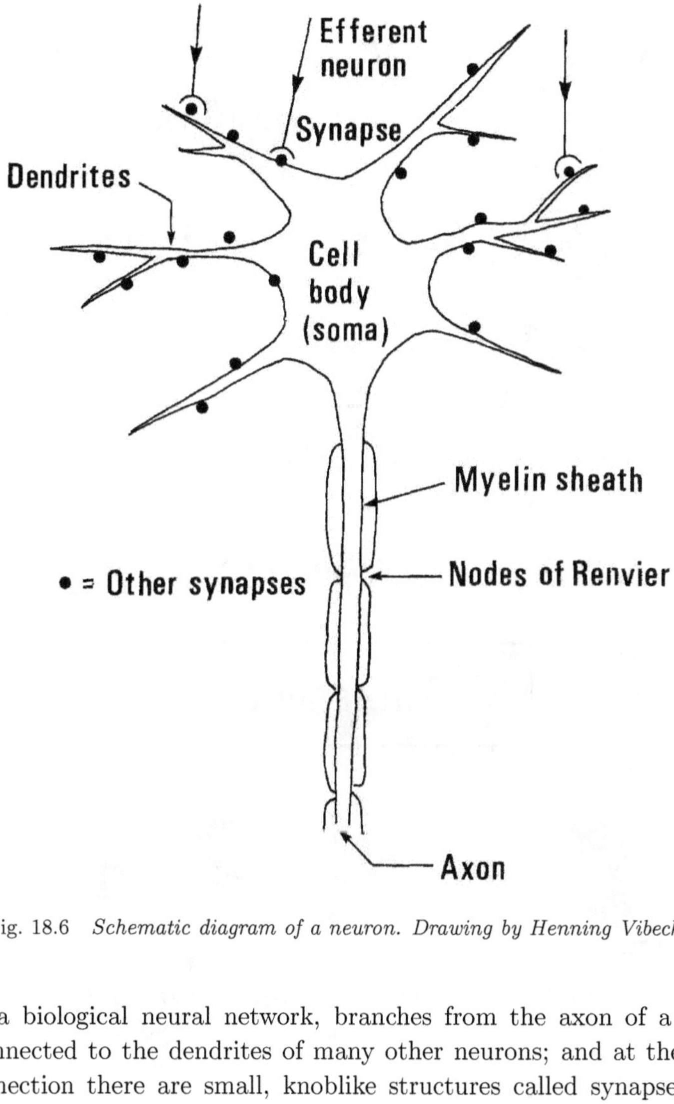

Fig. 18.6 *Schematic diagram of a neuron. Drawing by Henning Vibeck.*

In a biological neural network, branches from the axon of a neuron are connected to the dendrites of many other neurons; and at the points of connection there are small, knoblike structures called synapses. The "firing" of a neuron sends a wave of depolarization out along its axon. When the pulselike electrical and chemical disturbance associated with the wave of depolarization (the action potential) reaches a synapse, where the axon is connected with another neuron, transmitter molecules are released into the post-synaptic cleft. The neurotransmitter molecules travel across the post-synaptic cleft to receptors on a dendrite of the next neuron in the net, where they are bound to receptors.

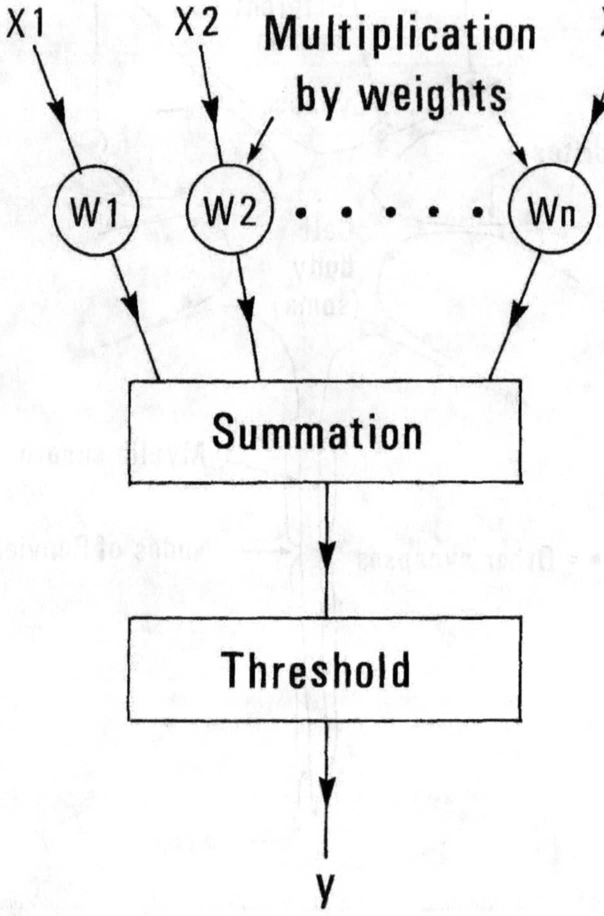

Fig. 18.7 *A Threshold Logic Unit (TLU) of the type proposed by McCulloch and Pitts. Drawing by Henning Vibeck.*

There are many kinds of neurotransmitter molecules, some of which tend to make the firing of the next neuron more probable, and others which tend to inhibit its firing. When the neurotransmitter molecules are bound to the receptors, they cause a change in the dendritic membrane potential, either increasing or decreasing its polarization. The post-synaptic potentials from the dendrites are propagated to the soma; and if their sum exceeds a threshold value, the neuron fires. The subtlety of biological neural networks derives from the fact that there are many kinds of neurotransmitters and synapses, and from the fact that synapses are modified by their past history.

Turning to Figure 17.8, we can compare the biological neuron with the Threshold Logic Unit of McCulloch and Pitts. Like the neuron, the TLU has many input channels. To each of the N channels there is assigned a weight, $w_1, w_2, ..., w_N$. The weights can be changed; and the set of weights gives the TLU its memory and learning capabilities. Modification of weights in the TLU is analogous to the modification of synapses in a neuron, depending on their history. In the most simple type of TLU, the input signals are either 0 or 1. These signals, multiplied by their appropriate weights, are summed, and if the sum exceeds a threshold value, θ the TLU "fires", i.e. a pulse of voltage is transmitted through the output channel to the next TLU in the artificial neural network.

Genetic algorithms

Genetic algorithms represent a second approach to machine learning and to computational problems involving optimization. Like neural network computation, this alternative approach has been inspired by biology, and it has also been inspired by the Darwinian concept of natural selection. In a genetic algorithm, the hardware is that of a conventional computer; but the software creates a population and allows it to evolve in a manner closely analogous to biological evolution.

One of the most important pioneers of genetic algorithms was John Henry Holland (1929-). After attending MIT, where he was influenced by Norbert Wiener, Holland worked for IBM, helping to develop the 701. He then continued his studies at the University of Michigan, obtaining the first Ph.D. in computer science ever granted in America. Between 1962 and 1965, Holland taught a graduate course at Michigan called "Theory of Adaptive Systems". His pioneering course became almost a cult, and together with his enthusiastic students he applied the genetic algorithm approach to a great variety of computational problems. One of Holland's students, David Goldberg, even applied a genetic algorithm program to the problem of allocating natural gas resources.

The programs developed by Holland and his students were modelled after the natural biological processes of reproduction, mutation, selection and evolution. In biology, the information passed between generations is contained in chromosomes - long strands of DNA where the genetic message is written in a four-letter language, the letters being adenine, thymine, guanine and cytosine. Analogously, in a genetic algorithm, the information is coded in a long string, but instead of a four-letter language, the code is binary: The chromosome-analogue is a long string of 0's and 1's, i.e., a long binary string. One starts with a population that has sufficient diversity so that natural selection can act.

The genotypes are then translated into phenotypes. In other words, the information contained in the long binary string (analogous to the genotype of each individual) corresponds to an entity, the phenotype, whose fitness for survival can be evaluated. The mapping from genotype to phenotype must be such that very small changes in the binary string will not produce radically different phenotypes.

From the initial population, the most promising individuals are selected to be the parents of the next generation, and of these, the fittest are allowed produce the largest number of offspring. Before reproduction takes place, however, random mutations and chromosome crossing can occur. For example, in chromosome crossing, the chromosomes of two individuals are broken after the nth binary digit, and two new chromosomes are formed, one with the head of the first old chromosome and the tail of the second, and another with the head of the second and the tail of the first. This process is analogous to the biological crossings which allowed Thomas Hunt Morgan and his "fly squad" to map the positions of genes on the chromosomes of fruit flies, while the mutations are analogous to those studied by Hugo de Vries and Hermann J. Muller.

After the new generation has been produced, the genetic algorithm advances the time parameter by a step, and the whole process is repeated: The phenotypes of the new generation are evaluated and the fittest selected to be parents of the next generation; mutation and crossings occur; and then fitness-proportional reproduction. Like neural networks, genetic algorithms are the subject of intensive research, and evolutionary computation is a rapidly growing field.

Evolutionary methods have been applied not only to software, but also to hardware. Some of the circuits designed in this way defy analysis using conventional techniques - and yet they work astonishingly well.

Like nuclear physics and genesplicing, artificial intelligence presents a challenge: Will society use its new powers wisely and humanely? The computer technology of the future can liberate us from dull and repetitive work, and allow us to use our energies creatively; or it can produce unemployment and misery, depending on how we organize our society. Which will we choose?

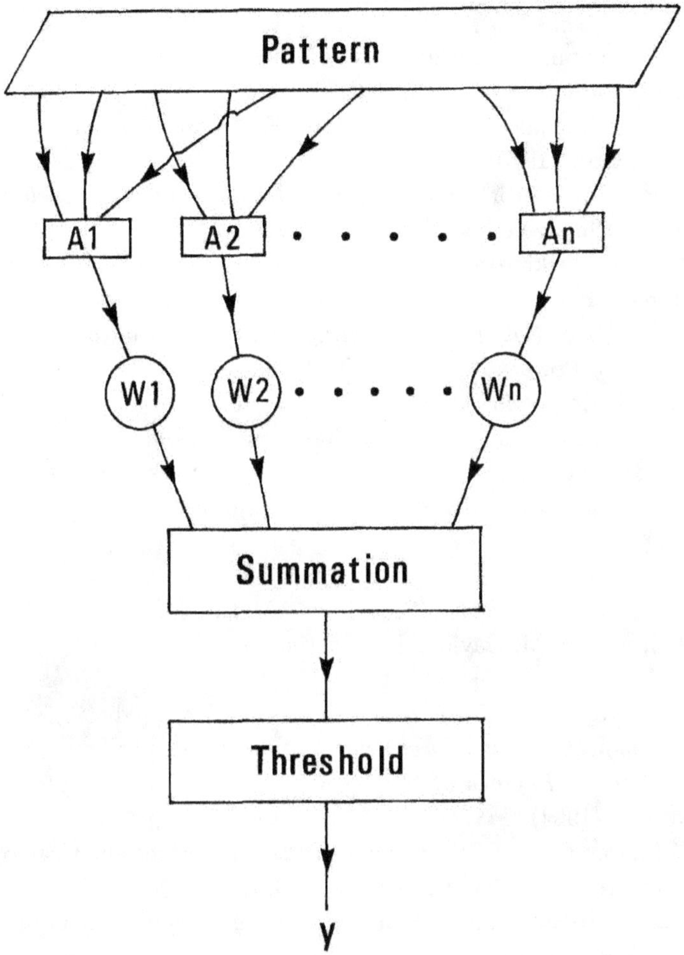

Fig. 18.8 *A perceptron, introduced by Rosenblatt in 1962. The perceptron is similar to a TLU, but its input is preprocessed by a set of association units (A-units). The A-units are not trained, but are assigned a fixed Boolean functionality. Drawing by Henning Vibeck.*

Suggestions for further reading

(1) N. Metropolis, J. Howlett, and Gian-Carlo Rota (editors), *A History of Computing in the Twentieth Century*, Academic Press (1980).
(2) S.H. Hollingdale and G.C. Tootil, *Electronic Computers*, Penguin Books Ltd. (1970).
(3) R. Randell (editor), *The Origins of Digital Computers, Selected Papers*, Springer-Verlag, New York (1973).
(4) Allan R. Mackintosh, *The First Electronic Computer*, Physics Today, March, (1987).
(5) H. Babbage, *Babbages Calculating Engines: A Collection of Papers by Henry Prevost Babbage*, MIT Press, (1984).
(6) A.M. Turing, *The Enigma of Intelligence*, Burnett, London (1983).
(7) Ft. Penrose, *The Emperor's New Mind: Concerning Computers, Minds, and the Laws of Physics*, Oxford University Press, (1989).
(8) S. Wolfram, *A New Kind of Science*, Wolfram Media, Champaign IL, (2002).
(9) A.M. Turing, *On computable numbers, with an application to the Entscheidungsproblem*, Proc. Lond. Math. Soc. Ser 2, **42**, (1937). Reprinted in M. David Ed., *The Undecidable*, Raven Press, Hewlett N.Y., (1965).
(10) N. Metropolis, J. Howlett, and Gian-Carlo Rota (editors), *A History of Computing in the Twentieth Century*, Academic Press (1980).
(11) J. Shurkin, *Engines of the Mind: A History of Computers*, W.W. Norten, (1984).
(12) J. Palfreman and D. Swade, *The Dream Machine: Exploring the Computer Age*, BBC Press (UK), (1991).
(13) T.J. Watson, Jr. and P. Petre, *Father, Son, and Co.*, Bantam Books, New York, (1991).
(14) A. Hodges, Alan Turing: *The Enegma*, Simon and Schuster, (1983).
(15) H.H. Goldstein, *The Computer from Pascal to Von Neumann*, Princeton University Press, (1972).
(16) C.J. Bashe, L.R. Johnson, J.H. Palmer, and E.W. Pugh, *IBM's Early Computers*, Vol. 3 in the History of Computing Series, MIT Press, (1986).
(17) K.D. Fishman, *The Computer Establishment*, McGraw-Hill, (1982).
(18) S. Levy, *Hackers*, Doubleday, (1984).
(19) S. Franklin, *Artificial Minds*, MIT Press, (1997).
(20) P. Freiberger and M. Swaine, *Fire in the Valley: The Making of the*

Personal Computer, Osborne/MeGraw-Hill, (1984).

(21) R.X. Cringely, *Accidental Empires*, Addison-Wesley, (1992).

(22) R. Randell editor, *The Origins of Digital Computers, Selected Papers*, Springer-Verlag, New York (1973).

(23) H. Lukoff, *From Dits to Bits*, Robotics Press, (1979).

(24) D.E. Lundstrom, *A Few Good Men from Univac*, MIT Press, (1987).

(25) D. Rutland, *Why Computers Are Computers (The SWAC and the PC)*, Wren Publishers, (1995).

(26) P.E. Ceruzzi, *Reckoners: The Prehistory of the Digital Computer, from Relays to the Stored Program Concept, 1935-1945*, Greenwood Press, Westport, (1983)

(27) S.G. Nash, *A History of Scientific Computing*, Adison-Wesley, Reading Mass., (1990).

(28) P.E. Ceruzzi, *Crossing the divide: Architectural issues and the emergence of stored programme computers*, 1935-1953, IEEE Annals of the History of Computing, **19**, 5-12, January-March (1997).

(29) P.E. Ceruzzi, *A History of Modern Computing*, MIT Press, Cambridge MA, (1998).

(30) K. Zuse, *Some remarks on the history of computing in Germany*, in *A History of Computing in the 20th Century*, N. Metropolis et al. editors, 611-627, Academic Press, New York, (1980).

(31) A.R. Mackintosh, *The First Electronic Computer*, Physics Today, March, (1987).

(32) S.H. Hollingdale and G.C. Tootil, *Electronic Computers*, Penguin Books Ltd. (1970).

(33) A. Hodges, *Alan Turing: The Enegma*, Simon and Schuster, New York, (1983).

(34) A. Turing, *On computable numbers with reference to the Entscheidungsproblem*, Journal of the London Mathematical Society, **II, 2. 42**, 230-265 (1937).

(35) J. von Neumann, *The Computer and the Brain*, Yale University Press, (1958).

(36) I.E. Sutherland, *Microelectronics and computer science*, Scientific American, 210-228, September (1977).

(37) W. Aspray, *John von Neumann and the Origins of Modern Computing*, M.I.T. Press, Cambridge MA, (1990, 2nd ed. 1992).

(38) W. Aspray, *The history of computing within the history of information technology*, History and Technology, **11**, 7-19 (1994).

(39) G.F. Luger, *Computation and Intelligence: Collected Readings*, MIT Press, (1995).

(40) Z.W. Pylyshyn, *Computation and Cognition: Towards a Foundation for Cognitive Science*, MIT Press, (1986).
(41) D.E. Shasha and C. Lazere, *Out of Their Minds: The Creators of Computer Science*, Copernicus, New York, (1995).
(42) W. Aspray, *An annotated bibliography of secondary sources on the history of software*, Annals of the History of Computing **9**, 291-243 (1988).
(43) R. Kurzweil, *The Age of Intelligent Machines*, MIT Press, (1992).
(44) S.L. Garfinkel and H. Abelson, eds., *Architects of the Information Society: Thirty-Five Years of the Laboratory for Computer Sciences at MIT*, MIT Press, (1999).
(45) J. Haugeland, *Artificial Intelligence: The Very Idea*, MIT Press, (1989).
(46) M.A. Boden, *Artificial Intelligence in Psychology: Interdisciplinary Essays*, MIT Press, (1989).
(47) J.W. Cortada, *A Bibliographic Guide to the History of Computer Applications, 1950-1990*, Greenwood Press, Westport Conn., (1996).
(48) M. Campbell-Kelly and W. Aspry, *Computer: A History of the Information Machine*, Basic Books, New York, (1996).
(49) B.I. Blum and K. Duncan, editors, *A History of Medical Informatics*, ACM Press, New York, (1990).
(50) J.-C. Guedon, La Planete Cyber, *Internet et Cyberspace*, Gallimard, (1996).
(51) S. Augarten, *Bit by Bit: An Illustrated History of Computers*, Unwin, London, (1985).
(52) N. Wiener, *Cybernetics; or Control and Communication in the Animal and the Machine*, The Technology Press, John Wiley and Sons, New York, (1948).
(53) W.R. Ashby, *An Introduction to Cybernetics*, Chapman and Hall, London, (1956).
(54) M.A. Arbib, *A partial survey of cybernetics in eastern Europe and the Soviet Union*, Behavioral Sci., **11**, 193-216, (1966).
(55) A. Rosenblueth, N. Weiner and J. Bigelow, *Behavior, purpose and teleology*, Phil. Soc. **10 (1)**, 18-24 (1943).
(56) N. Weiner and A. Rosenblueth, *Conduction of impulses in cardiac muscle*, Arch. Inst. Cardiol. Mex., **16**, 205-265 (1946).
(57) H. von Foerster, editor, *Cybernetics - circular, causal and feed-back mechanisms in biological and social systems. Transactions of sixth-tenth conferences*, Josiah J. Macy Jr. Foundation, New York, (1950-1954).

(58) W.S. McCulloch and W. Pitts, *A logical calculus of ideas immanent in nervous activity*, Bull. Math. Biophys., **5**, 115-133 (1943).
(59) W.S. McCulloch, *An Account of the First Three Conferences on Teleological Mechanisms*, Josiah Macy Jr. Foundation, (1947).
(60) G.A. Miller, *Languages and Communication*, McGraw-Hill, New York, (1951).
(61) G.A. Miller, *Statistical behavioristics and sequences of responses*, Psychol. Rev. **56**, 6 (1949).
(62) G. Bateson, *Bali - the value system of a steady state*, in M. Fortes, editor, *Social Structure Studies Presented to A.R. Radcliffe-Brown*, Clarendon Press, Oxford, (1949).
(63) G. Bateson, *Communication, the Social Matrix of Psychiatry*, Norton, (1951).
(64) G. Bateson, *Steps to an Ecology of Mind*, Chandler, San Francisco, (1972).
(65) G. Bateson, *Communication et Societe*, Seuil, Paris, (1988).
(66) S. Heims, *Gregory Bateson and the mathematicians: From interdisciplinary interactions to societal functions*, J. History Behavioral Sci., **13**, 141-159 (1977).
(67) S. Heims, *John von Neumann and Norbert Wiener. From Mathematics to the Technology of Life and Death*, MIT Press, Cambridge MA, (1980).
(68) S. Heims, *The Cybernetics Group*, MIT Press, Cambridge MA, (1991).
(69) G. van de Vijver, *New Perspectives on Cybernetics (Self-Organization, Autonomy and Connectionism)*, Kluwer, Dordrecht, (1992).
(70) A. Bavelas, *A mathematical model for group structures*, Appl. Anthrop. **7 (3)**, 16 (1948).
(71) P. de Latil, *La Pensee Artificielle - Introduction a la Cybernetique*, Gallimard, Paris, (1953).
(72) L.K. Frank, G.E. Hutchinson, W.K. Livingston, W.S. McCulloch and N. Wiener, *Teleological Mechanisms*, Ann. N.Y. Acad. Sci. **50**, 187-277 (1948).
(73) H. von Foerster, *Quantum theory of memory*, in H. von Foerster, editor, *Cybernetics - circular, causal and feed-back mechanisms in biological and social systems.* Transactions of the sixth conferences, Josiah J. Macy Jr. Foundation, New York, (1950).
(74) H. von Foerster, *Observing Systems*, Intersystems Publications, California, (1984).

(75) H. von Foerster, *Understanding Understanding: Essays on Cybernetics and Cognition*, Springer, New York, (2002).

(76) M. Newborn, *Kasparov vs. Deep Blue: Computer Chess Comes of age*, Springer Verlag, (1996).

(77) K.M. Colby, *Artificial Paranoia: A Computer Simulation of the Paranoid Process*, Pergamon Press, New York, (1975).

(78) J.Z. Young, *Discrimination and learning in the octopus*, in H. von Foerster, editor, *Cybernetics - circular, causal and feed-back mechanisms in biological and social systems.* Transactions of the ninth conference, Josiah J. Macy Jr. Foundation, New York, (1953).

(79) M.J. Apter and L. Wolpert, *Cybernetics and development.* I. Information theory, J. Theor. Biol. **8**, 244-257 (1965).

(80) H. Atlan, *L'Organization Biologique et la Theorie de I'Information*, Hermann, Paris, (1972).

(81) H. Atlan, *On a formal definition of organization*, J. Theor. Biol. 45, 295-304 (1974).

(82) H. Atlan, *Organization du vivant, information et auto-organization*, in Volume Symposium 1986 de l'Encylopediea Universalis, pp. 355-361, Paris, (1986).

(83) E.R. Kandel, Nerve cells and behavior, Scientific American, **223**, 57-70, July, (1970).

(84) E.R. Kandel, *Small systems of neurons*, Scientific American, **241 no.3**, 66-76 (1979).

(85) A.K. Katchalsky et al., *Dynamic patterns of brain cell assemblies*, Neurosciences Res. Prog. Bull., **12 no.1**, (1974).

(86) G.E. Moore, *Cramming more components onto integrated circuits*, Electronics, April 19, (1965).

(87) P. Gelsinger, P. Gargini, G. Parker and A. Yu, *Microprocessors circa 2000*, IEEE Spectrum, October, (1989).

(88) P. Baron, *On distributed communications networks*, IEEE Trans. Comm. Systems, March (1964).

(89) V.G. Cerf and R.E. Khan, *A protocol for packet network intercommunication*, Trans. Comm. Tech. **COM-22, V5**, 627-641, May (1974).

(90) L. Kleinrock, *Communication Nets: Stochastic Message Flow and Delay*, McGraw-Hill, New York, (1964).

(91) L. Kleinrock, *Queueing Systems: Vol. II, Computer Applications*, Wiley, New York, (1976).

(92) R. Kahn, editor, *Special Issue on Packet Communication Networks*,

Proc. IEEE, **66**, November, (1978).

(93) L.G. Roberts, *The evolution of packet switching*, Proc. of the IEEE **66**, 1307-13, (1978).

(94) J. Abbate, *The electrical century: Inventing the web*, Proc. IEEE **87**, November, (1999).

(95) J. Abbate, *Inventing the Internet*, MIT Press, Cambridge MA, (1999).

(96) J.C. McDonald, editor, *Fundamentals of Digital Switching, 2nd Edition*, Plenum, New York, (1990).

(97) B. Metcalfe, *Packet Communication*, Peer-to-Peer Communication, San Jose Calif, (1996).

(98) T. Berners-Lee, *The Original Design and Ultimate Destiny of the World Wide Web by its Inventor*, Harper San Francisco, (1999).

(99) J. Clark, *Netscape Time: The Making of the Billion-Dollar Start-Up That Took On Microsoft*, St. Martin's Press, New York, (1999).

(100) J. Wallace, *Overdrive: Bill Gates and the Race to Control Cyberspace*, Wiley, New York, (1997).

(101) P. Cunningham and F. Froschl, *The Electronic Business Revolution*, Springer Verlag, New York, (1999).

(102) J.L. McKenny, *Waves of Change: Business Evolution Through Information Technology*, Harvard Business School Press, (1995).

(103) M.A. Cosumano, *Competing on Internet Time: Lessons From Netscape and Its Battle with Microsoft*, Free Press, New York, (1998).

(104) F.J. Dyson, *The Sun, the Genome and the Internet: Tools of Scientific Revolutions*, Oxford University Press, (1999).

(105) L. Bruno, *Fiber Optimism: Nortel, Lucent and Cisco are battling to win the high-stakes fiber-optics game*, Red Herring, June (2000).

(106) N. Cochrane, *We're insatiable: Now it's 20 million million bytes a day*, Melbourne Age, January 15, (2001).

(107) K.G. Coffman and A.N. Odlyzko, The size and growth rate of the Internet, First Monday, October, (1998).

(108) C.A. Eldering, M.L. Sylla, and J.A. Eisenach, *Is there a Moore's law for bandwidth?*, IEEE Comm. Mag., 2-7, October, (1999).

(109) G. Gilder, *Fiber keeps its promise: Get ready, bandwidth will triple each year for the next 25 years*, Forbes, April 7, (1997).

(110) A.M. Noll, *Does data traffic exceed voice traffic?*, Comm. ACM, 121-124, June, (1999).

(111) B. St. Arnaud, J. Coulter, J. Fitchett, and S. Mokbel, *Architectural and engineering issues for building an optical Internet*, Proc. Soc. Optical Eng. (1998).

(112) M. Weisner, *The computer for the 21st century*, Scientific American, September, (1991).
(113) R. Wright, *Three Scientists and Their Gods*, Time Books, (1988).
(114) S. Nora and A. Mine, *The Computerization of Society*, MIT Press, (1981).
(115) T. Forester, *Computers in the Human Context: Information Theory, Productivity, and People*, MIT Press, (1989).

Chapter 19

CARING FOR THE EARTH

Exponential growth

Measured on the time scale of ordinary genetic evolution, the cultural evolution of our species has been astonishingly rapid. Humans have been living on the earth for roughly two million years (more or less, depending on where one draws the line between our human and pre-human ancestors). During almost all of this time, our ancestors lived by hunting and food-gathering. They were not at all numerous, and not conspicuously different from other animals.

Then, suddenly, during the brief space of ten thousand years, our species exploded in numbers from a few million to more than five billion, populating all parts of the earth, and even setting foot on the moon. This population explosion, which is still going on, has been the result of dramatic cultural changes. Genetically, we are almost identical with our hunter-gatherer ancestors who lived ten thousand years ago; but cultural evolution has changed our way of life beyond recognition.

In genetic evolution, a species changes through inherited variations in the DNA of its individual members. However, our species has another means of change - through additions to the inherited body of techniques, customs and knowledge which we call culture.

Beginning with the development of speech, human cultural evolution began to accelerate. It began to move faster with the agricultural revolution, and faster still with the invention of writing and the invention of printing. Finally, modern science has accelerated the rate of technical and social change to a completely unprecedented speed. There has been, in other words, an "information explosion", to which modern science has contributed.

The growth of modern science is accelerating because knowledge feeds on itself: A new idea or a new development may lead to several other innovations, which can in turn start an avalanche of change. For example, the quantum theory of atomic structure lead to the invention of transistors, which made the development of high-speed digital computers possible. Computers have not only produced further developments in quantum theory; they have also revolutionized many other fields.

The growth law which follows from this type of relationship is exponential; and in fact, the number of scientific articles published per year has for some time been increasing exponentially, doubling every fifteen years. The exponential growth of technology is the driving force behind the other exponentially increasing graphs which can be made, such as the graphs of population growth and the growth of international trade.

When the increase of a quantity is proportional to the amount already present, the resulting growth is exponential. The exponential growth of science follows from the fact that its increase is proportional to the amount already present; and the same is true for the growth of a population whose birth rate exceeds the death rate.

The doubling time for an exponentially-growing quantity is approximately equal to 70 years divided by the annual percentage of increase. Thus, a population growing at the rate of 2% per year will double in 35 years, while a population growing at 3% per year will double in 23 years.

Seen in one way, the phenomenal growth of human population and economic activity is a success story whose hero is technical progress. Almost everyone now living owes his or her life to modern techniques of agriculture, industry and medicine. If humans had remained hunter-gatherers, the total global population would have continued to be only a few millions; and under those conditions, almost everyone now living would never have been born, or would have died in childhood. Therefore most of us must thank the progress of society for the fact that we are alive at all.

However, if we compare the present growth rates of population and economic activity with the world's reserves of non-renewable resources and arable land, the picture changes: We can then see the beginnings of a tragedy, with growth and "progress" perhaps playing the roles of villains.

Population and food supply

In 1930, the population of the world reached two billion; in 1958 three billion; in 1974 four billion; and in 1988 five billion. Today, more than 90 million people are being added to the world's population every year. United Nations experts believe that by the year 2100, the population of the earth will have stabilized at between 10 and 15 billion - roughly double or triple today's population - most of the increase having been added to the less-developed parts of the world.

In 1983, the Secretary-General of the United Nations established a World Commission on Environment and Development, led by the Prime Minister of Norway, Gro Harlem Brundtland. The Commission's report, *"Our Common Future"* (published in 1987), examines the question of whether the earth can support a population of 10 billion people without the collapse of the ecological systems on which all life depends. With respect to food, the report has this to say:

"...Researchers have assessed the 'theoretical' potential for global food production. One study assumes that the area under food production can be around 1.5 billion hectares (3.7 billion acres - close to the present level), and that the average yields could go up to 5 tons of grain equivalent per hectare (as against the present average of 2 tons of grain equivalent). Allowing for production from rangelands and marine sources, the total 'potential' is placed at 8 billion tons of grain equivalent."

"How many people can this sustain? The present global average consumption of plant energy for food, seed, and animal feed amounts to about 6,000 calories daily, with a range among countries of 3,000-15,000 calories, depending on the level of meat consumption. On this basis, the potential production could sustain a little more than 11 billion people. But if the average consumption rises substantially - say, to 9,000 calories - the population carrying capacity of the Earth comes down under 7.5 billion."

"These figures could be substantially higher if the area under food production and the productivity of 3 billion hectares of permanent pasturage can be increased on a sustainable basis. Nevertheless, the data do suggest that meeting the food requirements of an ultimate world population of around 10 billion would require some changes in food habits, as well as greatly improving the efficiency of traditional agriculture."

Thus, the next doubling will bring the global population of humans near to or beyond the maximum number that the earth can support, even assuming greatly improved agricultural yields. The study quoted in the

Brundtland report assumes that the world average for agricultural yields per hectare can be doubled; but this assumption raises many problems.

Extremely high-yield varieties of rice and wheat have indeed been produced by "Green Revolution" plant geneticists, such as Norman Borlaug. However, these high-yield crop varieties require heavy use of chemical fertilizers and pesticides, as well as large amounts of water. Will the enormous quantities of fertilizer required be available globally?

According to a recent study (*Man's Impact on the Global Environment*, MIT Press, 1970), the world's food production rose by 34% between 1951 and 1966; but this required a 146% increase in the use of nitrate fertilizers, and a 300% increase in the use of pesticides. Between 1964 and 1987, the fertilizer consumption of Asia increased by a factor of 10, from 4 million metric tons to 40 million metric tons. Much greater increases will be needed if global agriculture is to double its productivity per hectare during the next half century. Assuming the availability of the needed amounts of fertilizer, we can anticipate that the runoff from fields, heavily saturated with nitrates and phosphates and pesticides, will contaminate the ground-water, lakes and oceans, thus reducing fish populations.

One can already observe a catastrophic depletion of oxygen in the bottom layers of such bodies of water as the Baltic Sea (which is surrounded by countries presently making heavy use of fertilizers in agriculture). This oxygen depletion is due to the growth of algae in layers near to the surface, stimulated by the presence of nitrates and phosphates. Bacterial decay of the algae at the bottom exhausts the oxygen; and in many parts of the Baltic, all bottom-living species have disappeared.

Pesticides and fertilizer in drinking water can cause a variety of human health problems, including cancer and methemoglobinemia. (Methemoglobinemia is sometimes called "blue baby syndrome", and it results from drinking water containing too large a concentration of nitrates.)

If a global population of 10 billion is to be supported, another alternative is open: More land can be exploited for agriculture. However, we may encounter as many problems in doubling the area of the world's agricultural land as in doubling its productivity per hectare.

The cost of roads, irrigation, clearance and fertilizer for new agricultural land averages more than a thousand U.S. dollars per hectare. During the next half century, hunger will strike the poorest parts of the world's population. Capital for opening new agricultural land cannot come from those who are threatened by famine. It must be found in some other way.

A Report by the United Nations Food and Agricultural Organization

(*Provisional Indicative World Plan for Agricultural Development*, FAO, Rome, 1970) makes the following statement concerning new agricultural lands:

"In Southern Asia,...in some countries in Eastern Asia, in the Near East, and North Africa...there is almost no scope for expanding the agricultural area... In the dryer regions, it will even be necessary to return to permanent pasture the land which is marginal or submarginal for cultivation. In most of Latin America and Africa south of the Sahara, there are still considerable possibilities for expanding cultivated areas; but the costs of development are high, and it will often be more economical to intensify the utilization of the areas already settled."

In the 1950's, both the U.S.S.R and Turkey attempted to convert arid grasslands into wheat farms. In both cases, the attempts were defeated by drought and wind erosion, just as the wheat farms of Oklahoma were overcome by drought and dust in the 1930's.

If irrigation of arid lands is not performed with care, salt may be deposited, so that the land is ruined for agriculture. This type of desertification can be seen, for example, in some parts of Pakistan. Another type of desertification can be seen in the Sahel region of Africa, south of the Sahara. Rapid population growth in the Sahel has led to overgrazing, destruction of trees, and wind erosion, so that the land has become unable to support even its original population.

The earth's tropical rain forests are also rapidly being destroyed for the sake of new agricultural land. Tropical rain forests are thought to be the habitat of more than half of the world's species of plants, animals and insects; and their destruction is accompanied by an alarming rate of extinction of species. The Harvard biologist, E.O. Wilson, estimates that the rate of extinction resulting from deforestation in the tropics may now exceed 4,000 species per year - 10,000 times the natural background rate (*Scientific American*, September, 1989).

The enormous biological diversity of tropical rain forests has resulted from their stability. Unlike northern forests, which have been affected by glacial epochs, tropical forests have existed undisturbed for millions of years. As a result, complex and fragile ecological systems have had a chance to develop. Professor Wilson expresses this in the following words:

"Fragile superstructures of species build up when the environment remains stable enough to support their evolution during long periods of time. Biologists now know that biotas, like houses of cards, can be brought tumbling down by relatively small perturbations in the physical environment. They are not robust at all."

The number of species which we have until now domesticated or used in medicine is very small compared with the number of potentially useful species still waiting in the world's tropical rain forests. When we destroy them, we damage our future. But we ought to regard the annual loss of thousands of species as a tragedy, not only because biological diversity is potential wealth for human society , but also because every form of life deserves our respect and protection.

Every year, more than 100,000 square kilometers of rain forest are cleared and burned, an area which corresponds to that of Switzerland and the Netherlands combined. Almost half of the world's tropical forests have already been destroyed. Ironically, the land thus cleared often becomes unsuitable for agriculture within a few years.

Tropical soils may seem to be fertile when covered with luxuriant vegetation, but they are usually very poor in nutrients because of leeching by heavy rains. The nutrients which remain are contained in the vegetation itself; and when the forest cover is cut and burned, they are rapidly leached away.

Often the remaining soil is rich in aluminum oxide and iron oxide. When such soils are exposed to oxygen and sun-baking, a rock-like substance called laterite is formed. The temples of Angkor Wat in Cambodia are built of laterite; and it is thought that the Khmer civilization, which built these temples a thousand years ago, disappeared because of laterization of the soil.

It can be seen from the facts which we have just discussed that increasing the world's food supply to accommodate the next doubling of population will be difficult. If this goal can be achieved at all, it will be achieved at the cost of severe damage to the global environment and the extinction of many thousands of species.

Added to the agricultural and environmental problems, are problems of finance and distribution. Famines can occur even when grain is available somewhere in the world, because those who are threatened with starvation may not be able to pay for the grain, or for its transportation. The economic laws of supply and demand are not able to solve this type of problem. One says that there is no "demand" for the food (meaning demand in the economic sense), even though people are in fact starving.

We can anticipate that as the earth's human population approaches 10 billion, severe famines will occur in many developing countries. The beginnings of this tragedy can already be seen. It is estimated that roughly 40,000 children now die every day from starvation, or from a combination of disease and malnutrition. This terrible suffering and loss of life is almost certain to become worse in the next few decades; and the fact that the problem of increasing the world's food supply is very difficult by no means decreases its urgency.

An analysis of the global ratio of population to cropland shows that we may already have exceeded the sustainable limit of population through our dependence on petroleum: Between 1950 and 1982, the use of cheap petroleum-derived fertilizers increased by a factor of 8, and much our present agricultural output depends their use. Furthermore, petroleum-derived synthetic fibers have reduced the amount of cropland needed for growing natural fibers, and petroleum-driven tractors have replaced draft animals which required cropland for pasturage. Also, petroleum fuels have replaced fuelwood and other fuels derived for biomass. The reverse transition, from fossil fuels back to renewable energy sources, will require a considerable diversion of land from food production to energy production. For example, 1.1 hectares are needed to grow the sugarcane required for each alcohol-driven Brazilian automobile. This figure may be compared with the steadily falling average area of cropland available to each person in the world - .24 hectares in 1950, .16 hectares in 1982.

As population increases, the cropland per person will continue to fall, and we will be forced to make still heavier use of fertilizers to increase output per hectare. Also marginal land will be used in agriculture, with the probable result that much land will be degraded through erosion and salination. Reserves of oil are likely to be exhausted by the middle of next century. Thus there is a danger that just as global population reaches the unprecedented level of 10 billion or more, the agricultural base for supporting it may suddenly collapse. The resulting ecological catastrophe, possibly compounded by war and other disorders, could produce famine and death on a scale unprecedented in history - a catastrophe of unimaginable proportions, involving billions rather than millions of people. The present tragic famine in Africa is to this possible future disaster what Hiroshima is to the threat of thermonuclear war - a tragedy of smaller scale, whose horrors should be sufficient, if we are wise, to make us take steps to avoid the larger catastrophe.

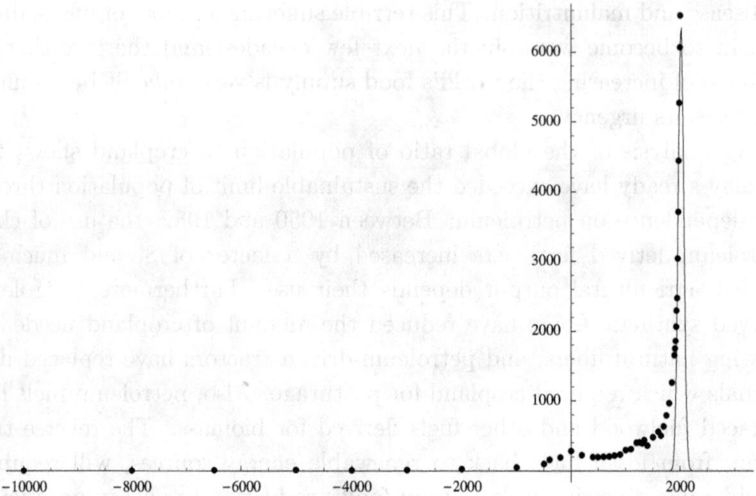

Fig. 19.1 *Population growth and fossil fuel use, seen on a time-scale of several thousand years. The dots are population estimates in millions from the US Census Bureau. Fossil fuel use appears as a spike-like curve, rising from almost nothing to a high value, and then falling again to almost nothing in the space of a few centuries. When the two curves are plotted together, the explosive rise of global population is seen to be simultaneous with, and perhaps partially driven by, the rise of fossil fuel use. This raises the question of whether the world's population is headed for a crash when the fossil fuel era has ended. (Author's own graph)*

Growth of cities

The global rate of population growth has slowed from 2.0% per year in 1972 to 1.7% per year in 1987; and one can hope that it will continue to fall. However, it is still very high in most developing countries. For example, in Kenya, the population growth rate is 4.0% per year, which means that the population of Kenya will double in seventeen years.

Because of increasing mechanization of agriculture, the extra millions added to the populations of developing countries are unable to find work on the land. They have no alternative except migration to overcrowded cities, where the infrastructure is unable to cope with so many new arrivals. Often the new migrants are forced to live in excrement-filled makeshift slums, where dysentery, hepatitis and typhoid are endemic, and where the conditions for human life sink to the lowest imaginable level.

During the 60 years between 1920 and 1980 the urban population of the developing countries increased by a factor of 10, from 100 million to almost a billion. In 1950, the population of Sao Paulo in Brazil was 2.7 million. By 1980, it had grown to 12.6 million; and it is expected to reach 24.0 million by the year 2000. Mexico City too has grown explosively to an unmanageable size. In 1950, the population of Mexico City was 3.05 million; in 1982 it was 16.0 million; and the projected population for 2000 is 26.3 million.

A similar explosive growth of cities can be seen in Africa and in Asia. In 1968, Lusaka, the capital of Zambia, and Lagos, the capital of Nigeria, were both growing at the rate of 14% per year, doubling in size every 5 years. In 1950, Nairobi, the capital of Kenya, had a population of 0.14 million. By 2000, it is expected to reach 5.3 million, having increased by a factor of almost 40.

In 1972, the population of Calcutta was 7.5 million, and it is expected to almost double in size by the turn of the century. This growth will produce a tragic increase in the poverty and pollution from which Calcutta already suffers. The Hoogly estuary near Calcutta is already choked with untreated industrial waste and sewage, and sixty% of Calcutta's population already suffer from respiratory diseases related to air pollution.

Governments in the third world, struggling to provide clean water, sanitation, roads, schools, medical help and jobs for all their citizens, are defeated by rapidly growing urban populations. Often the makeshift shantytowns inhabited by new arrivals have no piped water; or when water systems exist, the pressures may be so low that sewage seeps into the system.

Many homeless children, left to fend for themselves, sleep and forage in the streets of third world cities. These conditions have tended to become worse with time rather than better. Whatever gains governments can make are immediately canceled by growing populations.

The demographic transition

In discussing the Industrial Revolution, we noticed a general pattern in the social impact of technical change: Since technical changes can take place extremely rapidly, while social and political adjustments require more time, the first impact of new technology often throws society off balance, producing an initial period of suffering and social disruption. However, once society has made the needed adjustments, new techniques are usually beneficial.

In the case of the Industrial Revolution, great suffering resulted when an agricultural society, with traditional rights and duties, was replaced by a society functioning according to purely economic rules, where labor was regarded as a commodity to be bought and sold without regard for the needs of the humans involved. Later, however, after the appropriate social adjustments had been made, industrialization yielded great benefits.

We have just been discussing a more recent example of social dislocation and suffering produced by the initial impact of technical change: Advanced medical techniques transferred from industrialized countries to the third world have quickly lowered death rates without affecting basic social structures and traditions. The result has been overpopulation and poverty.

For example, in Sri Lanka (Ceylon), the death rate fell sharply, from 22 per thousand in 1945 to 10 per thousand in 1954, largely as the result of an antimalarial program. However, social customs remained the same: Girls continued to be married very early; and they continued to give their husbands large numbers of children, just as they had done when the death rate was high. The result was a population explosion which has produced almost as much suffering as the malaria which it replaced.

In the 1950's and 1960's there was great hope that transfer of technology from the industrialized countries would lead to development and prosperity in all parts of the world. President Kennedy proposed that the 1960's should be designated a "development decade", and this proposal was adopted by the United Nations.

The good intentions of the development decade were backed by substantial aid: According to official estimates, the industrialized nations contributed 8 billion U.S. dollars per year to the less developed parts of the world. However, in most third world countries, exploding populations blocked economic development, producing a trap of poverty. The gap between the rich and poor nations widened, rather than narrowed.

Rapidly-growing populations are both the cause and the effect of poverty: As we have seen, a rapidly-growing population makes economic development difficult or impossible. Furthermore, in an educated, prosperous, urban population, where women have high social status and jobs outside the home, the birth rate tends to be low. For example, in Denmark, each woman has, on the average, fewer than two children during her lifetime.

A recent study (conducted by Robert J. Lapham of the Demographic and Health Surveys and by W. Parker Mauldin of the Rockefeller Foundation) has shown that the use of birth control is correlated both with socio-economic setting and with the existence of strong family-planning programs. For example, in countries like Yemen, Burundi, Chad, Guinea, Malawi, Mali, Niger, Burkina Faso and Mauritania, where family-planning programs are weak or absent, only 1% of couples use birth control.

In Paraguay, where the socio-economic setting is high, but where a family-planning program is absent, 36% of couples use birth control. In Indonesia, with a lower-middle socio-economic setting but a strong government-supported family-planning program, the percentage is 48%. Finally, in Hong Kong, which has both a relatively high socio-economic status and a strong family-planning program, 80% of all couples use birth control.

China, the world's most populous nation, has adopted the policy of allowing only one child per family. This policy has, until now, been most effective in towns and cities, but with time it may also become effective in rural areas. Like other developing nations, China has a very young population, which will continue to grow even when fertility falls below the replacement level (because so many of its members will be contributing to the birth rate rather than to the death rate). China's present population is between 1.1 and 1.2 billion. Its projected population for the year 2025 is 1.5 billion.

Recent statistics show that the world can be divided into two demographic regions of roughly equal population. In the first region, which includes North America, Europe, the former Soviet Union, Australia, New Zealand and Eastern Asia, populations have completed or are completing the demographic transition from the old equilibrium where high birth rates were balanced by a high death rate to a new equilibrium with low birth rates balanced by a low death rate. In the second region, which includes Southeast Asia, Latin America, the Indian subcontinent, the Middle East and Africa, populations seem to be caught in a demographic trap, where high birth rates and low death rates lead to population growth so rapid that the development which could have slowed population growth is impossible.

The average population increase in the slow growth regions is 0.8% per year, with a range between 0.2% (Western Europe) and 1.0% (Eastern Asia). In the rapid growth regions, the average increase is 2.5% per year, with a range between 2.2% (Southeast Asia) and 2.8% (Africa). Thus there is a very marked division of the world into two demographic regions, and there seems to be no middle ground. Some individual countries in the rapid growth regions (such as Argentina, Cuba and Uruguay in Latin America) have completed or are completing the demographic transition, but their numbers are too small to influence the regional trends.

For countries caught in the demographic trap, government birth control programs are especially important, because one cannot rely on improved social conditions to slow birth rates. Since health and lowered birth rates should be linked, it is appropriate that family-planning should be an important part of programs for public health and economic development. In 1977, the World Health Organization resolved that during the coming decades its goal should be "the attainment by all citizens of the world by the year 2000 of a level of health that will permit them to lead a socially and economically productive life". Halfdan Mahler, who was then the Director General of the World Health Organization, has expressed the relationship between health, development and family planning in the following words:

"Country after country has seen painfully achieved increases in total output, food production, health and educational facilities and employment opportunities reduced or nullified by excessive population growth. Most underdeveloped countries therefore seek to limit their population growth."

"The lesson of recent years is that virtually wherever health-care facilities have been made available, women have demanded information and the necessary materials for spacing their children and limiting their families."

Non-renewable resources

Economists in the industrialized countries have long behaved as though growth were synonymous with economic health. If the gross national product of a country increases steadily by 4% per year, most economists express approval and say that the economy is healthy. If the economy could be made to grow still faster (they feel), it would be still more healthy. If the growth rate should fall, economic illness would be diagnosed.

Economics has been called "the impatient science of growth", and (with a few notable exceptions, such as the Club of Rome) economists seem to assume that growth can continue forever. This assumption, of course, cannot stand examination any better than the assumption that population can continue to grow forever. A "healthy" economic growth rate of 4% per year corresponds to increase by a factor of 50 in a century, by a factor of 2500 in two centuries, and by a factor of 125,000 in three centuries. No one can maintain that this type of growth is "sustainable" except by refusing to look more that a certain distance into the future.

It is obvious that on a finite earth, population cannot continue to grow indefinitely because of limits imposed by the food supply and because of limits to the ability of the environment to tolerate pollution. Exponential growth, where the population doubles in size every generation or every few generations, has brought us near to these limits with surprising rapidity. It is characteristic of exponential growth that one is surprised by the sudden approach of the limits, because one moves from a situation of plenty to one of scarcity in a single doubling time.

As we have seen above, global population will soon exceed the carrying capacity of the environment. Economic growth will encounter the same limit, as well as limits imposed by the depletion of non-renewable resources. Our failure to see this fact clearly is probably due to our unwillingness to look more than a few years ahead. We say to ourselves, "What happens fifty years from now is not our worry". However we owe it to our children to try look as far as possible into the future, since "we did not inherit the earth from our parents; we borrowed it from our children".

The total ultimately recoverable resources of fossil fuels amount to roughly 7300 terawatt-years of energy[1] Of this total amount, 6700 TWy is coal, while oil and natural gas each constitute roughly 300 TWy.[2] In 1890, global consumption of energy was 1 terawatt, but by 1990 this figure had grown to 13.2 TW, distributed as follows: oil, 4.6; coal, 3.2; natural gas, 2.4; hydropower, 0.8; nuclear, 0.7; fuelwood, 0.9; crop wastes, 0.4; and dung, 0.2. Thus, if we continue to use oil at the 1990 rate, it will last for 65 years, while natural gas will last for twice that long. The reserves of coal are much larger; and used at the 1990 rate, coal would last for 2000 years. However, it seems likely that as oil and natural gas become depleted, coal will be converted to liquid and gaseous fuels, and its rate of use will increase. Also, the total global energy consumption is likely to increase because of increasing population and rising standards of living in the developing countries.

It is easy to calculate that a global population of 10 billion, using oil and energy at the same rate as present-day Americans, could exhaust the world's supply of petroleum in seven years, and could burn all of the world's remaining reserves of fossil fuels in only 60 years, meanwhile producing a catastrophic change in the earth's climate through the release of greenhouse gases. It may be just as difficult for the developed countries to abandon their habit of encouraging economic growth as it will be for the developing countries to abandon their habit of encouraging large families; but both

[1] 1 terawatt $\equiv 10^{12}$ Watts is equivalent to 5 billion barrels of oil per year or 1 billion tons of coal per year

[2] British Petroleum, "B.P. Statistical Review of World Energy", London, 1991

these changes of attitude are necessary for the future of our planet.

The burning of coal and oil, and the burning of tropical rain forests, release so much carbon dioxide that its atmospheric concentration has increased from 290 parts per million in 1860 to 347 parts per million in 1985. At present 6 billion tons of carbon are released into the atmosphere every year by human activities; and if this continued at the same rate, the CO_2 concentration will reach 550 ppm by the end of the 21st century (double the preindustrial concentration) with a resulting global warming of between 3 and 5 °C. Although the exact climatic consequences of this warming are difficult to predict, there is a fear that some areas of the world which are now able to produce and export large quantities of grain may become arid.

Global warming of between 3 and 5 °C would also produce a rise in sea level of between 1 and 2 meters (because of the expansion of the water in the oceans and because of melting of the polar ice caps) with a resulting loss of fertile cropland in low-lying regions of the world. Thus, both because of limited reserves and because of the greenhouse effect, we will be forced to replace fossil fuels by renewable energy sources.

The industrialized countries use much more than their fair share of global resources. For example, with only a quarter of world's population they use more than two thirds of its energy; and in the U.S.A. and Canada the average per capita energy consumption is 12 kilowatts, compared with 0.1 kilowatts in Bangladesh. If we are to avoid severe damage to the global environment, the industrialized countries must rethink some of their economic ideas, especially the assumption that growth can continue forever.

The present use of resources by the industrialized countries is extremely wasteful. A growing national economy must, at some point, exceed the real needs of the citizens. It has been the habit of the developed countries to create artificial needs by means of advertising, in order to allow economies to grow even beyond the point where all real needs have been met; but this extra growth is wasteful, and in the future it will be important not to waste the earth's diminishing supply of non-renewable resources.

Thus, the times in which we live present a challenge: We need a revolution in economic thought. We must develop a new form of economics, taking into account the realities of the world's present situation - an economics based on real needs and on a sustainable equilibrium with the environment, not on the thoughtless assumption that growth can continue forever.

The resources of the earth and the techniques of modern science can support a global population of moderate size in comfort and security; but the optimum size is undoubtedly much smaller than the world's present population. Given a sufficiently small global population, renewable sources of energy can be found to replace disappearing fossil fuels. These include solar energy, wind energy, geothermal energy, hydroelectric power, and energy derived from biomass.

Technology may also be able to find renewable substitutes for many disappearing mineral resources for a global population of a moderate size. What technology cannot do, however, is to give a global population of 10 billion people the standard of living which the industrialized countries enjoy today.

Like a speeding truck headed for a brick wall, the earth's rapidly growing human population and its growing economic activity are headed for a collision with a very solid barrier - the carrying capacity of the environment. As in the case of the truck and the wall, the correct response is to apply the brakes in good time.

A global population of 10 billion people using energy at the present U.S.and Canadian rate would produce catastrophic environmental degradation; and for the developed countries to continue to use resources at the present rate while denying this privilege to the rest of the world would produce dangerous political tensions. The environmental crisis thus involves not only the problems of depletion of non-renewable resources, loss of cropland through erosion and salination, poisoning of the environment through fossil fuel emissions, destruction of forests through acid rain, eutrophication of rivers and lakes, threatened climatic change from the release of greenhouse gases, and a rate of extinction of species thousands of times the normal background rate. The crisis also involves problems of social injustice - a quarter of the world's population using almost three-fourths of its resources, and dying from overeating, overdrinking and oversmoking, while the remaining three quarters of humankind lives in near-poverty or absolute poverty, lacking safe water and sanitation, lacking elementary education and primary health care, with fourteen million children dying every year from diseases, most of which are preventable by simple means, such as vaccination, re-hydration therapy and proper nutrition.

In June, 1992, 35000 people from 172 countries met at Rio de Janero in Brazil for the United Nations Conference on Environment and Development. They included 118 heads of state or heads of governments, and before the meeting there were high hopes for international agreement on a new and equitable world order and for agreements which would address critical environmental problems. However, although some progress was made, the results of the meeting were disappointing because discussion of the two most important problems, overconsumption in the industrialized countries and the population explosion in the developing countries, was blocked respectively by the North and the South.

To avoid a North-South confrontation like that which blunted the effectiveness of the Rio meeting, a compromise is needed: Through a combination of increased energy efficiency and a more modest lifestyle (especially more modest transportation requirements) we should aim at a global society where both the developed and developing countries reach the same per capita energy consumption of between 1.5 and 3 kilowatts per person. This rate of energy consumption is near to the present global average. It is, however, considerably less than the present U.S. and Canadian level of 12 kilowatts per person and very much greater than the present figure for Bangladesh - 0.1 kilowatts per person!

The developed world must reduce its consumption of fossil fuels and other resources while aiming at a life which would have a high quality in other respects than purely material ones. The developing world should find its own way forward to the future, not imitating the wasteful and unsustainable lifestyle of the west, but evolving a way of life which is high in quality but low in resource consumption.

A more modest life-style need not be unpleasant. What is needed is a change in our system of values. We should recognize that a high quality of life is not synonymous with a high level of consumption. For example, the quality of life in our cities would be improved by a shift from private cars to bicycles and public transport, and this would at the same time reduce our consumption of energy. A less hectic and consumption-oriented life-style would also give us more leisure to enjoy our families.

In today's world, power and material goods are valued more highly than they deserve to be. "Civilized" life often degenerates into a struggle of all against all for power and possessions. However, the industrial complex on which the production of goods depends cannot be made to run faster and faster indefinitely, because we will soon encounter shortages of energy and raw materials.

Looking ahead to the distant future, we can hope that the values of society will change, and that non-material human qualities, such as kindness, politeness, knowledge, and musical, artistic or literary ability, will come to be valued more highly, and that people will derive a larger part of their pleasure from the appreciation of unspoiled nature.

Our power-worshiping industrial society can perhaps learn from the values of our hunter-gatherer ancestors, who lived in harmony with nature. We are now so numerous that we cannot return to a primitive way of life; but we can learn to respect nature as our ancestors did. Harmony is a better ideal than power. We must learn to live in harmony with other humans and with other species. We must learn to care for the earth.

Suggestions for further reading

(1) D.H. Meadows, D.L. Meadows, J. Randers and W.W. Behrens, *The Limits of Growth*, (Reports to the Club of Rome), The New American Library, New American Library, New York (1972).

(2) Berry Commoner, *The Closing Circle: Nature, Man and Technology*, Bantam Books, New York (1972).

(3) Barbara Ward and René Dubos, *Only One Earth*, Penguin Books Ltd. (1973).

(4) Gerald Foley, *The Energy Question*, Penguin Books Ltd. (1976).

(5) J. Holdren and P. Herrera, *Energy*, Sierra Club Books, New York (1971).

(6) J.R. Frisch (editor), *Energy 2000-2020: World Prospects and Regional Stresses*, World Energy Conference, Graham and Trotman (1983).

(7) T.R. Malthus, *An Essay on the Principle of Population*, J.M. Dent and Sons, London (1963).

(8) Elizabeth Draper, *Birth Control in the Modern World*, Penguin Books Ltd. (1972).

(9) Ernest Havemann, *Birth Control*, Time-Life Books (1967).

(10) Gordon Bridger and Maurice de Soissons, *Famine in Retreat?*, Dent, London (1970).

(11) Roland Pressat, *Population*, Penguin Books Ltd. (1970).

(12) Carlo M. Cipola, *The Economic History of World Population*, Penguin Books Ltd. (1974).

(13) Paul R. Ehrlich, *The Populatiuon Bomb*, Sierra/Ballentine, New York (1972).

(14) Paul R. Ehrlich, Anne H. Ehrlich and John P. Holdren, *Human Ecology* W.H. Freeman (1973).

(15) World Commission on Environment and Development, *Our Common Future*, Oxford University Press (1987).

(16) William C. Clark and others, *Managing Planet Earth*, Scientific American, Special Issue, September (1989).

(17) World Bank, *Poverty and Hunger; Issues and Options for Food Security in Developing Countries*, Washington D.C. (1986).

(18) G. Hagman and others, *Prevention is Better Than Cure*, Report on Human Environmental Disasters in the Third World, Swedish Red Cross, Stockholm (1986).

(19) P.W. Hemily and M.N. Ozdas (editors), *Science and Future Choice*, Clarendon Press, Oxford (1979).

(20) Council on Environmental Quality and U.S. Department of State, *The Global 2000 Report to the President: Entering the Twenty-First Century, The Technical Report, Volume 2*, U.S. Government Printing Office, Washington D.C. (1980).

(21) L. Timberlake, *Only One Earth: Living for the Future*, BBC/Earthscan, London (1987).

(22) L.R. Brown and others, *State of the World in 1987*, W.W. Norton, London (1987).

(23) UNESCO, *International Coordinating Council of Man and the Biosphere*, MAB Report Series No. 58, Paris (1985).

(24) Peter Donaldson, *Worlds Apart; The Economic Gulf Between Nations*, Penguin Books Ltd. (1973).

Chapter 20

LOOKING TOWARDS THE FUTURE

A sustainable global society

From the history sketched briefly in the preceding chapters, we can see that the impact of science on society has been profound and on the whole beneficial. However, we can also see that because the evolution of our social and political institutions is slow compared with the enormous speed of scientific and technological change, this change has frequently thrown society off balance. For example, the Industrial Revolution produced great suffering until social legislation, labor organization and birth control distributed its benefits more evenly. Similarly, the rapid lowering of death rates in the developing countries through the introduction of modern medicine has thrown these countries off balance: The resulting population explosion has produced terrible poverty and suffering and has blocked development. Equilibrium can only be restored by lowering birth rates; but when this is done the effects of medical progress will be purely beneficial.

Uncontrolled industrial expansion in the developed countries is now leading to a new situation where society will be thrown off balance: We now face environmental degradation and depletion of non-renewable resources. To prevent these negative effects of progress we must make the appropriate economic and social adjustments. Similarly, automation will lead to widespread unemployment unless we think carefully about the economic and political adjustments which will be needed to avoid it. Finally, science and technology have produced weapons of such destructiveness, and global communications and interdependence have increased to such an extent that our present international political system seems inadequate, characterized as it is by absolutely sovereign nation-states and an absence of international law.

Thus, the rapid growth of science-based technology has presented both dangers and opportunities: If we use science wisely - if we build a new global society where population is stabilized, where ecology and economics are merged to form a single discipline, and where our political and ethical development matches our technical progress, then we have the opportunity for a degree of widely shared happiness previously unknown in history. If not, technical progress presents us with dangers of catastrophe on a scale previously unknown.

It is interesting that the word for "crisis", when written in Japanese, consists of two characters, one meaning "danger" and the other "opportunity"; and this Japanese double word is very appropriate to describe our present situation. It is up to us to build a future world where the opportunities will be utilized and the dangers avoided. Our responsibility to future generations calls us to give our best efforts to this cause.

The end of the Cold War provides us with a unique opportunity, because there is now a general consensus that war is unacceptable as a means of settling international disputes, and because of the enormous amounts of money which a reduction in military spending can release for constructive uses - the "peace dividend". If properly used, the peace dividend can help us to take the steps needed to build a sustainable global society, and at the same time re-employ young people thrown out of work by automation.

What are the necessary steps towards sustainability? The Worldwatch Institute, Washington D.C., lists the following: 1. Stabilizing population; 2. Shifting to renewable energy; 3. Increasing energy efficiency 4. Recycling resources; 5. Reforestation; 6. Soil conservation.[1] All of these measures are labor-intensive, and they can therefore help us to solve the problem of technological unemployment. Especially the shift to renewable energy sources will be an enormous, labor-intensive task.

The transition from fossil fuel use (at present 77 percent of total energy consumption) to renewable energy sources should begin immediately. This transition will be difficult and time-consuming because of the immense capital investment in our present energy-production system - roughly 8 trillion dollars - and because of the long lifetimes of installations - typically 40 years.

Renewable energy sources include wind energy, hydroelectric power, energy from tides, geothermal energy, biomass and solar energy. Power from nuclear fission is not renewable, since uranium is needed for fuel. Further-

[1]Lester R. Brown and Pamela Shaw, Worldwatch Paper 48, March 1982.

more, widespread use of fission for power generation would carry a severe danger of nuclear weapon proliferation because plutonium is produced as a byproduct. Fusion does not have these drawbacks, but it is difficult to predict when or whether it will become an economically viable energy source.

Several forms of renewable energy technology have reached or are nearing the stage where they can compete in price with fossil fuels. For example, in Brazil a highly efficient technology has been developed for producing ethanol from sugar cane. Anhydrous ethanol is combined with 20% gasoline and used as a motor fuel. In 1981, Brazil produced 4 billion liters of ethanol for fuel at costs as low as 18.5 U.S. cents per liter. Vehicles driven on the ethanol-gasoline mixture produce very little local pollution, and no net CO_2 is released into the atmosphere by the burning of ethanol derived from photosynthesis.

Fig. 20.1 *Mityaevo Solar Station in Ukraine. By Active Solar [CC BY-SA 4.0], Wikimedia Commons.*

Another promising renewable energy technology uses thermal or photovoltaic solar energy devices to split water into hydrogen and oxygen. It is estimated that solar installations covering 500,000 square kilometers (2% of the world's desert area) could produce hydrogen equivalent to the world's total fossil fuel consumption. The hydrogen would then be compressed and distributed by pipeline to centers of population and industry. Fuel cell technologies are being developed for the direct conversion of hydrogen's energy into electricity. In one design, H_2 molecules are converted to H^+ ions and free electrons at a permeable anode. The electrons flow through an external circuit, providing power. Meanwhile, the H^+ ions migrate through a phosphoric acid solution to the cathode, where they combine with the electrons and molecular oxygen, producing steam. If the energy of the steam is utilized, the efficiency of such fuel cells can be as high as 60%.

In 2012, the World Bank issued a report warning that without quick action to curb CO_2 emissions, global warming is likely to reach 4 °C during the 21st century. This is dangerously close to the temperature which initiated the Permian-Triassic extinction event: 6 °C above normal.[2] The Permian-Triassic extinction event occurred 252 million years ago. In this event, 96% of all marine species were wiped out, as well as 70% of all terrestrial vertebrates.

New hope that such a disaster can be avoided comes from the inspiring and beautiful encyclical *Laudato si'* by Pope Francis. We can hope that other world leaders will speak out equally strongly on the need for a rapid end to the fossil fuel era. Quick governmental action is also needed. The enormous subsidies presently given to the fossil fuel and nuclear industries must be ended, or better yet used instead to support the development of renewable energy infrastructure.

Another reason for hope can be found in the extremely high present growth rate of renewable energy, and in the remarkable properties of exponential growth. According to figures recently released by the Earth Policy Institute,the global installed photovoltaic capacity is currently (in 2015) able to deliver 240,000 megawatts, and it is increasing at the rate of 27.8% per year. Wind energy can deliver 370,000 megawatts, and it is increasing at the rate of 20% per year. Because of the remarkable properties of exponential growth, we can calculate that if these growth rates are continued, renewable energy can give us 28,000,000 megawatts within only 15 years. This is more than the present rate of use of all forms of energy.

The need for a system of international law

It is extremely important that research funds be used to develop renewable energy sources and to solve other urgent problems now facing humankind, rather than for developing new and more dangerous weapons systems. In

[2]http://www.worldbank.org/en/news/feature/2012/11/18/Climate-change-report-warns-dramatically-warmer-world-this-century

spite of the end of the Cold War, the world still spends more than a *trillion* U.S. dollars per year on armaments. At present, more than 40% of all research funds are used for projects related to the arms industry.

Since the Second World War, there have been over 150 armed conflicts; and on any given day, there are an average of 12 wars somewhere in the world. While in earlier epochs it may have been possible to confine the effects of war mainly to combatants, in recent decades the victims of war have increasingly been civilians, and especially children.

Civilian casualties often occur through malnutrition and through diseases which would be preventable in normal circumstances. Because of the social disruption caused by war, normal supplies of food, safe water and medicine are interrupted, so that populations become vulnerable to famine and epidemics. In the event of a nuclear war, starvation and disease would add greatly to the loss of life caused by the direct effects of nuclear weapons.

The indirect effects of war and the threat of war are also enormous. For example, the World Health Organization lacks funds to carry through an anti-malarial programme on as large a scale as would be desirable; but the entire programme could be financed for less than the world spends on armaments in a single day. Five hours of world arms' spending is equivalent to the total cost of the 20-year WHO programme which resulted, in 1979, in the eradication of smallpox. With a diversion of funds consumed by three weeks of the military expenditures, the world could create a sanitary water supply for all its people, thus eliminating the cause of more than half of all human illness.

It is often said that we are economically dependent on war-related industries; but if this is so, it is a most unhealthy dependence, analogous to drug-dependence or alcoholism. From a purely economic point of view, it is clearly better to invest in education, roads, railways, reforestation, retooling of factories, development of disease-resistant high-yield wheat varieties, industrial research, research on utilization of solar and geothermal energy, and other elements of future-oriented economic infrastructure, rather than building enormously costly warplanes and other weapons. At worst, the weapons will contribute to the destruction of civilization. At best, they will become obsolete in a few years and will be scrapped. By contrast, investment in future-oriented infrastructure can be expected to yield economic benefits over a long period of time.

It is instructive to consider the example of Japan and of Germany, whose military expenditures were severely restricted after World War II. The impressive post-war development of these two nations can very probably be attributed to the restrictions on military spending which were imposed on them by the peace treaty.

As bad as conventional arms and conventional weapons may be, it is the possibility of a nuclear war that still poses the greatest threat to humanity. One argument that has been used in favor of nuclear weapons is that no sane political leader would employ them. However, the concept of deterrence ignores the possibility of war by accident or miscalculation, a danger that has been increased by nuclear proliferation and by the use of computers with very quick reaction times to control weapons systems.

With the end of the Cold War, the danger of a nuclear war between superpowers has faded; but because of nuclear proliferation, there is still a danger of such a war in the Middle East or in the India-Pakistan dispute, as well as the danger of nuclear blackmail by terrorists or political fanatics.

Recent nuclear power plant accidents remind us that accidents frequently happen through human and technical failure, even for systems which are considered to be very "safe". We must also remember the time scale of the problem. To assure the future of humanity, nuclear catastrophe must be avoided year after year and decade after decade. In the long run, the safety of civilization cannot be achieved except by the abolition of nuclear weapons, and ultimately the abolition of the institution of war.

In the long run, because of the terrible weapons which have been produced through the misuse of science, and because of the even more destructive weapons which are likely to be devised in the future, the only way that we can insure the survival of civilization is to abolish war as an institution. It seems likely that achievement of this goal will require revision and strengthening of the United Nations Charter. The Charter should not be thought of as cast in concrete for all time. It needs instead to grow with the requirements of our increasingly interdependent global society. We should remember that the Charter was drafted and signed before the first nuclear bomb was dropped on Hiroshima; and it also could not anticipate the extraordinary development of international trade and communication which characterizes the world today.

Among the weaknesses of the present U.N. Charter is the fact that it does not give the United Nations the power to make laws which are binding on individuals. At present, in international law, we treat nations as though they were persons: We punish entire nations by sanctions when the law is broken, even when only the leaders are guilty, even though the burdens of the sanctions fall most heavily on the poorest and least guilty of the citizens, and even though sanctions often have the effect of uniting the citizens of a country behind the guilty leaders. To be effective, the United Nations needs a legislature with the power to make laws which are binding on individuals, and the power to to arrest individual political leaders for flagrant violations of international law.

Another weakness of the present United Nations Charter is the principle of "one nation one vote" in the General Assembly. This principle seems to establish equality between nations, but in fact it is very unfair: For example it gives a citizen of China or India less than a thousandth the voting power of a citizen of Malta or Iceland. A reform of the voting system is clearly needed.

The present United Nations Charter contains guarantees of human rights, but there is no effective mechanism for enforcing these guarantees. In fact there is a conflict between the parts of the Charter protecting human rights and the concept of absolute national sovereignty. Recent history has given us many examples of atrocities committed against ethnic minorities by leaders of nation-states, who claim that sovereignty gives them the right to run their internal affairs as they wish, free from outside interference.

One feels that it ought to be the responsibility of the international community to prevent gross violations of human rights, such as the use of poison gas against civilians (to mention only one of the more recent political crimes); and if this is in conflict with the notion of absolute national sovereignty, then sovereignty must yield. In fact, the concept of the absolutely sovereign nation-state as the the supreme political entity is already being eroded by the overriding need for international law. Recently, for example, the Parliament of Great Britain, one of the oldest national parliaments, acknowledged that laws made by the European Community take precedence over English common law.

Today the development of technology has made global communication almost instantaneous. We sit in our living rooms and watch, via satellite, events taking place on the opposite side of the globe. Likewise the growth of world trade has brought distant countries into close economic contact with each other: Financial tremors in Tokyo can shake New York. The impact of contemporary science and technology on transportation and communication has effectively abolished distance in relations between nations. This close contact and interdependence will increasingly require effective international law to prevent conflicts. However, the need for international law must be balanced against the desirability of local self-government. Like biological diversity, the cultural diversity of humankind is a treasure to be carefully guarded. A balance or compromise between these two desirable goals could be achieved by granting only a few carefully chosen powers to a strengthened United Nations with sovereignty over all other issues retained by the member states.

Fig. 20.2 *A Hungarian stamp showing a mother giving bread to her children and the FAO Emblem. Public domain, Wikimedia Commons.*

The United Nations has a number of agencies, such as the World Health Organization, the Food and Agricultural Organization, and UNESCO, whose global services give the UN considerable prestige and *de facto* power. The effectiveness of the UN as a global authority could be further increased by giving these agencies much larger budgets. In order to do this, and at the same time to promote the shift from fossil fuels to renewable energy sources, it has been proposed that the UN be given the power to tax CO_2 emissions. The amount of money which could thus be made available for constructive purposes is very large; and a slight increase in the prices of fossil fuels could make a number of renewable energy technologies economically competitive.

The task of building a global political system which is in harmony with modern science will require our best efforts, but it is not impossible. We can perhaps gain the courage needed for this task by thinking of the history of slavery. The institution of slavery was a part of human culture for so long that it was considered to be an inevitable consequence of human nature; but today slavery has been abolished almost everywhere in the world. The example of the dedicated men and women who worked to abolish slavery can give us courage to approach the even more important task which faces us today - the abolition of war.

Ethics in a technological age

Modern science has, for the first time in history, offered humankind the possibility of a life of comfort, free from hunger and cold, and free from the constant threat of death through infectious disease. At the same time, science has given humans the power to obliterate their civilization with nuclear weapons, or to make the earth uninhabitable through overpopulation and pollution. The question of which of these paths we choose is literally a matter of life or death for ourselves and our children.

Will we use the discoveries of modern science constructively, and thus choose the path leading towards life? Or will we use science to produce more and more lethal weapons, which sooner or later, through a technical or human failure, may result in a catastrophic nuclear war? Will we thoughtlessly destroy our beautiful planet through unlimited growth of population and industry? The choice between these alternatives is ours to make, and it is an ethical choice.

Ethical considerations have traditionally been excluded from scientific discussions. This tradition perhaps has its roots in the desire of the scientific community to avoid the bitter religious controversies which divided Europe following the Reformation. Whatever the historical reason may be, it has certainly become customary to speak of scientific problems in a dehumanized language, as though science had nothing to do with ethics or politics.

The great power of science is derived from an enormous concentration of attention and resources on the understanding of a tiny fragment of nature; but this concentration is at the same time a distortion of values. To be effective, a scientist must believe, at least temporarily, that the problem on which he or she is working is more important than anything else in the world, which is of course untrue. Thus a scientist, while seeing a fragment of reality better than anyone else, becomes blind to the larger whole. For example, when one looks into a microscope, one sees the tiny scene on the slide in tremendous detail, but that is all one sees. The remainder of the universe is blotted out by this concentration of attention.

The system of rewards and punishments in the training of scientists produces researchers who are highly competent when it comes to finding solutions to technical problems, but whose training has by no means encouraged them to think about the ethical or political consequences of their work.

Scientists may, in fact, be tempted to escape from the intractable moral and political difficulties of the world by immersing themselves in their work. Enrico Fermi, (whose research as much as that of any other person made nuclear weapons possible), spoke of science as "soma" - the escapist drug of Aldous Huxley's *Brave New World.* Fermi perhaps used his scientific preoccupations as an escape from the worrying political problems of the '30's and '40's.

The education of a scientist often produces a person with a strong feeling of loyalty to a particular research discipline, but perhaps without sufficient concern for the way in which progress in that discipline is related to the general welfare of humankind. To remedy this lack, it would be very desirable if the education of scientists could include some discussion of ethics, as well as a review of the history of modern science and its impact on society.

The explosive growth of science-driven technology during the last two centuries has changed the world completely; and our social and political institutions have adjusted much too slowly to the change. The great problem of our times is to keep society from being shaken to pieces by the headlong progress of science - the problem of harmonizing our social and political institutions with technological change. Because of the great importance of this problem, it is perhaps legitimate to ask whether anyone today can be considered to be educated without having studied the impact of science on society. Should we not include this topic in the education of both scientists and non-scientists?

Science has given us great power over the forces of nature. If wisely used, this power will contribute greatly to human happiness; if wrongly used, it will result in misery. In the words of the Spanish writer, Ortega y Gasset, "We live at a time when man, lord of all things, is not lord of himself"; or as Arthur Koestler has remarked, "We can control the movements of a spaceship orbiting about a distant planet, but we cannot control the situation in Northern Ireland."

Thus, far from being obsolete in a technological age, wisdom and ethics are needed now, more than ever before. We need the ethical insights of the great religions and philosophies of humankind - especially the insight which tells us that all humans belong to a single family, that in fact all living creatures are related, and that even inanimate nature deserves our care and respect.

Modern biology has given us the power to create new species and to exert a drastic influence on the course of evolution; but we must use this power with great caution, and with a profound sense of responsibility. There is a possibility that human activities may cause 20% of all species to become extinct within a few decades if we do not act with restraint. The beautiful and complex living organisms on our planet are the product of more than three billion years of evolution. The delicately balanced and intricately interrelated communities of living things on earth must not be destroyed by human greed and thoughtlessness. We need a sense of evolutionary responsibility - a non-anthropocentric component in our system of ethics.

Science and human values

In many ways, the scientific community is very well qualified to help in the task of building a more unified world. Science is, after all, essentially international. The great expense of scientific research can best be justified when the results are freely available to the entire international community. Furthermore, the laws of nature have a universal validity which scientists from every nation can agree upon. Almost every important scientific meeting is international, and not only international, but also characterized by a spirit of close friendship and cooperation. Also, certain human values seem to grow naturally out of the results of scientific research:

Relativity theory reminds us that the laws of nature are independent of the observer. Albert Einstein, the founder of relativity, was always unwilling to accept the prejudices of a particular time or place as representing absolute truth. Both in his scientific work, and in his moral and political judgements, he freed himself from the narrow prejudices of a particular frame of reference. Respect for objective truth and freedom from personal bias thus seem natural to anyone who has worked with relativity.

Not only relativity theory, but also thermodynamics, ought to give scientists special insight. Knowledge of the second law of thermodynamics, the statistical law favoring disorder over order, ought to make scientists especially aware of the danger of our present situation. The second law of thermodynamics reminds us that life itself is always balanced on a tightrope above an abyss of disorder: Destruction is always easier than construction.

It is easier to burn down a house than to build one - easier to kill a human than to raise and educate one. It might take only hours to destroy our civilization, but it has taken millions of dedicated hands millennia to build it.

Biology at the molecular level has shown us the complexity and beauty of even the most humble living organisms. Looking through the eyes of contemporary biochemistry, we can see that even the single cell of an amoeba is a structure of miraculous complexity and precision, worthy of our respect and wonder. This knowledge should lead us to a reverence for the order and beauty of all life, underlining the importance of a principle which religion has always taught.

The basic biochemistry of all life on earth has been shown to be the same. Thus, the insight of St. Francis, who called birds and animals his brothers and sisters, has been confirmed by modern biology. The unity of all life is a theme common to the great religions of humankind; and the truth of this theme has been confirmed by twentieth century research.

Modern astronomy has revealed the majestic dimensions of the universe, with its myriads of galaxies, each containing billions of stars; and humans have even voyaged out into space. The beauty and majesty of the fathomless universe, which men and women of our time have been privileged to see through the eyes of science, should make us not arrogant, but humble. We should recognize the vastness of what we do not know, and the smallness of what we know.

What kind of world do we want for the future? We want a world where war is abolished as an institution, and where the enormous resources now wasted on war are used constructively. We want a world where a stable population of moderate size lives in comfort and security, free from fear of hunger or unemployment. We want a world where peoples of all countries have equal access to resources, and an equal quality of life. We want a world with a new economic system where the prices of resources are not merely the prices of the burglar's tools needed to crack the safes of nature, a system which is not designed to produce unlimited growth, but which aims instead at meeting the real needs of the human community in equilibrium with the environment. We want a world of changed values, where extravagance and waste are regarded as morally wrong; where kindness, wisdom and beauty are admired; and where the survival of other species than our own is regarded as an end in itself, not just a means to our own ends. In our reverence for the intricate beauty and majesty of nature, and our respect for the dignity and rights of other humans, we can feel united with the great religious and philosophical traditions of mankind, and with the traditional wisdom of our ancestors.

Pictures sent back by the astronauts show the earth as it really is - a small, fragile, beautiful planet, drifting on through the dark immensity of space - our home, where we must learn to live in harmony with nature and with each other.

Suggestions for further reading

(1) W. Brandt, *World Armament and World Hunger: A Call for Action*, Victor Gollanz Ltd., London (1986).
(2) Olof Palme and others, *Common Security, A Programme for Disarmament*, Pan Books, London (1982).
(3) E. Chivian and others (editors), *Last Aid: The Medical Dimensions of Nuclear War*, W.H. Freeman, San Francisco (1982).
(4) Medical Association's Board of Science and Education, *The Medical Effects of Nuclear War*, Chichester, Wiley (1983).
(5) Leonard S. Spector, *The New Nuclear Nations*, Vantage Books, Random House, New York (1985).
(6) M. Khanert and others (editors), *Children and War*, Peace Union of Finland, Helsinki (1983).
(7) Konrad Lorenz, *On Aggression*, Bantam Books, New York (1977).
(8) Irenäus Eibl-Eibesfeldt, *The Biology of Peace and War*, Thames and Hudson, New York (1979).
(9) R.A. Hinde, *Biological Bases of Human Social Behaviour*, McGraw-Hill, New York (1977).
(10) R.A. Hinde, *Towards Understanding Relationships*, Academic Press, London (1979).
(11) Albert Szent-Györgyi, *The Crazy Ape*, Philosophical Library, New York (1970).
(12) E.O. Wilson, *Sociobiology*, Harvard University Press (1975).
(13) C. Zhan-Waxler, *Altruism and Aggression: Biological and Social Origins*, Cambridge University Press (1986).
(14) R. Axelrod, *The Evolution of Cooperation*, Basic Books, New York (1984).
(15) Pope Francis I, *Laudato si'*, https://laudatosi.com/watch
(16) Lester R. Brown, *Eco-Economy: Building an Economy for the Earth*, (2001), http://www.earth-policy.org/books/eco

(17) Lester R. Brown, *Plan B, 4.0: Mobilizing to Save Civilization*, (2009), http://www.earth-policy.org/books/pb4
(18) Lester R. Brown *World on the Edge*, (2011), http://www.earth-policy.org/books/wote
(19) Lester R. Brown, *Full Planet, Empty Plates*, (released in 2012). http://www.earth-policy.org/books/fpep
(20) John S. Avery, *The Need for a New Economic System*, Irene Publishing, Sparsnäs Sweden, (2016).
(21) John S. Avery, *Collected Essays, Volumes 1-3*, Irene Publishing, Sparsnäs Sweden, (2016).
(22) John S. Avery, *Space-Age Science and Stone-Age Politics*, Irene Publishing, Sparsnäs Sweden, (2016).

Index

www.ingramcontent.com/pod-product-compliance
Lightning Source LLC
Chambersburg PA
CBHW061231220326
41599CB00028B/5391

* 9 7 8 9 8 1 3 1 4 7 7 1 3 *